C0-BUT-990

Geobotany Studies

Basics, Methods and Case Studies

Editor
Franco Pedrotti
University of Camerino
Via Pontoni 5
62032 Camerino
Italy

Editorial Board:
S. Bartha, Vacratot, Hungary
F. Bioret, University of Brest, France
E.O. Box, University of Georgia, Athens, USA
A. Čarni, Slovenian Academy of Sciences, Ljubljana (Slovenia)
K. Fujiwara, University of Yokohama, Japan
D. Gafta, University "Babes-Bolyai" of Cluj-Napoca (Romania)
J.-M. Géhu, Inter-Phyto, Nouvion sur Ponthieux, France
J. Loidi, University of Bilbao, Spain
L. Mucina, University of Perth, Australia
S. Pignatti, University of Rome, Italy
R. Pott, University of Hannover, Germany
A. Velasquez, Centro de Investigacion en Sciencias Ambientales, Morelia, Mexico
R. Venanzoni, University of Perugia, Italy

For further volumes:
http://www.springer.com/series/10526

About the Series

The series includes outstanding monographs and collections of papers on a given topic in the following fields: Phytogeography, Phytosociology, Plant Community Ecology, Biocoenology, Vegetation Science, Eco-informatics, Landscape Ecology, Vegetation Mapping, Plant Conservation Biology and Plant Diversity. Contributions are expected to reflect the latest theoretical and methodological developments or to present new applications at large spatial or temporal scales that could reinforce our understanding of ecological processes acting at the phytocoenosis and vegetation landscape level. Case studies based on large data sets are also considered, provided they support habitat classification refinement, plant diversity conservation or vegetation change prediction. Geobotany Studies: Basics, Methods and Case Studies is the successor to Braun-Blanquetia published by the University of Camerino between 1984 and 2011 with cooperation of Station Internationale de Phytosociologie (Bailleul-France) and Dipartimento di Botanica ed Ecologia (Université de Camerino - Italia) and under the aegis of Societé Amicale Francophone de Phytosociologie, Societé Francaise de Phytosociologie, Rheinold Tuexen Gesell-schaft and the Eastern Alpine and Dinaric Society for Vegetation Ecology. This series aims to promote the expansion, evolution and application of the invaluable scientific legacy of the Braun-Blanquetia school.

Franco Pedrotti

Plant and Vegetation Mapping

Springer

Franco Pedrotti
University of Camerino
Camerino
Italy

QK
101
.P43
2013

ISBN 978-3-642-30234-3   ISBN 978-3-642-30235-0 (eBook)
DOI 10.1007/978-3-642-30235-0
Springer Heidelberg New York Dordrecht London

Library of Congress Control Number: 2012943936

The Italian edition published in 2004 by Pitagora Editrice S.r.l.

© Springer-Verlag Berlin Heidelberg 2013
This work is subject to copyright. All rights are reserved by the Publisher, whether the whole or part of the material is concerned, specifically the rights of translation, reprinting, reuse of illustrations, recitation, broadcasting, reproduction on microfilms or in any other physical way, and transmission or information storage and retrieval, electronic adaptation, computer software, or by similar or dissimilar methodology now known or hereafter developed. Exempted from this legal reservation are brief excerpts in connection with reviews or scholarly analysis or material supplied specifically for the purpose of being entered and executed on a computer system, for exclusive use by the purchaser of the work. Duplication of this publication or parts thereof is permitted only under the provisions of the Copyright Law of the Publisher's location, in its current version, and permission for use must always be obtained from Springer. Permissions for use may be obtained through RightsLink at the Copyright Clearance Center. Violations are liable to prosecution under the respective Copyright Law.
The use of general descriptive names, registered names, trademarks, service marks, etc. in this publication does not imply, even in the absence of a specific statement, that such names are exempt from the relevant protective laws and regulations and therefore free for general use.
While the advice and information in this book are believed to be true and accurate at the date of publication, neither the authors nor the editors nor the publisher can accept any legal responsibility for any errors or omissions that may be made. The publisher makes no warranty, express or implied, with respect to the material contained herein.

Printed on acid-free paper

Springer is part of Springer Science+Business Media (www.springer.com)

B333465

Foreword

Mapping plants and vegetation has a long history in Europe, where it is known as geobotanical mapping, i.e. mapping of botany in the landscape or in relation to other geographic factors. Much of the methodology and results, however, is documented only in its original Italian, Polish, Russian or other language, and so remains largely unknown to those who read only the English, French or German scientific literature. I first became aware of this mapping tradition in 1990, at the annual meeting of the International Association for Vegetation Science in Warszawa (Poland), hosted by Janusz B. Faliński, a foremost specialist in vegetation mapping. Many detailed maps were on display, some made by quite unfamiliar methods and some a bit difficult to read but nevertheless compelling. Maps depicting vegetation dynamics or naturalness seemed especially original. Altogether, these maps demonstrated a wealth of information that was new to many of us from the 'West'.

I was reminded of this in the summer of 2010 upon first seeing the original Italian version of this book, *Cartografia Geobotanica*, by Professor Franco Pedrotti (University of Camerino). Immediately I thought that this book should be available in English. Learning to speak and write a foreign language is difficult and takes time; learning to read, on the other hand, is not difficult, especially for scientific material with its more limited and somewhat more international vocabulary. So, even though I had never studied Italian formally, it was indeed possible to translate the book by working in Camerino with the author for about 5 weeks the next summer.

This book presents mainly Italian geobotanical mapping as it developed among specialists and their students there. Most examples involve maps of Italy, and little attempt was made to include maps from wider areas because the Italian tradition provides so much already. There was also little attempt to include newer examples. In particular, the recent explosion of computer maps based on automated data sources (but often with little quality control) is not covered in great detail.

Many original terms, such as 'synphytosociological' or 'phytogeoceneoses', come across in English as very 'European' and hard to remember, even if initially understood. So, for some words, I have tried to create shorter, more analytical equivalents, using basic English words but retaining the more synthetic European constructions in parentheses. Map names in the original language are also retained, as far as possible, with English translations in parentheses. Finally, some Italian terms appear to have no English equivalents, such as *fotolimiti* (the edges or boundaries that can be discerned on an aerial photo); for these I have used familiar English words like, indeed, edge or boundary.

I hope the reader will find this book useful, understanding that it is not an attempt to be comprehensive but rather to present a mapping tradition that may otherwise not be well known. I would also like to take this opportunity to thank Franco Pedrotti for his wonderful Italian hospitality in Camerino.

Athens (Georgia)   Elgene Owen Box
27 August 2011

Preface

Geobotanical maps do not represent the reality of the world but rather what we know about it.

(Janusz Bogdan Faliński)

Geobotanical research finds its syntheses in the production of cartographic documents (maps) which constitute a privileged medium of presentation of information and scientific data on plants in relation to environmental conditions. The significance of new geobotanical maps for already many years has been greatly augmented above all by the contribution that these may make to understanding and solution of environmental problems, such as protection of the flora, zonation of protected areas, management of plant resources and urban planning.

For such reasons, the Faculty of Natural Sciences and Technology of the University of Camerino instituted in 1998 the course *Geobotanical Cartography* for members of the *Scuola di Specializzazione in Gestione dell'Ambiente Naturale e delle Aree Protette* (School of Management of the Natural Environment and Protected Areas) and for students in the Natural and Biological Sciences as a general degree course.

In this book, born of the necessity to provide students a textbook for geobotanical cartography, we try to present the fundamental concepts in this field, which are otherwise sparse in truly scientific publications which often have limited circulation and are difficult to find. This has theoretical as well as practical value, and tries to give the students both an instructional aid for preparation for their exams and theses, and a reference text. Given the high degree of specialization in the material treated, this text is filled with many bibliographic references from various authors, not only to give the reader the greatest possibility for research from original sources but also as a cultural basis.

The book is concerned principally with geobotanical mapping but also contains a chapter dedicated to environmental mapping, because of the contribution that this has received from the former and because of its own current trends and contributions.

In drafting the text, I have tried to improve and expand on what I already wrote about geobotanical mapping in earlier contributions. The text is dedicated predominantly to botanical aspects of cartography and only in part to those techniques (photogrammetry, use of satellite data, etc.) for which numerous, more specialized manuals already exist.

Beginning in 1962, in the Institute of Botany of the University of Camerino (from 1986 the Department of Botany and Ecology), I have had the possibility to collect data and produce geobotanical maps of various kinds, with a group of students and collaborators whom I would like to mention here: Ettore Orsomando, Edoardo Biondi (now at the University of Ancona), Roberto Venanzoni (now University of Perugia), Dan Gafta (now University of Cluj-Napoca, Romania), Krunica Hruska, Andrea Catorci, Paolo Minghetti, Aurelio Manzi, Renato Gerdol (now University of Ferrara), Fabio Taffetani (now University of Ancona), Claudio Chemini (now at the Centro di Ecologia Alpina in Viotte del Monte Bondone of Trento) and Rainer Buchwald (now Universität Vechta, Germany) for vegetation mapping; Carmela Cortini, Michele Aleffi, Sandro Ballelli and Fabio Conti for floristic mapping; and Roberto Canullo and Giandiego Campetella for mapping of populations. To these must be added researchers from other institutes, in particular Francesco Maria Raimondo (Palermo) and Filippo Piccoli (Ferrara). More recently, there has been a group of students of the School of Management of the Natural Environment and Protected Areas of Camerino that is dedicated, under my guidance, to various aspects of cartography: Luciana Carotenuto (Pavia), Anna Maria Castellaneta (Martina Franca), Wilcka Fanesi (Osimo), Renzo Feliziani (Acquasanta Terme), Simone Galassi (Macerata), Jessica Mazzarelli (Foce di Montemonaco), Stefania Menini (Livorno), Bruno Petriccione (Roma), Donatella Rosi (Visso), Sergio Ruggieri (Vieste), Roberta Tacchi (San Severino Marche) and Rosella Vallozzi (Ascoli Piceno).

Finally, I am particularly happy to acknowledge some botanists and ecologists with whom I have been able to study various aspects of geobotanical and environmental mapping, in the laboratory and in the field: the late Janusz Bogdan Faliński (Białowieza and Warsaw), Maximo Liberman Cruz (La Paz, Bolivia), Marcello Martinelli (São Paulo, Brazil), Paul Ozenda (Grenoble), Udo Bohn (Bad Godesberg) and Vasile Cristea (Cluj-Napoca). Janusz Bogdan Faliński, who is the author of the three-volume manual *Kartografia Geobotaniczna* (1990–1991), and I have together described the vegetation of the Gargano (promontory) of Adriatic Italy, of the Białowieza forest in Poland and of Pikhtovka in Siberia. To Dan Gafta (Cluj-Napoca), I am indebted for advice on the content of the book and for critical reading of the text.

I thank Marco Mogetta (Camerino) for producing the layout and Massimo Maccari (Camerino) for preparing printed illustrations and for correcting the proofs.

Particular thanks go to Augusto Persico and to the SELCA of Firenze (Florence), which has a grand tradition in cartographic publication, for supplying specific contributions.

The English edition of my book *Cartografia Geobotanica* was made possible thanks to Prof. Elgene O. Box of the University of Georgia, Athens, USA, who as a friend made himself available for the task of translating the book text from Italian into English.

The structure of the book and the division of the themes into 14 chapters was left the same as in the original, but many chapters were expanded with new information and bibliographic references. Some illustrations were also modified.

The discussions with Prof. Box during the translation work (which was done in Camerino during July 2011) were very useful for improving the original text. Together we have travelled to various parts of the world for geobotanic purposes, beginning with the international phytosociological excursion to Argentina in 1983 and followed by field excursions to Japan, USA, South Africa, Mexico and various countries of Europe, including Italy, France, Poland and Romania. Also, the symposia of the *International Association of Vegetation Science* and of the *Association Amicale Francophone de Phytosociologie* (organized at Rinteln by Rheinold Tüxen and at Bailleul by Jean-Marie Géhu and then in other cities) have provided opportunities to meet and discuss with many botanists interested in vegetation mapping, including Kazue Fujiwara (Yokohama), Frédéric Bioret (Brest), Guillaume Decocq (Amiens), Richard Pott (Hannover), Rüdiger Wittig (Frankfurt am Mein), Xavier Loidi (Bilbao), Laco Mucina (Perth) and others (in addition to the many mentioned already in the introduction to the Italian version of the book).

I am very grateful to Prof. Box for his challenging translation work and thank him very deeply. I thank also Massimo Maccari (Camerino), who adapted the figures for the English edition and prepared the new figures.

Camerino
21 September 2011   Franco Pedrotti

Contents

Geobotanical Mapping and Its Levels of Study

Definition of Geobotanical Mapping

Geobotany is a broad science that deals with the study of species and of vegetation communities in relation to the environment; it includes other, perhaps more familiar sciences, such as plant geography, plant ecology, and chorology, and phytosociology (plant sociology).

Geobotanical cartography is a field of *thematic cartography* that deals with the interpretation and representation, in the form of maps, of those spatial and temporal phenomena that pertain to flora, vegetation, vegetated landscapes, vegetation zones, and phytogeographical units.

The production of a geobotanical map represents the last stage in a cognitive process that begins with observations in the field and continues with the collection of sample data, interpretation of the phenomena observed, and their appropriate cartographic representation; geobotanical cartography is closely tied to the concepts and scope of *geobotany* in general.

It was Rübel (1912a) who first used the term 'geobotanical cartography', but only for the mapping of vegetation. Nevertheless, its meaning was amplified later to include the mapping of all phytogeographic, phytocoenotic and phytosociological aspects, as emphasized by Faliński (1990–1991, 1999). A *phytocoenosis* is a concrete vegetation stand in the field, as opposed to a *community*, which is theoretical; and *phytosociology* is the analytical classification of plant communities based on full-floristic vegetation samples called *relevés*. From this it is evident that the field of research called *geobotanical cartography* is in fact broader than that of *vegetation mapping*.

F. Pedrotti, *Plant and Vegetation Mapping*, Geobotany Studies,
DOI 10.1007/978-3-642-30235-0_1, © Springer-Verlag Berlin Heidelberg 2013

A Bit of History

Geobotanical mapping began relatively recently, about the middle of the 1800s, but developed rapidly during the 1900s. Its beginning was preceded by a lengthy formative period, during which there were various floristic, phytogeographical and vegetation investigations, with the development of elaborate theories, which in turn led to the production of true cartographic documents. Some international congresses pointed out goals attained by geobotanical mapping and objectives to pursue. The most recent such congresses were held in Stolzenau in 1959 (Tüxen 1963), in Tolosa in 1961 (Gaussen 1961a), in St. Petersburg in 1975 (Sochava and Isachenko 1976), in Klagenfurt in 1979 (Ozenda 1980–1982), in Grenoble in 1980 (Ozenda 1981), in Warsaw in 1990 (Faliński 1991), again in Grenoble in 1996 (Michalet and Pautou 1998) and in České Budějovice in 1997 (Bredenkamp et al. 1998). Specialists from all over the world participated in these congresses, which contributed ideas in various areas of cartography.

Journals dedicated to cartography were also published, such as the *Bulletin du Service de la Carte Phytogéographique* edited by Louis Emberger (Montpellier) beginning in 1956 (now discontinued); the *Documents pour la Carte de la végétation des Alpes,* then *Documents de Cartographie Écologique* (1963–1987) edited by Paul Ozenda (Grenoble); *Geobotaniceskoe Kartografirovanie (Geobotanical Mapping)* (from 1963) edited by V.B. Sochava and E.M. Lavrenko (Saint Petersburg); and the *Supplementum Cartographiae Geobotanicae* (from 1988) edited by Janusz Bogdan Faliński (Białowieza-Warsaw).

During all this time the production of vegetation maps increased and improved in both form and content, culminating in 2000 in the publication of the *Map of the Natural Vegetation of Europe* directed by Udo Bohn (Bad Godesberg) and involving coordinating work by Robert Neuhäusl (Prague-Průhonice), Wladyslaw Matuszkiewicz (Warsaw), Paul Ozenda (Grenoble), Tatiana Yurkovskaya (Saint Petersburg) and others (Bohn et al. 2000a,b, 2003).

Numerous specialized publications on vegetation mapping, illustrating both theoretical and practical aspects, were also produced (e.g. Sochava 1954, 1962, 1979; Küchler 1967; Ozenda 1986; Küchler and Zonneveld 1988; Faliński 1990–1991; Pirola and Vianello 1992; Alexander and Millington 2000), as well as geobotanical publications that contain chapters on cartography (Braun-Blanquet 1928, 1951, 1964; Tomaselli 1956a; Ozenda 1964, 1982; Borza and Boşcaiu 1965; Puscaru-Soroceanu and Popova-Cucu 1966; Ivan and Doniţă 1975; Ivan 1979; Dierschke 1994; Cristea et al. 2004; and others). A quite readable and less specialized summary of vegetation mapping, including its earlier history, was provided by de Laubenfels (1975).

Levels of Synthesis in Geobotanical Mapping

The themes and other subjects that can be represented on geobotanical maps are quite diverse. These correspond to the level of geobotanical study and can be listed as follow (see also Fig. 1.1):

I. *Level of individual plants* (species): the mapping of plant populations in great detail, as may be used in studies of competition, facilitation, vegetative regeneration, dynamic processes, etc., which manifest themselves in small areas; the scale of such maps varies according to the dimensions of the individual plants.

II. *Level of populations* (species): distributions of populations of plant species in a particular territory.

III. *Level of synusiae*: mapping of the synusiae present in a definite phytocoenosis. A synusia (see Chap. 3) is a group of structurally and functionally similar plant species in a vegetation stand. According to Gams (1918), synusiae can be distinguished at a first order (populations of a single species in a phytocoenosis), at a second order (populations of more species all belonging to the same life form), and at a third order (populations of diverse species that differ structurally by each occupying a particular microhabitat). In fact, in accepted current usage, the concept of a synusia of order I coincides with that of a population and that of a synusia of order III coincides with that of a microphytocoenosis (for example, epixylous assemblages on fallen trunks or epilithic assemblages on large rocks in the understorey). Normally the objective of mapping at this level is a synusia of order III, which has a wider connotation, especially in ecology, than that of order I or II. The term synusia, on the other hand, has been used by many authors with sometimes quite different meanings, as in the cases of Barkman (1973), Guinochet (1973), Vigo (1998) and others. Referring to synusiae, some authors have proposed a new concept, that of coeno-association, which will be examined briefly in Chap. 6 (section on Maps of Coeno-Associations) (Gillet 1986, 1988; Gillet et al. 1991; Gillet and Gallandat 1996).

IV. *Level of phytocoenosis*: mapping of vegetation types of higher floristic homoteneity[1] found in specific stable conditions; corresponds to classic phytosociological mapping at the level of associations or subassociations (Braun-Blanquet 1964) and to mapping based on classifications that refer to other definite vegetation units from other schools, such as phytogeocoenoses (Sukachev and Zonn 1961).

---

[1] Homoteneity is a synthetic concept based on all the relevés collected for diverse phytocoenoses belonging to the same type of community (van der Maarel and Westhoff 1973). An association has high homoteneity if the expression $(S_{IV} + S_V)/(S_{II} + S_{III}) > 1$, which is to say that the proportion of dominant species (constancy $> 60\%$) is greater than in the neutral model of a random distribution of species in the plant community (S_i = number of species that re-enter in classes of constancy i).

V. *Level of ecotopes (teselas)*: the mapping of vegetation series, or synphytoso-ciological units, based on the concepts of sigmeta or vegetation series or sigma-associations according to Rivas-Martínez (1985) and Géhu (1991a).
VI. *Level of catenas* (vegetated landscapes): at the scale being used, this coincides with geo-synphytosociological or catenal mapping, which is based on the concept of geoseries or geosigmeta or catenas of vegetation series (Géhu 1991a).
VII. *Level of lower phytogeographical units*: land areas that are distinct in terms of their distributions of species, genera and families and in particular their endemics; one can distinguish also lower and higher phytogeographical units (see level VIII); the lower phytogeographical units were called districts, sectors and provinces (or dominions) by Rivas-Martínez (1985) and, respectively, meso-chorocomplexes, macro-chorocomplexes and mega-chorocomplexes by Theurillat (1992, 1994).
VIII. *Level of higher phytogeographical units* (regions and kingdoms) *and biomes*: this level can involve very different kinds of maps due to the two possible bases for subdivision of the vegetation cover of the world, namely according to the flora or the vegetation type. Maps of floristic subdivisions (phytogeo-graphic maps) are based on qualitative and quantitative characteristics of the flora, as opposed to maps of vegetation zonation, which are based on qualita-tive and quantitative characteristics of the vegetation formations. Maps that represent phytogeographic regions and kingdoms use, as the fundamental criterion, the floristic uniqueness of a particular territory, including its endemics. Maps of vegetation zones are based on the physiognomy of the vegetation formation and on the climate in which it developed.

Levels I and II are for mapping populations, and level II also for mapping chorological (or floristical) areas; level III is for mapping synusiae; levels IV and

Level	Schematic	Abstract unit (theoretical model)	Map type	
I. Individual plant		Species	Phytoecological	Floristic cartography
II. Population		Species	Populations and chorology	

Fig. 1.1 (continued)

Level	Schematic	Abstract unit (theoretical model)	Map type	
III. Synusia		"Association"	Synusial	Vegetation cartography
IV. Vegetation stand (phytocoenosis)		Plant association	Phytosociological	Vegetation cartography
V. "Tesela"	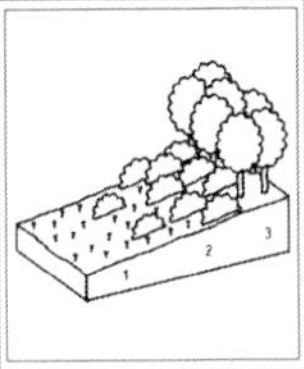	Sigmetum (series)	Synphytosociological or Dynamic-phytosociological or Integrated-phytosociological	Vegetation cartography
VI. Catena (vegetated landscape)	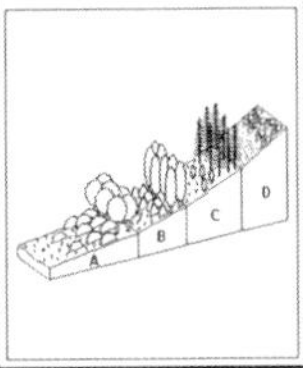	Geosigmetum (geoseries)	Geo-synphytosociological	Vegetation cartography
VII. Lower plant-geographic units		District, Sector, Province	Regional-phytogeographical	Plant-geographical cartography
VIIIa. Higher plant-geographic units		Region, Kingdom	Plant-geographical	Plant-geographical cartography
VIIIb. Biomes		Zone and belt vegetation	Global-phytoclimatical	Plant-geographical cartography

Fig. 1.1 Levels of synthesis in geobotanical mapping

V are for mapping vegetation distributions; level VI is for plant or vegetation landscapes (geo-synphytosociological mapping); and levels VII and VIII are for phytogeographical mapping and the mapping of vegetation formations.

Types of Geobotanical Maps

Based on the above, it is possible to distinguish the following types of maps, with the levels of synthesis (above) shown in parentheses:

Population maps (levels I and II)

Chorological maps (level II)

Synusial maps (level III)

Vegetation maps (phytosociological maps of actual or potential vegetation, etc.) (levels IV and V)

Landscape maps (geo-synphytosociological maps) (level VI)

Phytogeographical maps at lower level (level VII)

Phytogeographical maps at higher level and maps of vegetation distribution (vegetation formations) (level VIII).

These cartographies are very different from each other, by map content or by techniques of representation. Nevertheless they all integrate knowledge of complex biological phenomena, as are found in the plant world.

To these map types already listed can be added the phytoecological maps, which may treat species, including individuals and populations (autoecology), or vegetation (synecology), according to the particular goals.

Finally, one must also mention the maps of plant biodiversity.

A plant population is the set of the individuals (organisms, phytoindividuals) of the same species that live in a given place at a given moment and interact with each other. A mapping of individuals coincides in part with that of populations, but it is still preferable to keep the two types (phytoindividuals and populations) distinct because they have different meanings and purposes, and because the methodologies used may be different.

Maps at this level, at fine (large) scale, are today a useful basis for conservation biology, since threatened species have greater probability of surviving due to the *rescue effect* involving migration between neighboring populations. On the other hand, maps at broader (smaller) scale become chorological, in that they may contain entire populations or, more generically, the absolute range of a species.

Maps of Individual Plants

The purpose of maps of individual plants is to represent their distributions in localized areas. The mapping is done at a fine scale, recording also the composition in terms of organisms (units) of diverse types and functions (Falińska 1984). Beyond just the spatial representation of complex individuals, including genotypes identified by molecular methods (Figs. 2.1 and 2.2), this kind of the mapping permits monitoring all individuals in permanent areas, generally small but adequate, from 1/4 m^2 to a few tens of m^2 (Fig. 2.3). Naturally the mapping of herbaceous species is always more complex than that of woody species, as can be seen in Fig. 2.4, which shows a plot of 50 $\times$ 50 m inside which all individuals of the herbaceous species present have been mapped.

Repeated mapping over time permits gathering information on the manner and rate of growth of the individuals, hence of the population that they compose (Fig. 2.5), and on the duration and disappearance of the individuals (Fig. 2.6), etc.

Use of sophisticated geopositioning systems (from theodolites to GPS) does not always result in the adequate required level of detail, as opposed to simpler and more immediate topographic positioning by means of corner coordinates, which

F. Pedrotti, *Plant and Vegetation Mapping*, Geobotany Studies,
DOI 10.1007/978-3-642-30235-0_2, © Springer-Verlag Berlin Heidelberg 2013

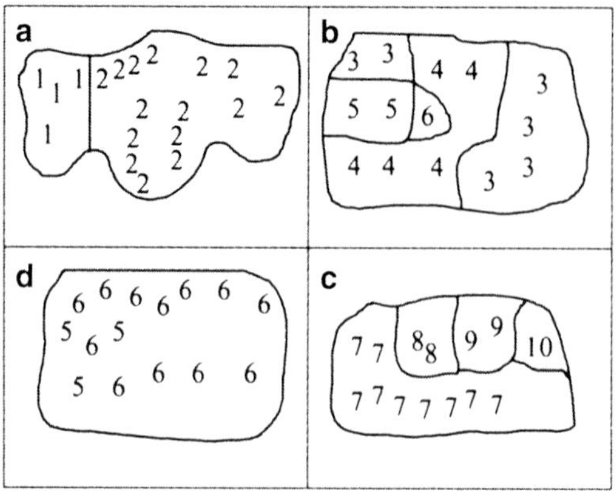

Fig. 2.1 Distribution of four populations genotypes of *Festuca rubra* (**a**, **b**, **c**, **d**) identified by electrophoresis; the *numbers* shows different genotypes (From Falińska 1998)

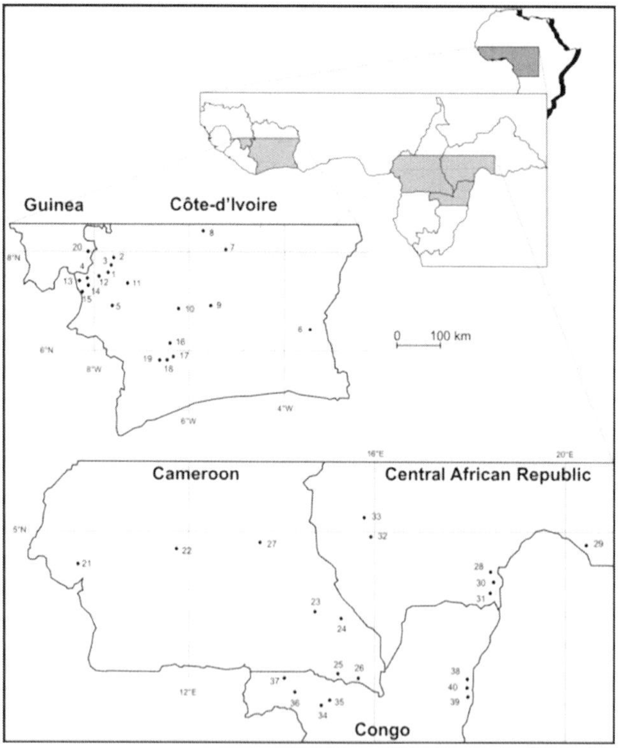

Fig. 2.2 Geographic and genetic origins of the wild *Coffea canephora* genotypes in West and Central Africa; the *numbers* corresponds to different genotypes names and origin (From Gomez et al. 2009)

Fig. 2.3 Spatial distribution
of *Centaurea diomedea*
individuals between those of
Thymelaea hirsuta, Tremiti
Islands, southern Italy (From
Falińska 1999)

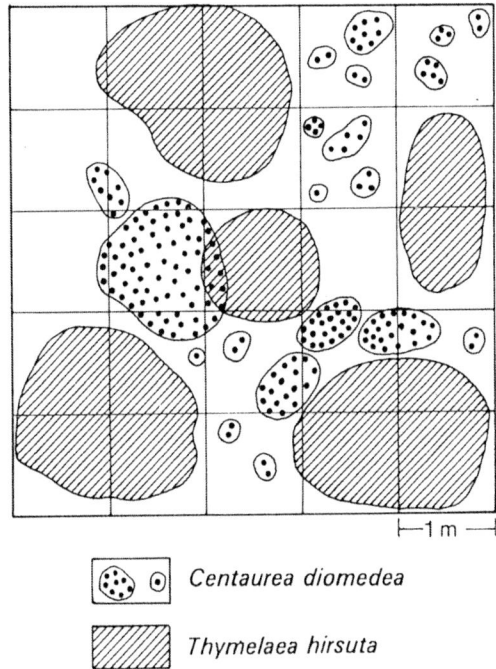

Centaurea diomedea

Thymelaea hirsuta

permit finding the mapped unit also at less phenologically optimum moments and with minimal impact. In these cases, the geographic coordinates of individuals are measured directly in the field, as for two populations of *Polylepis tarapacana* from the Nevado Sajama (Andes of Bolivia) in areas of about 4,000 m^2, with the following distinct functional age classes: seedling, young individual, fertile adult, sterile adult and dead individual (Fig. 2.7).

Naturally these considerations depend on the dimensions of the organisms: woody species require more space and benefit from more modern techniques for topographic positioning, such as photogrammetry (Fig. 2.8) or remote sensing (Fig. 2.9).

Generally, maps of the distribution of individual trees are made for rare species, as in the case of the Nebrodi fir (*Abies nebrodensis*), a species which exists in nature only as 30 individuals and as such has been mapped repeatedly, three times over the past 30 years (Fig. 2.10) (Morandini 1969; Morandini et al. 1994; Virgilio et al. 2000); and the case of the *bagolaro* (*Celtis tournefortii*), with a range limited to a few sites on the southwest slope of Mt. Etna (Fig. 2.11). Mapping of this type has also been done for the century-old individuals of larch pine (*Pinus laricio*), noted as the "giants of Sila" in the forest Bosco Fallistro (Avolio and Ciancio 1985), and for the *Quercus pubescens* trees of the zone Abbadia di Fiastra in the Marche (region) of Adriatic Italy, in which pollution damage suffered by the trees is indicated distinctly in four classes: no damage, plus light, moderate and severe damage (Fig. 2.12).

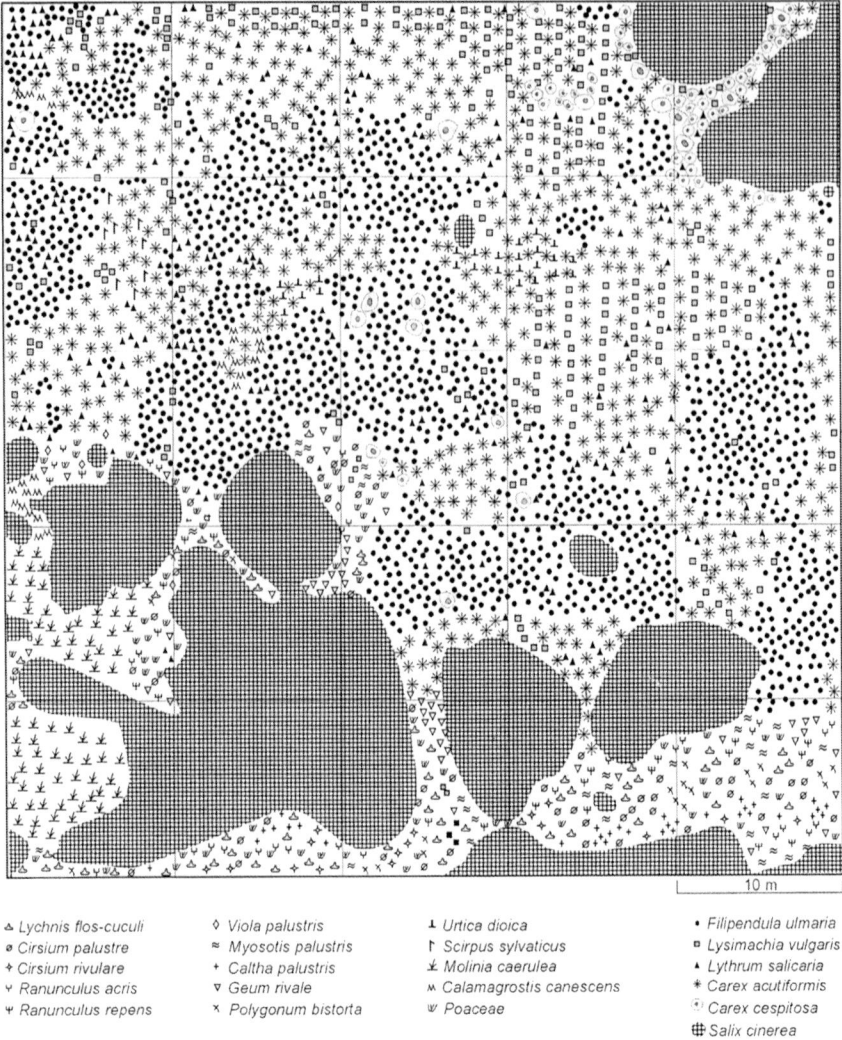

⚘ Lychnis flos-cuculi	◊ Viola palustris	⊥ Urtica dioica	• Filipendula ulmaria
◠ Cirsium palustre	≈ Myosotis palustris	↾ Scirpus sylvaticus	▫ Lysimachia vulgaris
✦ Cirsium rivulare	+ Caltha palustris	⊥ Molinia caerulea	▲ Lythrum salicaria
Ⅎ Ranunculus acris	▽ Geum rivale	⋀ Calamagrostis canescens	✳ Carex acutiformis
ⱳ Ranunculus repens	× Polygonum bistorta	ⱳ Poaceae	⊙ Carex cespitosa
			⊞ Salix cinerea

Fig. 2.4 Spatial distribution of perennial herbaceous species in abandoned meadow in the Remski wetland, Narewka Valley, Białowieza (Poland) after 25 years, partially invaded by *Salix cinerea* (From Falińska 2003)

Trees are also mapped as a function of their ecology. Kondō and Sakai (2011) surveyed and mapped some tree species in relation to microtopography in a mountain area of central Japan. The species there occur mainly in particular geomorphological situations, such as *Salix cardiophylla* on debris cones; *Tsuga diversifolia*, *Abies mariesii* and *Abies veitchii* on mountain slopes and terraces; and *Alnus matsumurae* on valley and lower slopes (Fig. 2.13).

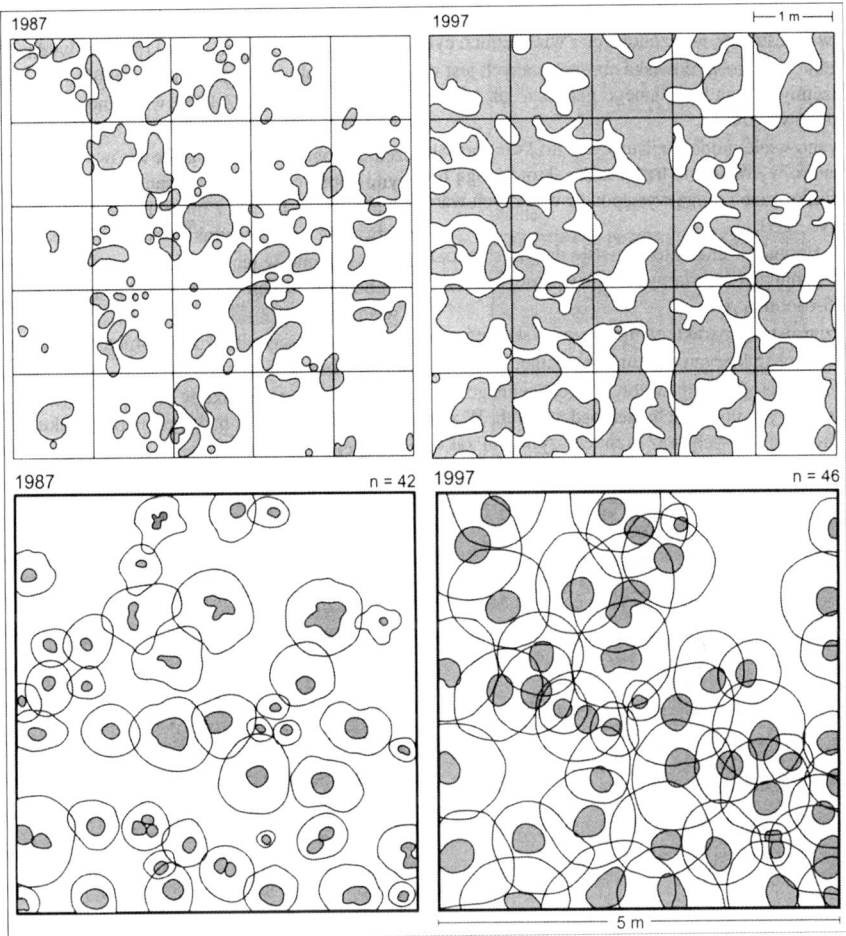

Fig. 2.5 Spatial dynamic of a population of *Filipendula ulmaria* and another of *Carex caespitosa*, from 1987 to 1997 (From Falińska 2002)

Individual trees have also been mapped for the remains of two forests left standing under water when the two lakes in Trentino were created by a landslide around 1000BC. At the end of the Lago di Molveno lake, which was drained completely for construction of a hydroelectric impoundment, 142 individuals belonging to 7 species have been found and mapped, among them *Abies alba*, *Taxus baccata* and *Fagus sylvatica* (Marchesoni 1954). At the Lago di Tenno lake, 74 individuals belonging to 7 species were found and mapped, including especially *Fagus sylvatica* (Fig. 2.14). This mapping was done by a diver who marked the trunks found by means of cables tied to floats. Using a theodolite along the shoreline, it was possible to map the positions of the individual floats, each of which corresponded to a tree trunk on the

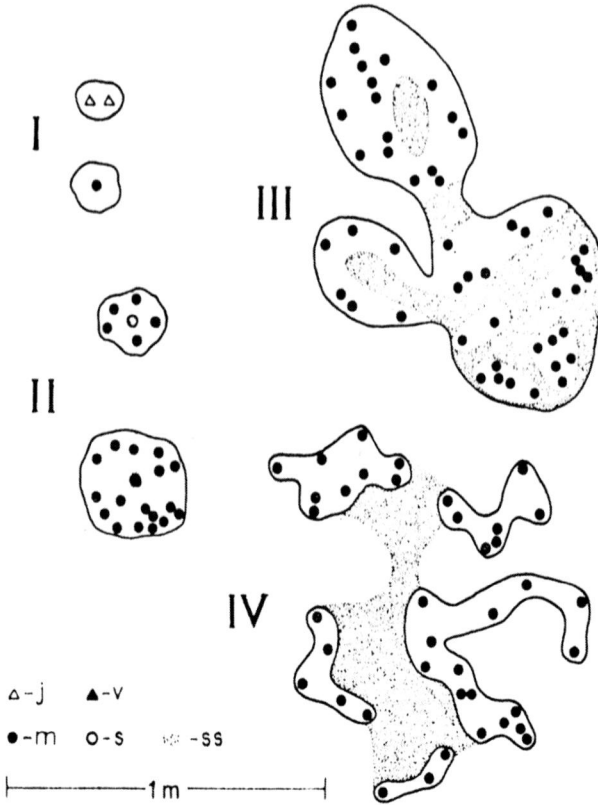

Fig. 2.6 Development and shape variation of *Filipendula ulmaria* polycormon. Development phases: *I* – beginning, *II* – juvenile; *III* – mature; *IV* – senescent (over 10 years); *j* – young, *v* – virginale (physiologically ready for reproduction; but still without differentiated reproductive organs), *m* – fertile; *s* – old, *ss* – dead parts of the underground organs (From Falińska 1984)

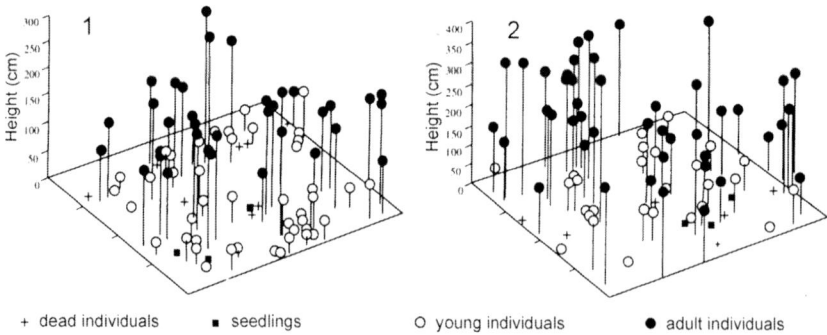

+ dead individuals   ▪ seedlings   ○ young individuals   ● adult individuals

Fig. 2.7 Spatial distribution of two populations of *Polylepis tarapacana*, Nevado Sajama, Bolivia, number 1 at 4,720 m and number 2 at 4,370 m; dimensions and age were surveyed for each individual (From Liberman Cruz et al. 1997)

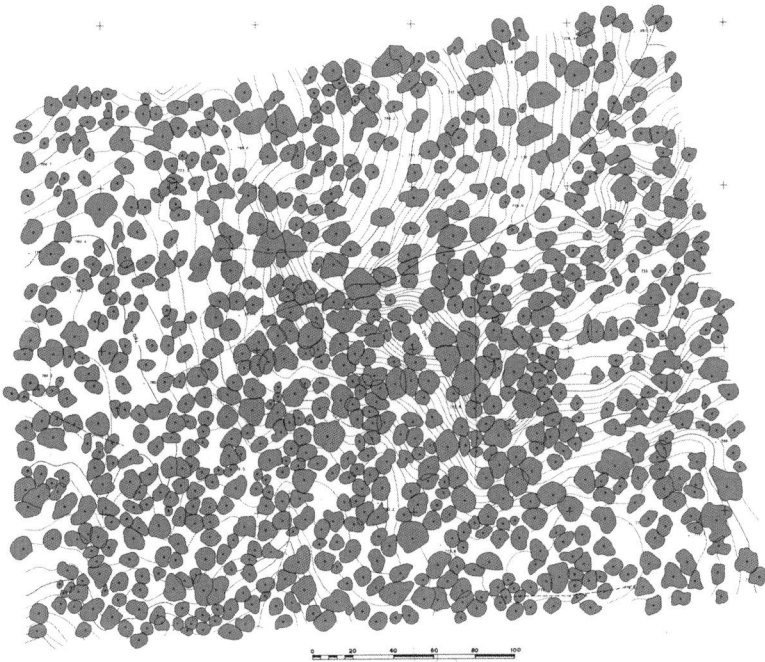

Fig. 2.8 Distribution of *Fagus sylvatica* trees in the canopy of forest in the Riserva Pavari, Foresta Umbra, Mt. Gargano, Apulia region of Adriatic; southern Italy; the distribution was obtained by photorestitution

lake bottom. With such research it is possible to understand the composition of the forest at the time the lakes were formed.

Maps of Populations

Cartographic representation of one population may involve:
- One species in a phytocoenosis (coenopopulation or coenotic population, i.e. all individuals of the same species in a phytocoenosis), at the location of a given ecosystem; in this case it is possible to show the dynamic and spatial relations of subpopulations and ecotonal or synusial populations (Fig. 2.15); or
- One species in a territory (administrative, geomorphological, landscape, etc.) in which different coenotic populations are distinguishable; in this case it is possible to show the functional relations within the metapopulations, i.e. a set of populations connected by a flow of individuals but separated by environmental heterogeneity and various dynamic processes (including phenology).

Mapping the dynamics of these populations permits expression of the relations among ecological and dynamic processes in populations, interpretable as spatial process patterns that can occur simultaneously or sequentially in a single "dynamic

Fig. 2.9 IKONOS image Carterra PAN – MS (1 and 4 m resolution) from 22 June, 2000, on the Kassandra peninsula, Chalkidiki, Greece. The image show the number of *Pinus halepensis* trees and the density of tree biomass, as represented by *crosses* of different sizes (*Software developed by Advanced Computer Systems A.C.S. S.p.A., Rome*)

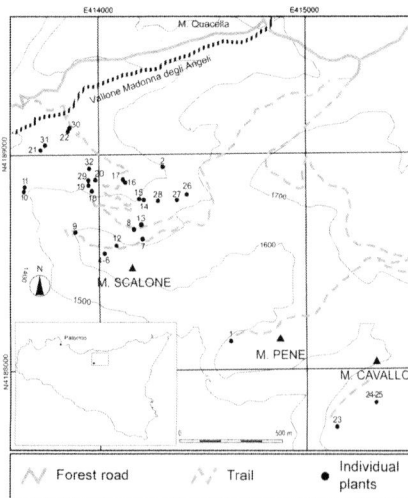

Fig. 2.10 On the *left*, the distribution of individual *Abies nebrodensis*; trees, numbered from 1 to 32, Mt. Scalone, Sicily; on the *right*, the tree no. 27 (From Virgilio et al. 2000; photo Rosario Schicchi)

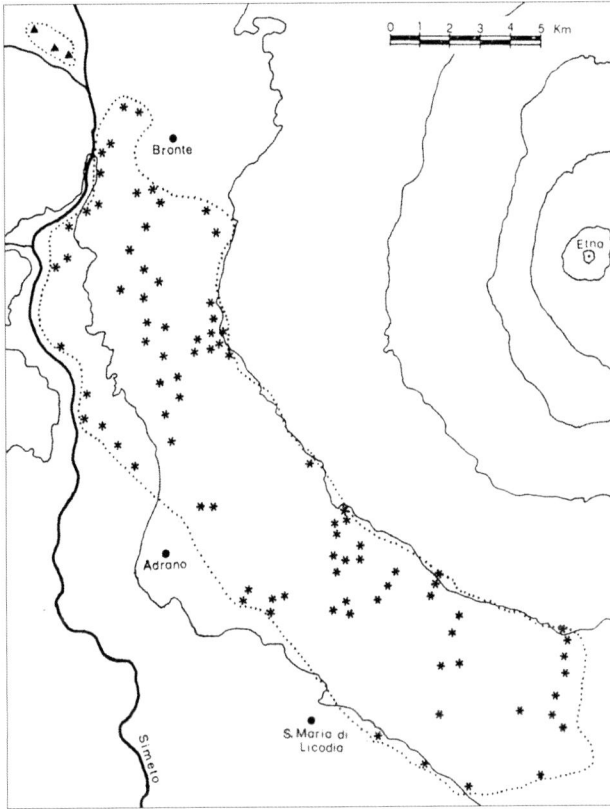

Fig. 2.11 Distribution of *Celtis tournefortii* on Mt. Etna, Sicily (From Poli et al. 1974)

landscape". The expansion of a species on a territory can be described as a complex balance among distinct populations in various dynamic states (pioneers, expanding, regressing, etc.) (Canullo 1991b, 1993a, b), as can be seen in the case of *Anemone nemorosa* at Białowieza (Fig. 2.16) and of *Cytisus sessilifolius* at Torricchio (Fig. 2.17). The variations in the populations can be observed and mapped also in relation to the various successional stages of the communities in which they occur, for example during the secondary succession of a marsh and progressive formation of a shrubby stand of *Salix cinerea* (Fig. 2.18) or of *Alnus glutinosa* (Fig. 2.19), or of *Cyclamen hederifolium* in the beechwood of Mt. Gargano (Fig. 2.20). In the wetland area of Reski (Białowieza forest), Falińska (2003) followed variations in populations in the association Cirsietum rivularis from 1980 to 1995, producing four maps, one every 5 years. The wetland was invaded progressively by shrub species, in particular *Salix cinerea*, which replaced the pre-existing populations of herbaceous species completely over the course of the 15 years (Fig. 2.18).

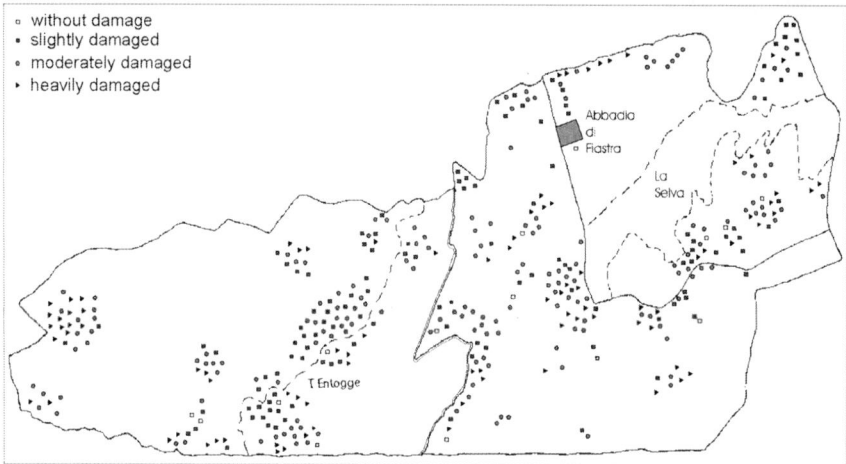

Fig. 2.12 Distribution of field oaks (*Quercus pubescens*) from the hills of Abbadia di Fiastra, Marche Region, Adriatic central Italy, with degrees of pollution damage (From Campetella et al. 2002)

In particular cases it is possible to map *biogroups*, i.e. aggregations of several species of trees and bushes of various ages, of which one is older, is found at the center of the aggregation and grows taller than the others, assuming a promoting role (Canullo and Falińska 2003). In the Torricchio nature reserve (near Camerino, see list of frequently cited locations), *Juniperus communis* and other woody species frequently grow according to this model of development (Fig. 2.21).

Accurate boundary mapping of individual populations can be done by standard survey equipment, such as a Laser Electronic Theodolite, Electronic Distance Measurer or a Global Positioning System. The different accuracies of these

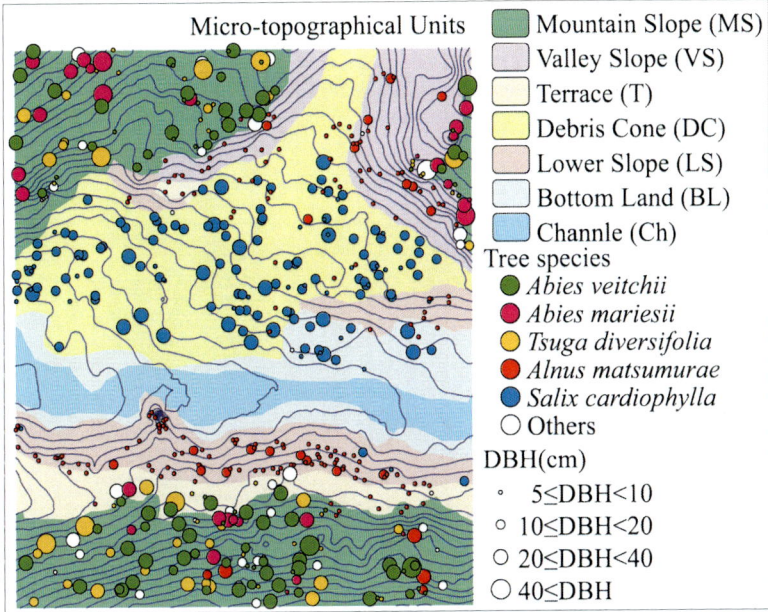

Fig. 2.13 Micro-topographical units and tree distribution in central Japan (From Kondō and Sakai 2011)

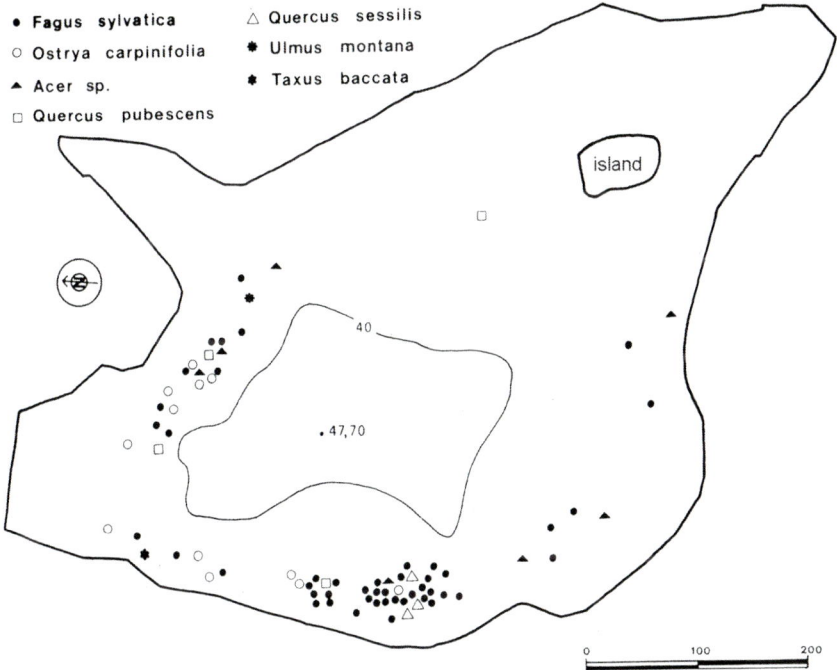

Fig. 2.14 Distribution of tree trunks rooted on the bottom of the Lago di Tenno (lake), Trentino-Alto Adige Region, northern Italy (From Biondi et al. 1981)

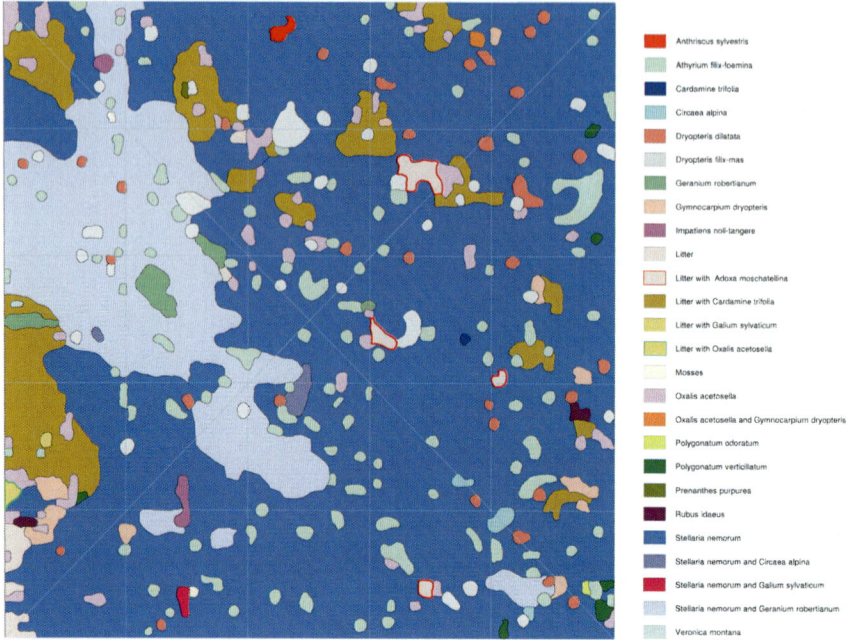

Legend:
- Anthriscus sylvestris
- Athyrium filix-foemina
- Cardamine trifolia
- Circaea alpina
- Dryopteris dilatata
- Dryopteris filix-mas
- Geranium robertianum
- Gymnocarpium dryopteris
- Impatiens noli-tangere
- Litter
- Litter with Adoxa moschatellina
- Litter with Cardamine trifolia
- Litter with Galium sylvaticum
- Litter with Oxalis acetosella
- Mosses
- Oxalis acetosella
- Oxalis acetosella and Gymnocarpium dryopteris
- Polygonatum odoratum
- Polygonatum verticillatum
- Prenanthes purpurea
- Rubus idaeus
- Stellaria nemorum
- Stellaria nemorum and Circaea alpina
- Stellaria nemorum and Galium sylvaticum
- Stellaria nemorum and Geranium robertianum
- Veronica montana

Fig. 2.15 Dominant plant species (populations) in the understory of a beech forest belonging to the *Galio odorati-Fagetum*. Vallone Vallor, Piano del Cansiglio, Veneto Region, northern Italy. The map was produced during summer 2005, based on a visual record 1:100, in a 50 × 50 m permanent monitoring plot of the ICP forests, CONECOFOR national network. The dominant species units refer to the understory up to 2 m high, with 50 cm accuracy. *Stellaria nemorum* dominates the ground layer (more than 85 %), sometimes in co-dominance with other species. Other species are represented by a number of small patches (*Athyrium filix-foemina* and *Oxalis acetosella*, represented respectively by 113 and 50 units), often single individuals, polichormones or clones (e.g. *Cardamine trifolia* or *Polygonatum odoratum*). Scattered populations occur on litter-covered areas (*Field surveys by Leonardo Ghirelli and Maria Cristina Villani; elaboration by Monica Foglia and Roberto Canullo*)

instruments define the scale at which they can be used profitably, depending on the local population dimension, grain and extent (Elzinga et al. 1998).

Maps of Vegetation Structure

The horizontal structure of vegetation is the (horizontal) spatial distribution pattern of plant populations, synusiae or phytocoenoses; the vertical structure is the way in which vegetation is organized into strata (Ivan 1979). Horizontal structure involves the distributions, in a certain area, of the components of the phytocoenoses, i.e. the individuals of the various species. This is done separately for the herbaceous, shrub and tree strata (layers). The methodology is similar to that used for populations

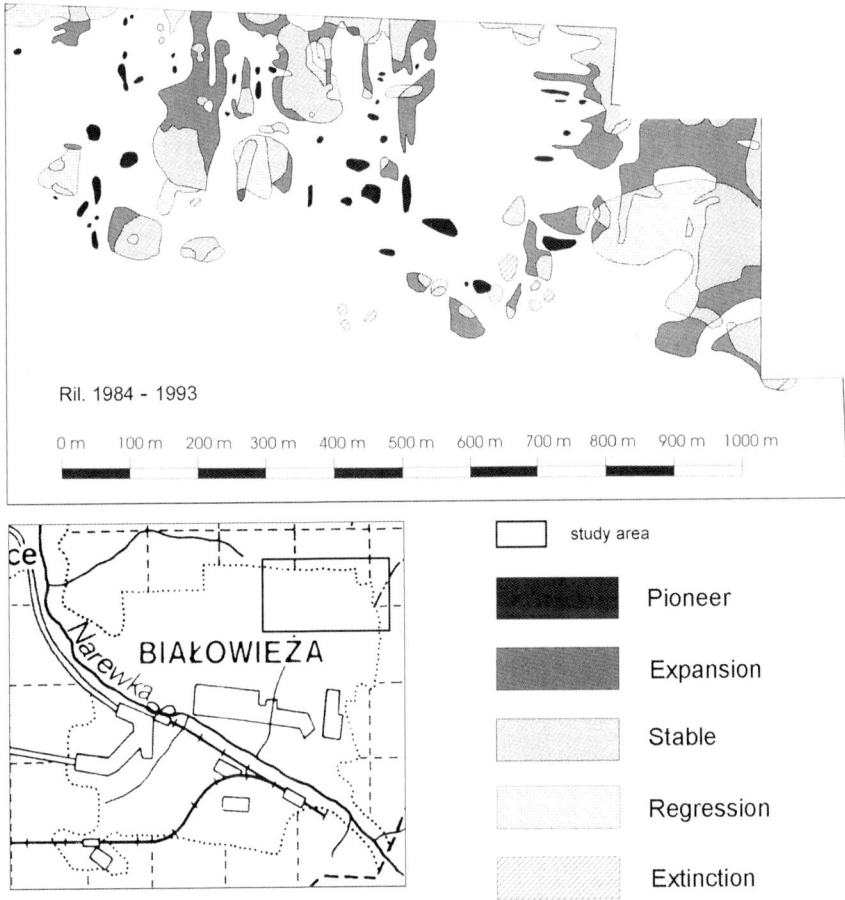

Fig. 2.16 Map of populations of *Anemone nemorosa* in different dynamic states in *Tilio-Carpinetum* forest, from 1984 to 1993, in a clearing in the Białowieza primeval forest, Poland (From Canullo and Tavolini 1999)

(Fig. 2.22). Vertical structure, on the other hand, considers the height of individual plants as they appear in distinct strata (Fig. 2.22). The demographic structure of vegetation (Canullo and Falińska 2003) considers also the age of the individuals, as was done for two populations of *Polylepis tarapacana* from the Andes of Bolivia (Fig. 2.7).

Maps of vegetation structure can be analytical or synthetic. Analytical maps are for individuals, as already said; synthetic maps are for homogeneous vegetation units with the same type of structure, as can be observed in the map of the vegetation structure of the Fiastra forest (Fig. 2.23).

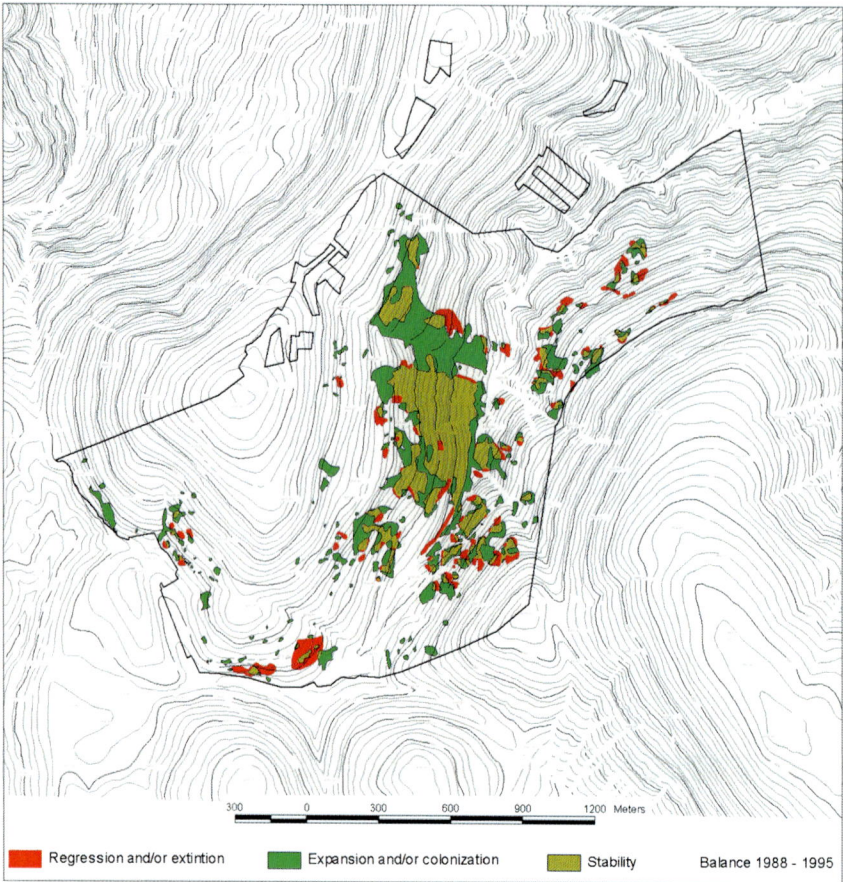

Fig. 2.17 Variations in the populations of *Cytisus sessilifolius* (1988–1995) in Torricchio Natural Reserve, Marche Region, Adriatic central Italy (From Canullo and Campetella 2010)

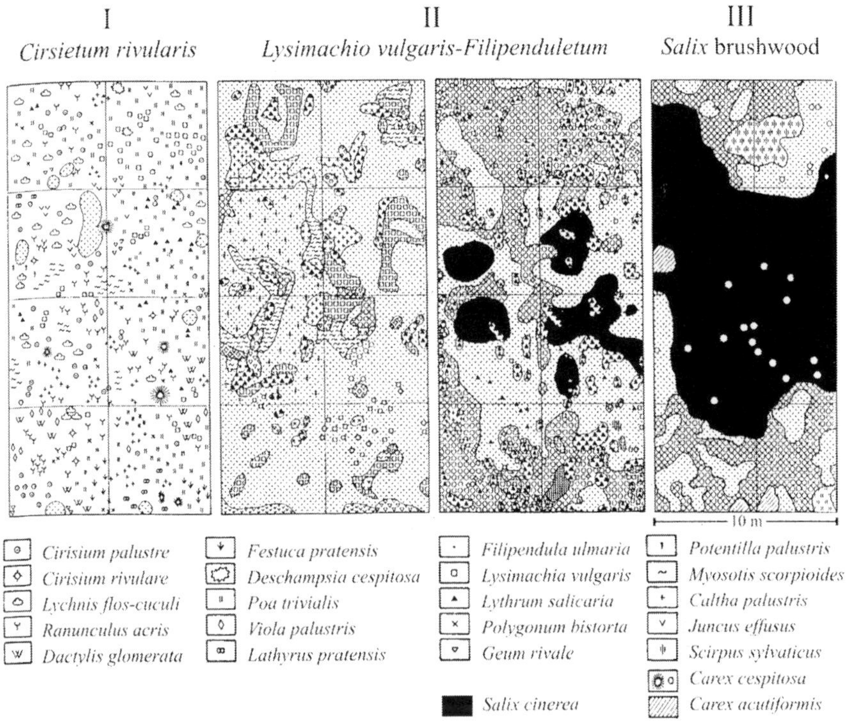

Fig. 2.18 Dynamics of spatial structure of populations during secondary succession. *Stage I*: beginning (meadow of *Cirsietum rivularis*); *stage II*: temporary (mega-forbs of *Lysimachio-Filipenduletum*); *stage III*: final (*Salix cinerea* brushwood) (From Falińska 1998)

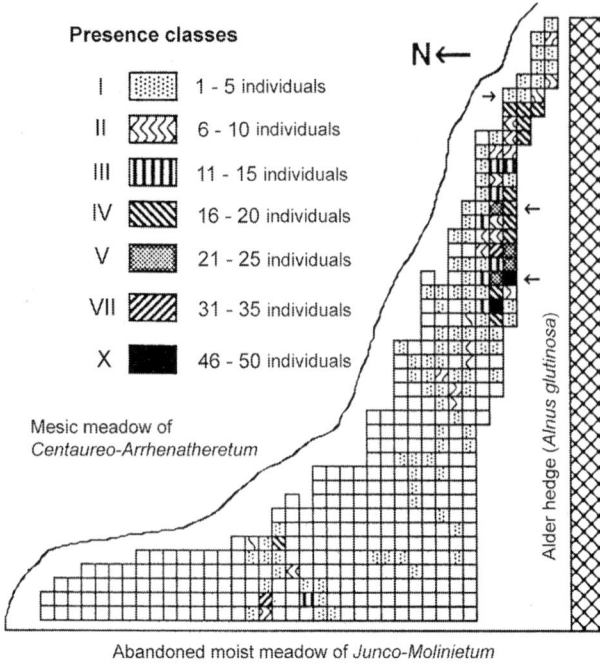

Fig. 2.19 Occurrence of individuals of *Alnus glutinosa* in abandoned meadow of the association *Junco-Molinietum*, Piné, Trentino-Alto Adige Region, northern Italy; the occurrence is evaluated and mapped in 0.5 × 0.5 m areas (From Gafta and Canullo 1992)

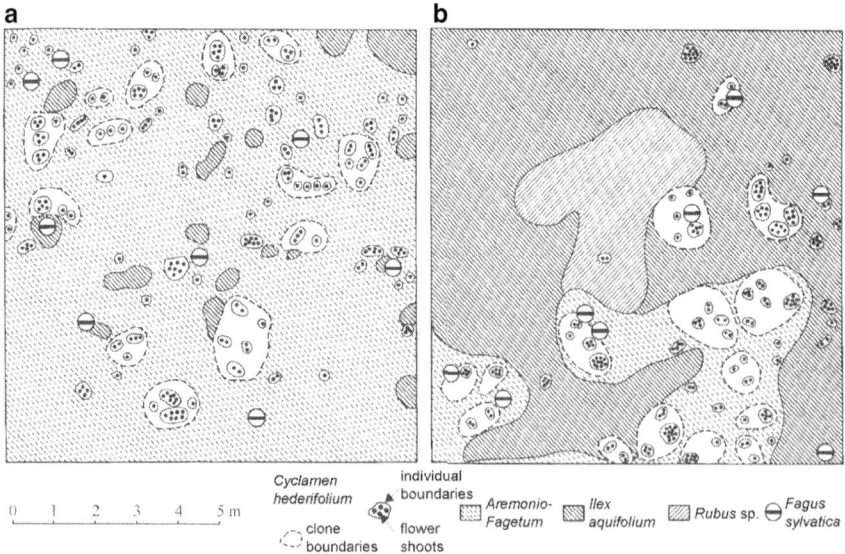

Fig. 2.20 Changes in the spatial structure of *Cyclamen hederifolium* population in some dynamical stages of *Aremonio-Fagetum* forest, Mt. Gargano, Apulia Region, Adriatic southern Italy; (**a**) forest in fluctuation stage; (**b**) forest in degeneration stage because of invasion of *Ilex aquifolium* (From Falińska 1993)

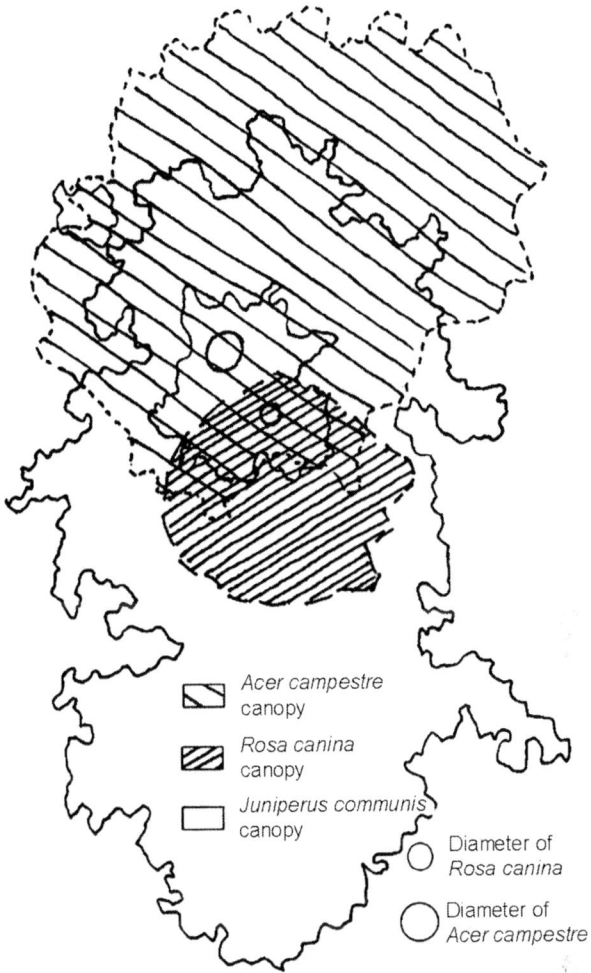

Fig. 2.21 Canopy projection and trunk positions of species in the biogroup formed by *Juniperus communis*, *Acer campestre* and *Rosa canina* in the Torricchio Natural Reserve, Marche Region, Adriatic central Italy (*Field survey of Simone Galassi, Camerino*)

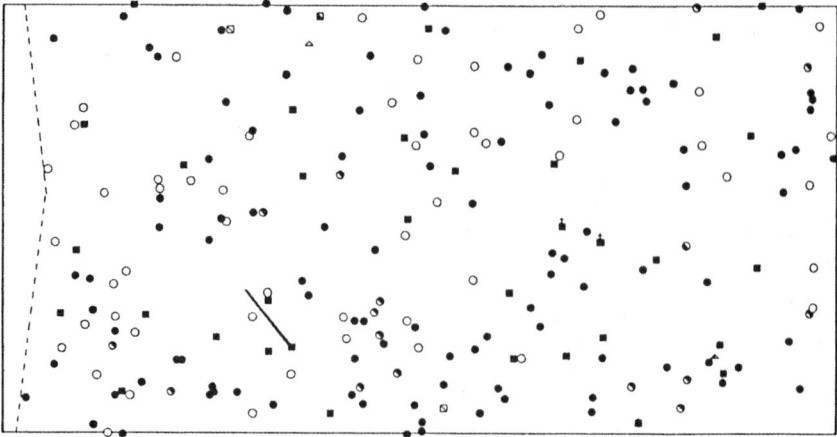

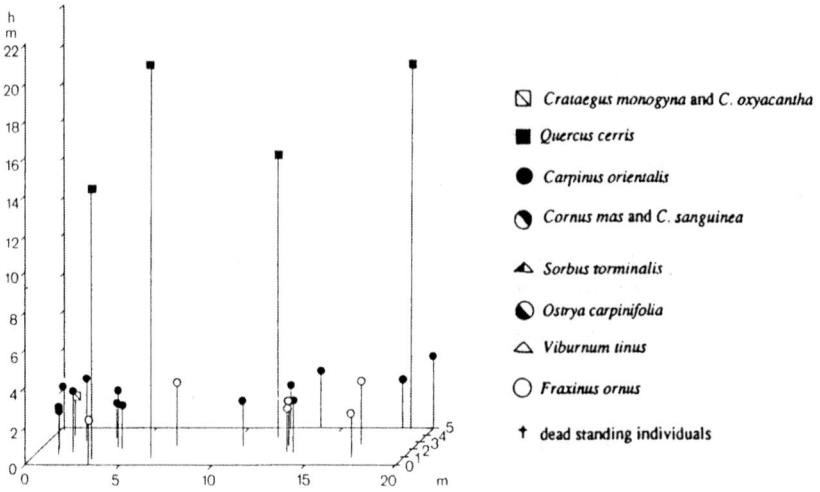

Fig. 2.22 *Above*: horizontal species distribution of individuals or stumps in tree study plots in mixed deciduous coppice; *below*: the vertical structure of the tree study plots, as simplified by a representative transect of the tree and shrub layers (From Canullo 1991b)

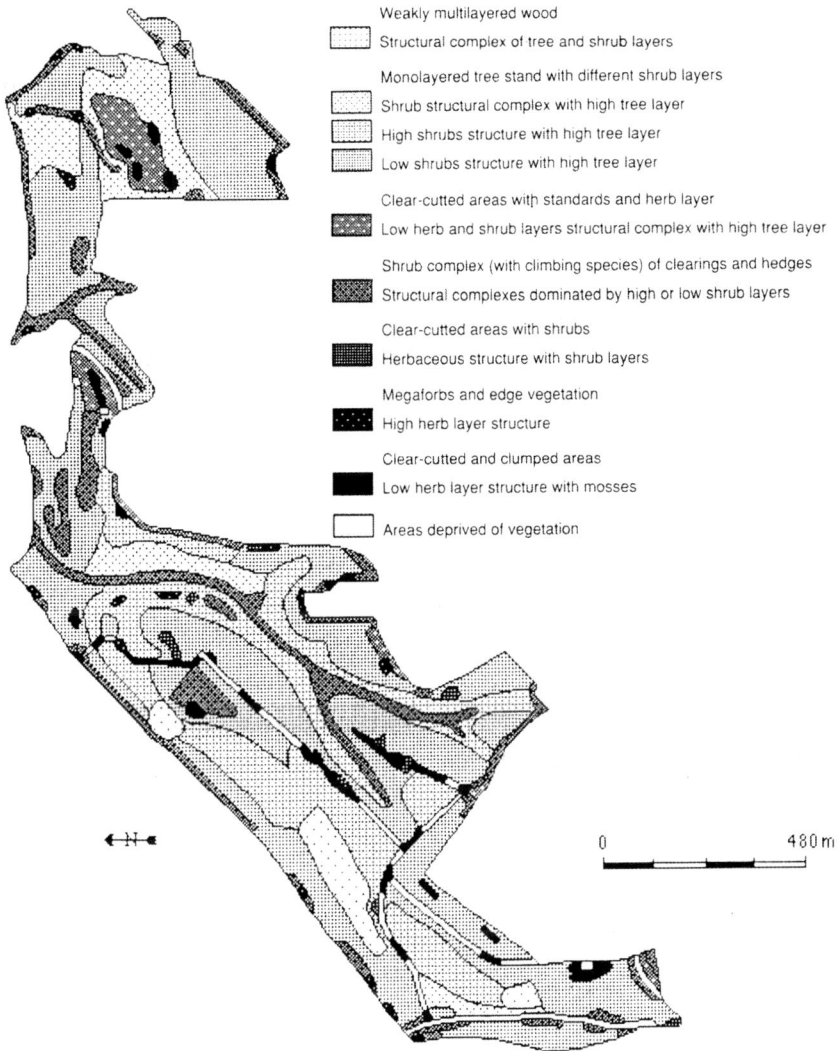

Fig. 2.23 Vegetation structure of the Fiastra Abbey forest, a relict oak-mixed coppice wood in the submediterranean hilly belt of the Marche Region, Adriatic central Italy (From Canullo 1991a)

Mapping Synusiae

The term *synusia* (plural synusiae) is Latinized Greek and refers to groups of structurally and functionally similar plant species in a vegetation stand, such as all the spring-ephemeral geophytes in a nemoral forest. (The analog for animals is guilds.) Synusiae possess a double significance: structural, as a concrete part of a phytocoenosis, and adaptive or functional, since they unite species with similar adaptations. This means species that forage for resources in a similar manner through convergence in morphology, ecophysiology and phenology (that have a comparable functional-ecological role) and that participate as common structural elements (Ivan 1979).

The synusial approach is useful especially for mapping the spatial heterogeneity of a phytocoenosis by means of the co-occurrence of species in the same microsite (Kirkpatrick 1990; Canullo and Falińska 2003).

A map of the synusial structure of the taiga at Pikhtovka in Siberia, formed by *Abies sibirica*, *Picea obovata* and *Pinus cembra* ssp. *sibirica*, at scale 1:250, is shown in Fig. 3.1. On this map it is possible to recognize seven synusiae (of order III, *sensu* Gams 1918) forming three groups: synusiae that occupy submerged depressions, flat terrain and small rises. The synusiae of the depressions are characterized by moss species (*Calliergon cordifolium*) and by tall forbs (*Filipendula ulmaria*, *Cacalia hastata*, *Geum rivale*, etc.); the synusiae of flat terrain have only tall forbs; and the synusiae of the rises are characterized by *Linnaea borealis*, *Vaccinium vitis-idaea*, *Majanthemum bifolium* and other species (Faliński and Venanzoni 1991).

The set of multiple synusiae surveyed separately constitutes a coeno-association, i.e. a complex, multi-layered association defined during the integrated synusial analysis of the vegetation by integrating the different synusiae at a site (Gillet 1988). For the mapping of coeno-associations, see Chap. 6.

F. Pedrotti, *Plant and Vegetation Mapping*, Geobotany Studies,
DOI 10.1007/978-3-642-30235-0_3, © Springer-Verlag Berlin Heidelberg 2013

O Picea obovata   ● Abies sibirica   ◎ Pinus cembra ssp. sibirica   0 ____ 5 ____ 10 m   ● Betula pubescens   o· 5–10 cm;   O·11–30 cm;   ◯·>30 cm ∅

I. Plant synusia occupying the submerged hollows:

   Hollows without plants or with simple pioneer bryophytes

   Synusium of *Calliergon cordifolium* and other hygrophilous bryophytes

   Synusium of hygrophilous megaforbs: *Caltha palustris, Geum rivale & Filipendula ulmaria* and bryophytes: *Calliergon cordifolium* etc.

II. Plant synusia occupying the flat terrain:

   Synusium of hygrophilous megaforbs: *Geum rivale, Filipendula ulmaria & Cacalia hastata*

III. Plant synusia occupying the flat terrain and hummocks:

   Synusium of *Carex macroura, Oxalis acetosella & Calamagrostis obtusata*

   Synusium of *Linnaea borealis, Vaccinium vitis-idaea, Pyrolaceae, Oxalis acetosella & Maianthemum bifolium.*

Fig. 3.1 Map of synusiae in taiga of *Abies sibirica*, *Picea obovata* and *Pinus cembra* ssp. *sibirica* at Piktovka, Novosibirsk, Siberia (From Faliński and Venanzoni 1991)

Chorological Mapping

4

Definition of Chorological Mapping

Chorological maps show the ranges (distribution areas) of species and of other taxonomic units. The purpose of chorology is to study the spatial distributions (ranges) of taxa, in isolation, that is to say without their interactions with other species. This may involve both lower taxonomic units (subspecies, varieties, etc.) and higher (genera, families, etc.).

One understands the range (*areal*) of a species to be the territory in which it is present at all locations with environmental conditions suitable for it, such as lakes and wetlands in the case of water lilies (*Nymphaea alba* L.). The definition of Pignatti (1988a) improves on this: "the dispersal range (*areal*) is the limit to which the surface occupied by the individuals of a plant species with a given dispersal power tends, in presence of the existing limiting factors". It is good to stress that "location" of a species means the geographical location where it is found, while "station" or "ecotope" is understood to mean the environment in which it grows (Pop 1977–1979; Ozenda 1982).

Ranges, or areas, can be divided into various types according to their form and extent: virtual, effective and progressive (also called actual and possible), continuous and disjunct, cosmopolitan (Fig. 4.1) and endemic (Figs. 4.2 and 4.3), etc. The study of the relations between the distribution of a species and environmental factors, between extent and form of ranges and geological history of the earth, of modifications of ranges, etc., is the specific task of the relevant phytogeography (see, for example, Zenari 1950; Cain 1951; Walter 1954; Ozenda 1982; Polunin 1967; Pignatti 1988a). Here we treat only methods of data collection and of the cartographic representation of ranges, as well as fundamental tools for analysis of phytogeographical phenomena at a local and general level.

Still it must be remembered that the ranges of different species are not independent. The different species, in most cases, have evolved together under similar changing conditions, over geological time, such that in a given territory species with very similar areas can coexist. This constitutes the premise for the

F. Pedrotti, *Plant and Vegetation Mapping*, Geobotany Studies,
DOI 10.1007/978-3-642-30235-0_4, © Springer-Verlag Berlin Heidelberg 2013

Fig. 4.1 *Poa annua*, a
cosmopolitan species

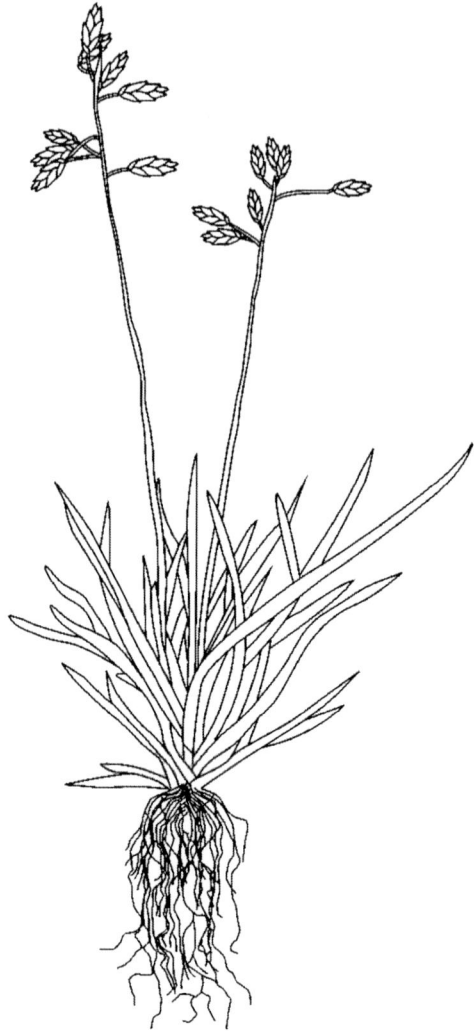

production of other geobotanical maps (phytogeographical maps or maps of
floristic subdivisions).

General Characteristics and Current Trends in Chorological Mapping

Chorological maps are topographic maps when at fine scale or geographic maps
when at broad scale, on which the distribution of species is indicated by various
graphical techniques. As will be seen below, these can be very diverse according to
the means employed.

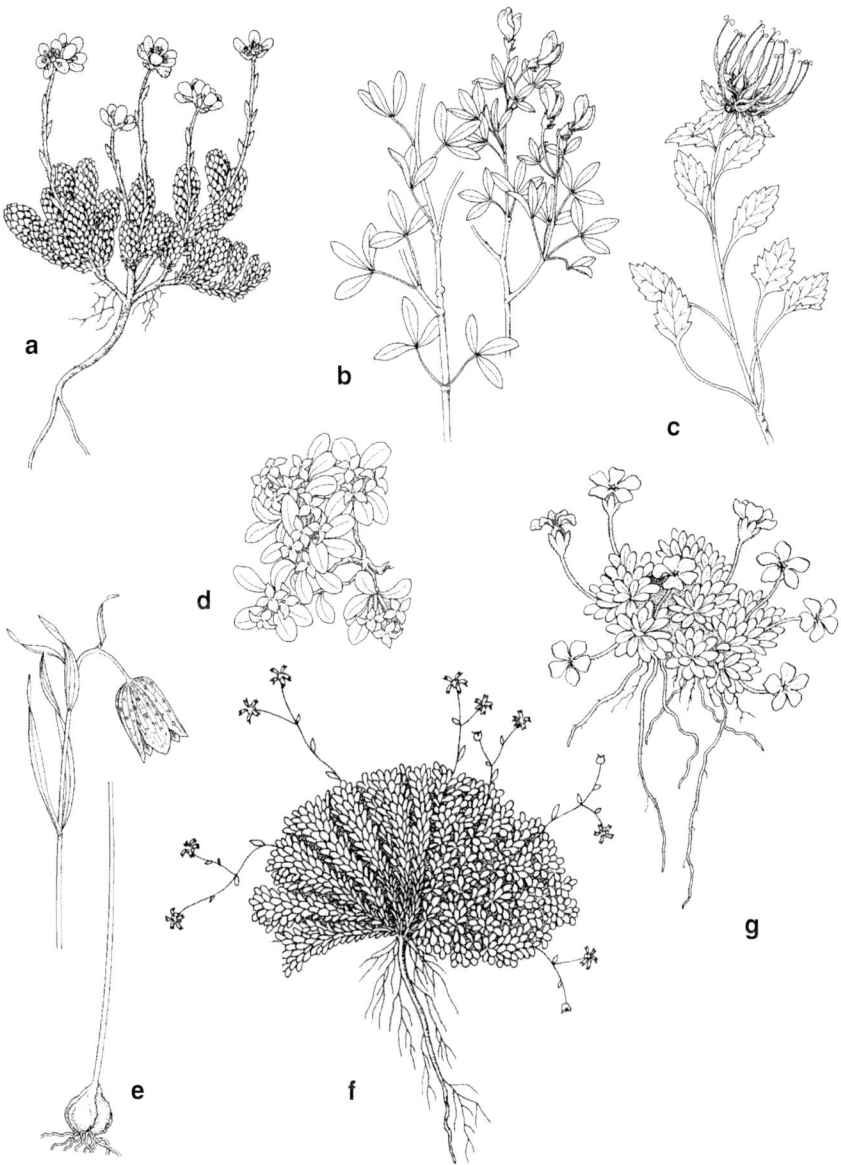

Fig. 4.2 Endemic species of the Italian Pre-Alps: (**a**) *Saxifraga tombeanensis*; (**b**) *Cytisus emeriflorus*; (**c**) *Physoplexis comosa*; (**d**) *Daphne reichsteinii*; (**e**) *Fritillaria tubaeformis*; (**f**) *Saxifraga presolanensis*; (**g**) *Androsace brevis* (From Conti et al. 1992)

Fig. 4.3 Endemic species of the Apennines: (**a**) *Aquilegia magellensis*; (**b**) *Iris marsica*; (**c**) *Centaurea scannensis*; (**d**) *Jonopsidium savianum*; (**e**) *Goniolimon italicum*; (**f**) *Astragalus aquilanus*; (**g**) *Athamanta cortiana*; (**h**) *Achillea lucana* (From Conti et al. 1992)

Fig. 4.4 Herbarium sample
of *Iris pseudo-acorus*, Piani di
Montelago, 888–916 m, lg.
Pedrotti, 14 July 1966
(CAME)

Chorological mapping involves the following main steps: verification and control of species data taken from the specialized literature of particular regions (local and general floras, floristic field guides, catalogs, contributions, other reports, etc.); verification and control of herbarium samples of the species studied (Fig. 4.4); and field sampling.

Beyond scientific interest, current methods and trends in chorological mapping come today from conservation problems: knowledge of the locations of species is essential for protection, including establishment of protected areas. Chorological data are also essential for compilation of the "red books", for whole countries or more regionally, and for evaluating plant biodiversity.

Types of Chorological Maps

Chorological maps may show part or all of the geographic range of a species. According to the detail (precision), method of representation and scale, chorological maps can be classified as follows:

(a) Location maps, indicating by symbols on a topographic map the locations where the species occurs;

(b) Grid maps, referring the locations to a determinate grid that subdivides the territory and noting in which cells the species is found at least once; normally the reference grid is the UTMG (Universal Transverse Mercator Grid);

(c) Maps of homogeneous territories, referring the species locations not to grid cells but to particular "study areas" corresponding to concrete zones of the territory that have homogeneous morphological and botanical characteristics;

(d) (choropleth) maps of territories with fixed internal boundaries, such as administrative units, indicating the chorology of the species in reference to the fixed sub-areas or regions;

(e) Absolute range maps, on which the range is shown by a closed curve that represents the absolute limit of distribution of the species (Pop 1977–1979); the area delimited can be left white or colored;

(f) Quantified chorological maps, based not only on presence/absence or other a qualitative criteria but also on a quantitative valuation of one or more characters, such as chorological elements, growth form, or ecological indices.

Location Maps

Location maps (or floristic maps) are analytical maps that show the locations where a given species is present. These may be made for documentation of other research, such as systematic revisions of species and genera, or karyological or cytotaxonomical research; revisions of the presence of a species in a given geographical area, or new findings; for compilation of chorological atlases of an entire country or of more limited territories, such as regions or provinces; for compilation of red books and atlases of rare or threatened species; for producing atlases of protected areas; or perhaps many other reasons.

Floristic maps are made by indicating with a symbol the locations where a species occurs on a topographic map. The territory studied may be that of a region (Fig. 4.5), a state (Fig. 4.6) or other territory under consideration. The chorological maps included in red books of the plants of Poland, Bulgaria, China and other countries have been produced using this criterion (Bulgarian Academy Science 1984; Fu Li-kuo 1992; Polish Academy Science 2001, etc.). When the data are available, it is possible to enrich these maps with the addition of other kinds of information, such as historical characters (distinguishing the locations

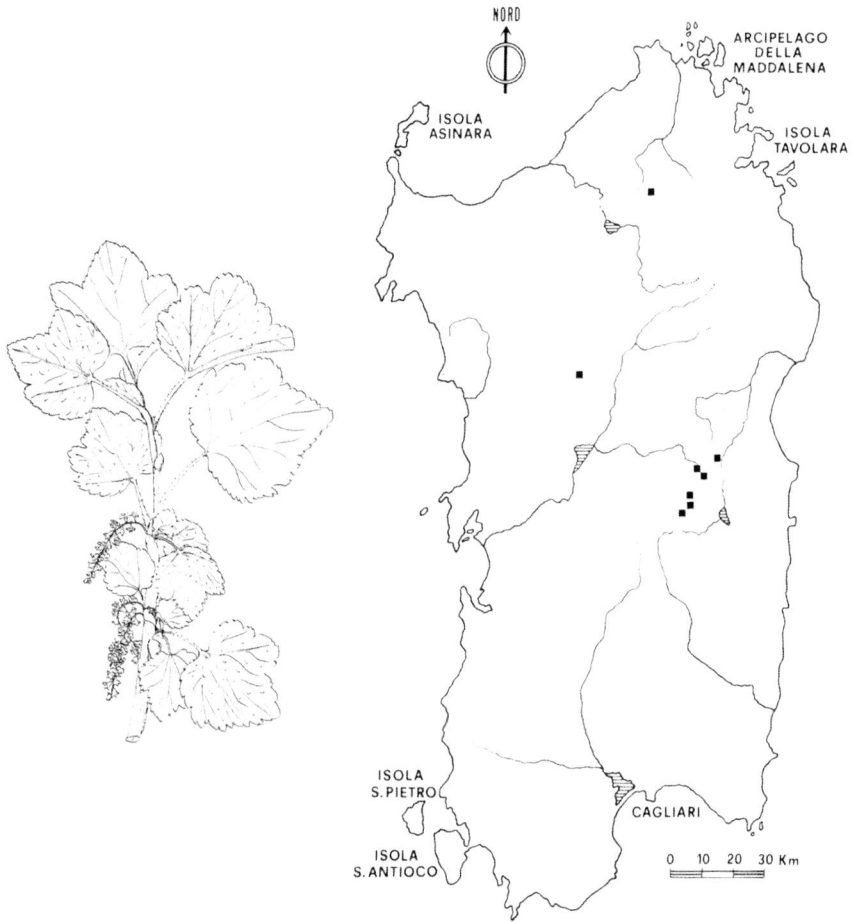

Fig. 4.5 Distribution of *Ribes sandalioticum*, an endemic species of Sardinia (From Arrigoni 1981)

according to the years when the species was found), ecological characters (site type and type of vegetation that the species characterizes), quantitative characters (number of individuals present), biological characters (fertile individuals, only growing vegetatively, etc.). It is also possible to map variations of the ranges of certain species due to anthropogenic causes, for example the progressive expansion of ranges through invasions into new territories (neophytism) (Fig. 4.7). In the same manner it is also possible to map range contraction for species undergoing progressive decline caused by changes in environmental conditions (Figs. 4.8 and 4.9).

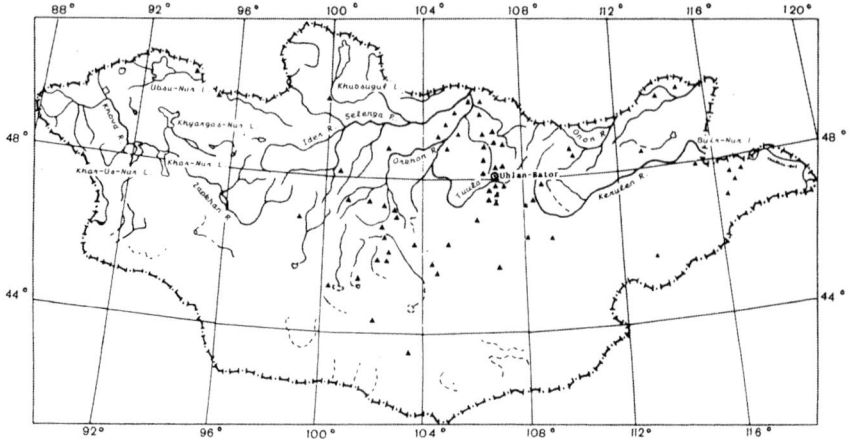

Fig. 4.6 Distribution of *Cymbaria dahurica* in Mongolia (From Karamysheva and Khramtsov 1995)

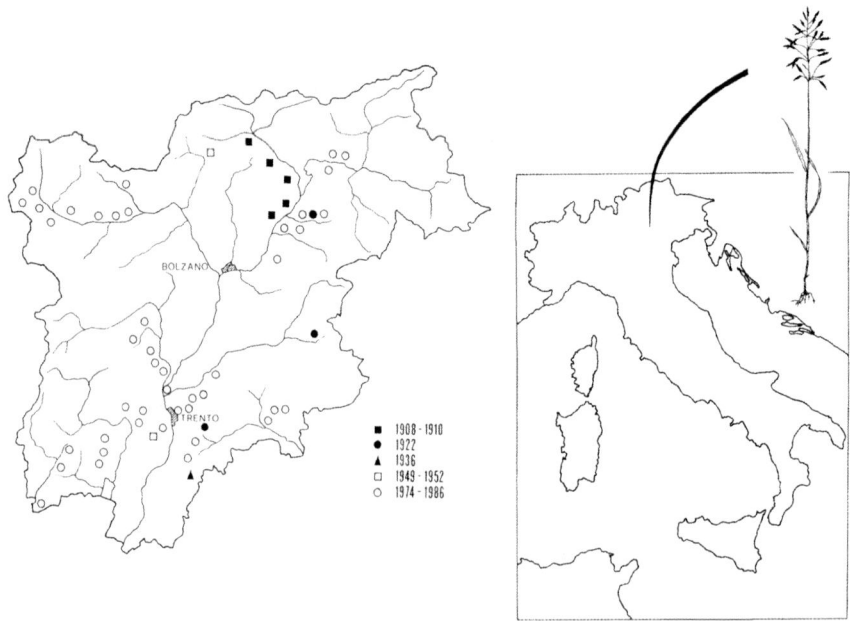

Fig. 4.7 Distribution of *Bromus inermis* in the Trentino-Alto Adige Region, northern Italy, with years of surveys; at *left* the arrival path of *Bromus inermis*, from eastern Europe at the beginning of the twentieth century (From Pedrotti 1987, 1999b)

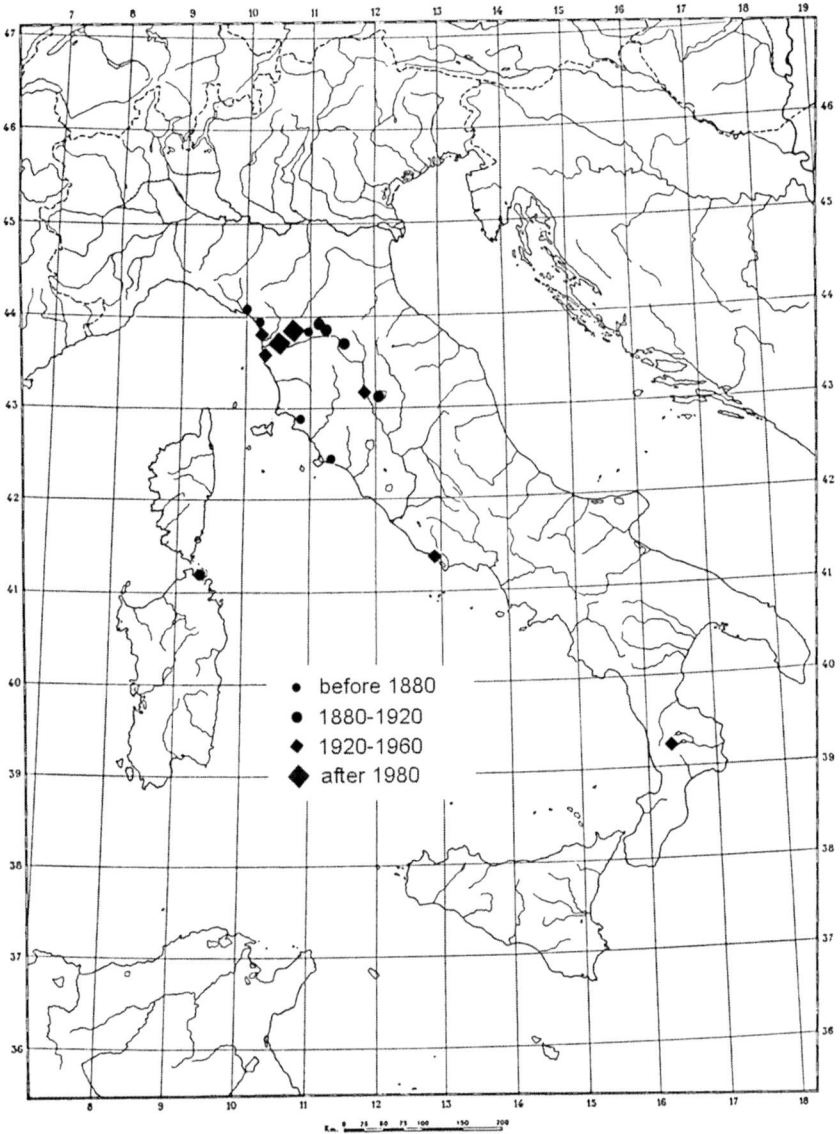

Fig. 4.8 Distribution of *Ludwigia palustris* in peninsular Italy in different years (From Moggi 1992)

Grid Maps

Grid maps are more synthetic than the previous, and it is possible to make general maps of complete or only partial species ranges. The atlases produced by this methodology may be for a continent, such as the atlas of the flora of Europe

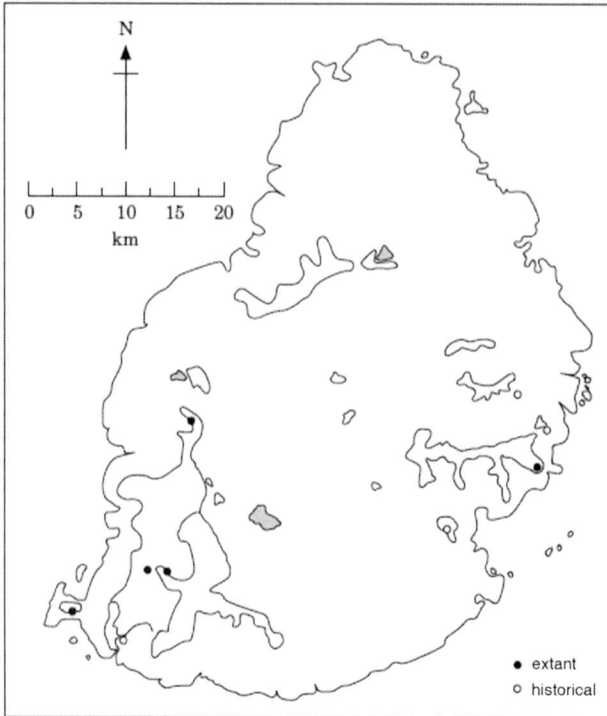

Fig. 4.9 Map of Mauritius Island showing the current and historical distributions of *Coffea myrtifolia*. This species grow in evergreen dry forests (From Dulloo et al. 1999)

(Fig. 4.10); for an entire country, such as the atlas of Grat Britain (Fig. 4.11); for an administrative region, such as the atlas of Friuli (Poldini 1991, 2002), which gives maps of 2,780 species (Fig. 4.12); for a particular geographic region, such as the River Bug in eastern Poland, with smaller maps of 1,123 species (Fig. 4.13); or for a protected area, such as the nature reserve of Zingaro in Sicily (Fig. 4.14) or for the city of Monte Sant'Angelo on the Gargano Promontory (Fig. 4.15). There is also an atlas of France, but it is limited to species from the red book (Olivier et al. 1995).

One must also note immediately that use of a grid results in a loss of information, because multiple locations can sometimes fall into the same grid cell, as can be seen by comparing maps of the distribution of *Ephedra major* Host in Italy, by individual locations and by UTM grid cells (Fig. 4.16). In some cases it is possible to remedy this by progressively subdividing the grid cells. Of course the precision of cartographic representation increases with the decreased size of the grid cells.

The use of a grid also has the advantage that it becomes possible to process the chorological data by numerical analysis of the spatial distribution, as for example

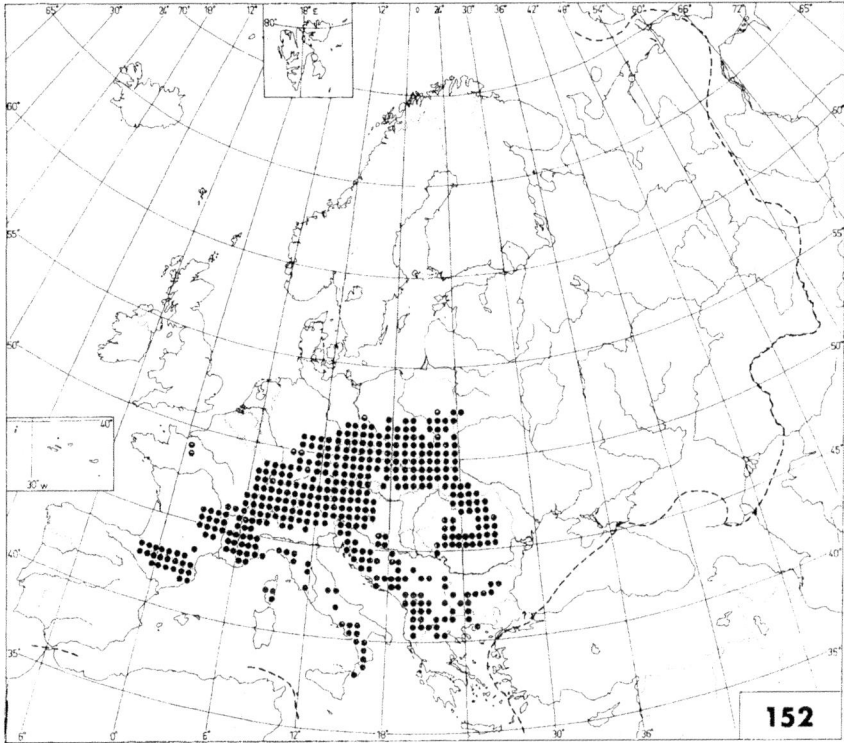

Fig. 4.10 Distribution of *Abies alba* in Europe, based on the UTM network (From Jalas and Suominen 1972)

with the study of floristic gradients along a profile, the counting of different species in different grid cells (relative and absolute floristic richness), the distribution of species and analysis of vegetation (Pignatti 1978; Géhu 1984) and so on.

Maps of Homogeneous Areas

This is the methodology used for the atlas of the Swiss flora (Welten and Sutter 1982), in which 2,855 species were mapped, each also with indication, by various symbols, of abundance in two levels and the elevation and type of environment according to the following typology: calcareous rocks and rubble, rocks and rubble poor or lacking in carbonates, broad-leaved forests (*sensu lato*), needle-leaved forests (*sensu lato*), dwarf shrubs, floodplain forests, raised bogs, low bogs, non-manured natural meadows, manured and mowed meadows plus artificial meadows, gardens, fields and rural environments (Fig. 4.17). These maps also have a more synthetic character than do location maps alone.

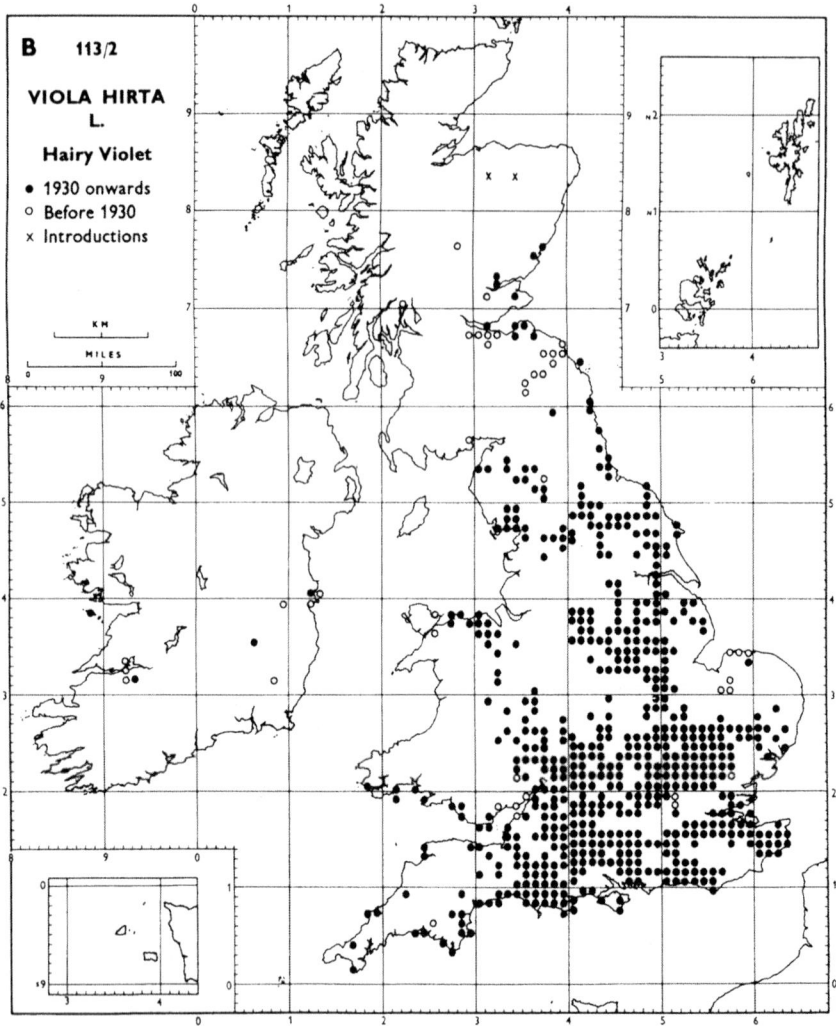

Fig. 4.11 Distribution of *Viola hirta* in Great Britain (From Perring and Walters 1962)

Maps with Fixed Internal Boundaries (e.g. Administrative Units)

Maps with fixed internal boundaries, each unit of which has a single value, are called choropleth maps. These are chorological maps on which the occurrence locations of a species are shown relative to fixed internal boundaries (choropleth maps), such as administrative units, as in the distribution maps included with the *Flora d'Italia* of Pignatti (1982a) which were constructed to indicate the presence of individual species in the administrative regions of Italy (Fig. 4.18). Analogous is the criterion adopted by Cortini Pedrotti (2001) for the distribution of mosses in

Huperzia selago

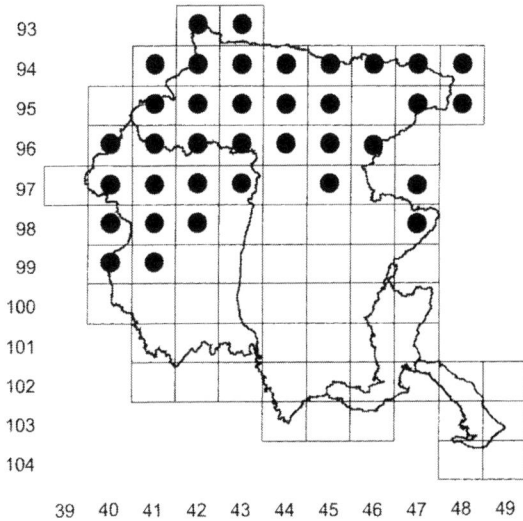

Fig. 4.12 Distribution of *Huperzia selago* in the Friuli-Venezia Giulia Region, northeastern Italy (From Poldini 1991)

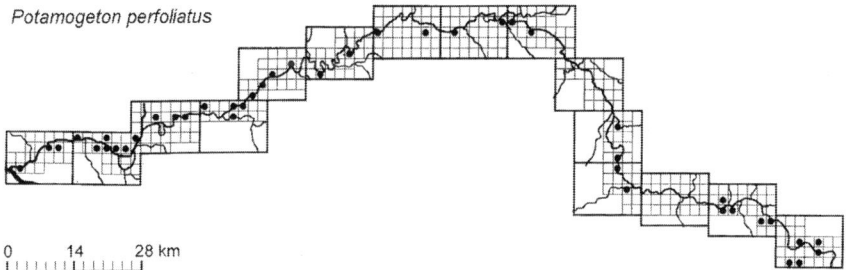

Fig. 4.13 Distribution of *Ranunculus lingua* and *Potamogeton perfoliatus* along the Bug River, Poland; occurrences ware observed in 363 square plots of 4 km^2 (2 × 2 km) (From Cwiklinski and Glowacki 2000)

Fig. 4.14 Distribution of
Limonium flagellare in the
Zingaro Natural Reserve,
Sicily (From Raimondo and
Schicchi 1998)

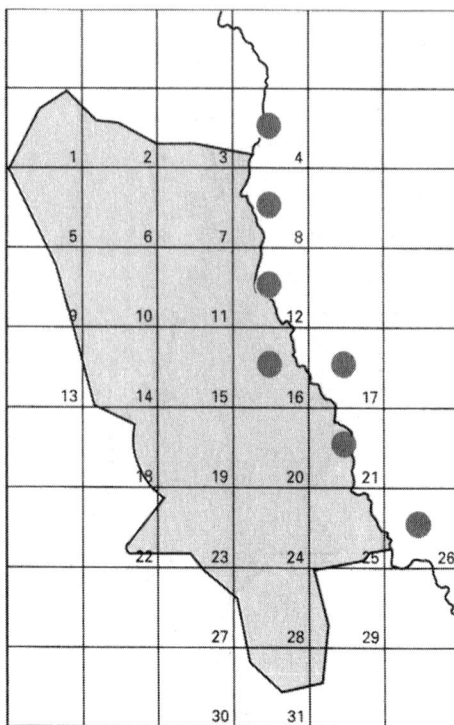

Italy. In the Flora of the Carolinas (USA) (Radford et al. 1968), the distribution of
the species is indicated for the individual counties (Fig. 4.19).

The range maps of species of the Alpine flora by Aeschimann et al. (2004) are
also based on administrative subdivisions; these maps were made by coloring the
subdivisions within which a given species occurs (Fig. 4.20).

Range Maps

Maps of absolute distributional limits (also called phytogeographical maps) are
generally synthetic maps at broad scale that show the ranges of species and related
biogeographical phenomena (disjunctions, vicariances, geographic relicts, etc.).
The floristic (location) maps first described furnish the data for producing range
maps. The atlas of comparative chorology by Meusel et al. (1965, 1978, 1992)
shows typical examples of range maps (Fig. 4.21). In some cases it is possible to
study in detail only a part of a species range, as Marchesoni (1958, 1959) did for
Pinus cembra L.; the distributional limits were analyzed in the basin of the Adige
River (Fig. 4.22), and a general schematic map was also provided, showing the full
range of the species and the range of a related, vicariant species, the Siberian pine

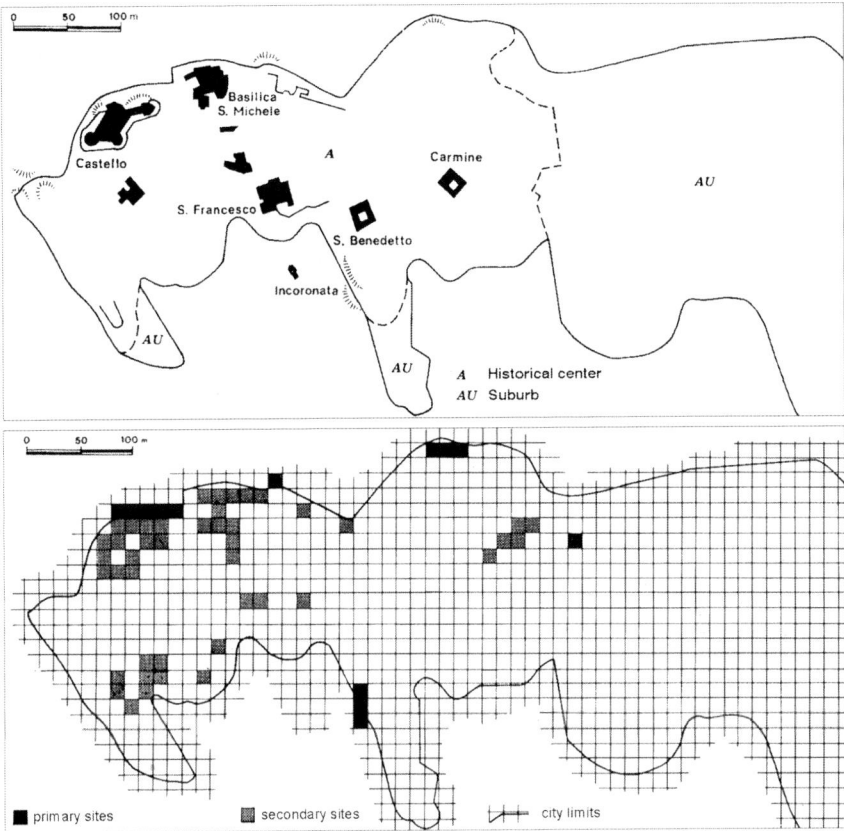

Fig. 4.15 Distribution of *Campanula garganica* in the city Monte San Angelo, at Mt Gargano, southern Adriatic Italy. The main occurrences are on cliffs inside the city, secondary occurrences are on ancient buildings and walls in the historic center; the species has not yet arrived in the suburban area with new buildings (From Pedrotti 1988d)

(*Pinus sibirica*). Distribution maps of other tree species of Trentino-Alto Adige Region of particular phytogeographical and ecological significance (Fig. 4.23) were also made with the same criteria adopted for *Pinus cembra* (Gafta and Pedrotti 1998).

It is also possible to make chorological maps with mixed characteristics, i.e. that show both the species range by a continuous closed curve and the actual locations where the species occurs, as on the map of *Primula tyrolensis*, a species endemic to the Dolomiti mountains (Fig. 4.24). In other cases, a mixed criterion has been adopted, with points used when the species occurrence is concentrated spatially and areas when the species occurs commonly over larger areas (Fig. 4.25), the latter made distinct through different intensities of hatching for the two degrees of frequency (Hultén 1964, 1971).

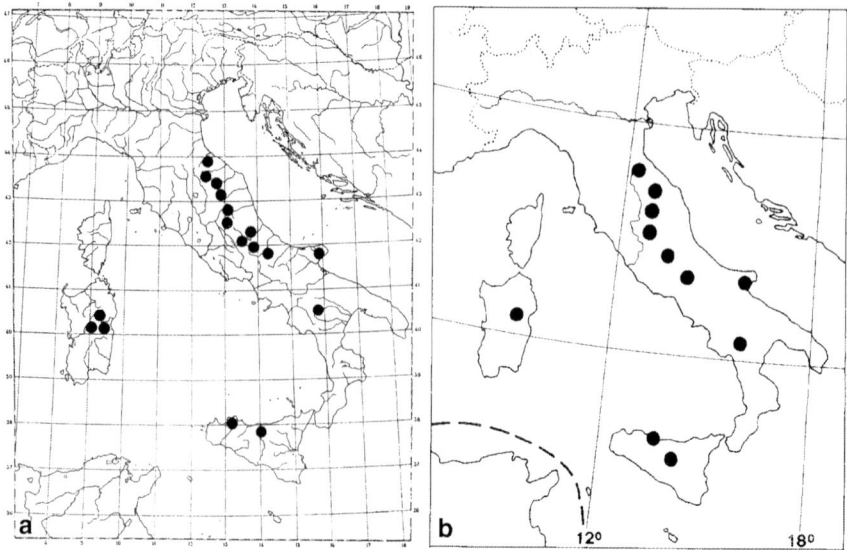

Fig. 4.16 (**a**) Distribution in Italy of *Ephedra major*; each *black point* showing one occurrence location; (**b**) distribution with UTM network, with loss of information because some sites fall into the same cell of the UTM grid (From Orsomando 1969 and Pedrotti 1983)

Quantified Chorological Maps

These are synthetic maps showing both chorological and phytoecological characteristics.

Quantified chorological maps that show the joint distribution of a set of characters (taxa, chorological elements, growth form, ecological indices, etc.) attributed to fixed areas into which the study area has been subdivided. For example the map by Zangheri (1959) shows the density per unit area of east-Balkan species in Italy.

These areas were called Unità Geografiche Operazionali (OGUs) by Nimis and Bolognini (1990, 1993) and by Bolognini and Nimis (1993). The maps (quantified chorological or chorogram maps) produced by these authors derive from matrices of OGUs and some characters, and thus may be submitted to multivariate analysis. By using an "area index" it is possible to express the degree of territorial expansion of a given group of species, for example the species of a beechwood, a deciduous oak forest, or an evergreen oakwood (*Quercus ilex*). The chorograms obtained give a synthetic view of the phytogeographical affinity of each local flora, and through further processing it is possible to show the existence of groups of species with similar distributions (Fig. 4.26).

The maps by Cortini Pedrotti (1996) of the mosses of Italy are also quantified chorological maps, because they take into account for each individual region a group of species defined by temperate, boreal, arctic-alpine and Mediterranean

Mapped areas and numbers

Huperzia selago (L.) Bernh. ex Schrank & Mart.
Lycopodium selago L.

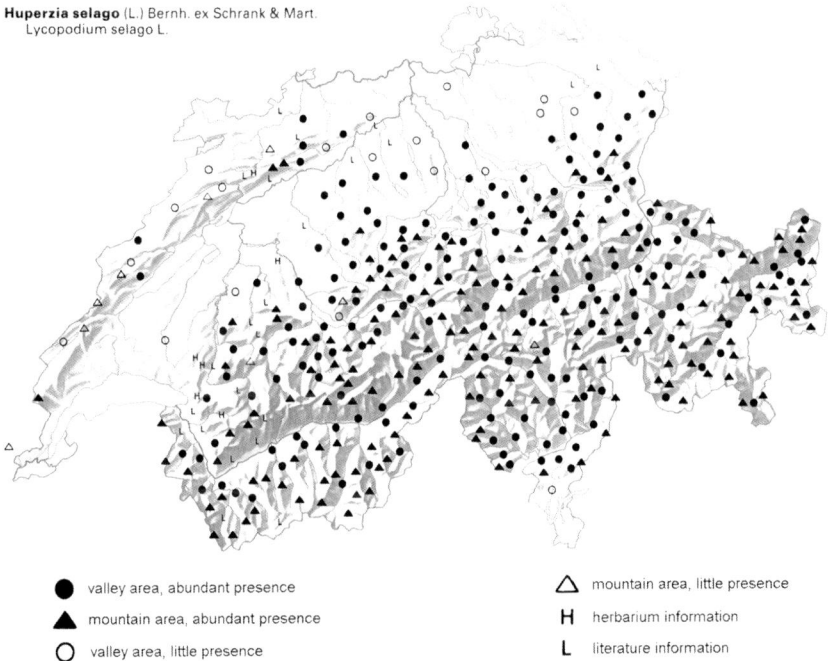

●	valley area, abundant presence	△	mountain area, little presence
▲	mountain area, abundant presence	H	herbarium information
○	valley area, little presence	L	literature information

Fig. 4.17 At the *top*: survey surfaces employed for the atlas of the flora of Switzerland; *below*: distribution of *Huperzia selago* (From Welten and Sutter 1982)

geoelements. These are quantified based on their proportions in the total moss flora of the region (Fig. 4.27).

Fig. 4.18 Distributions in Italy of *Oxyria digyna*, *Thalictrum minus* and *Saxifraga tombeanensis* (From Pignatti 1982a)

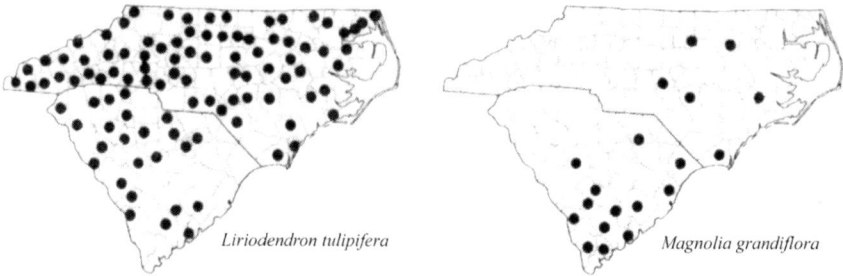

Fig. 4.19 Distributions of *Liriodendron tulipifera* and *Magnolia grandiflora* in North and South Carolina, U.S.A. (From Radford et al. 1968)

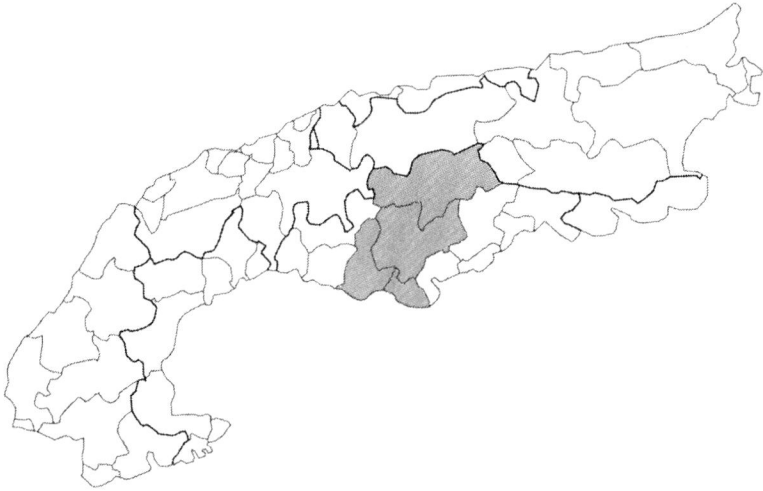

Fig. 4.20 Distribution in the Alps of *Saxifraga tombeanensis*, an endemic species of the Trentino and Lombardia Pre-Alps (From Aeschimann et al. 2004)

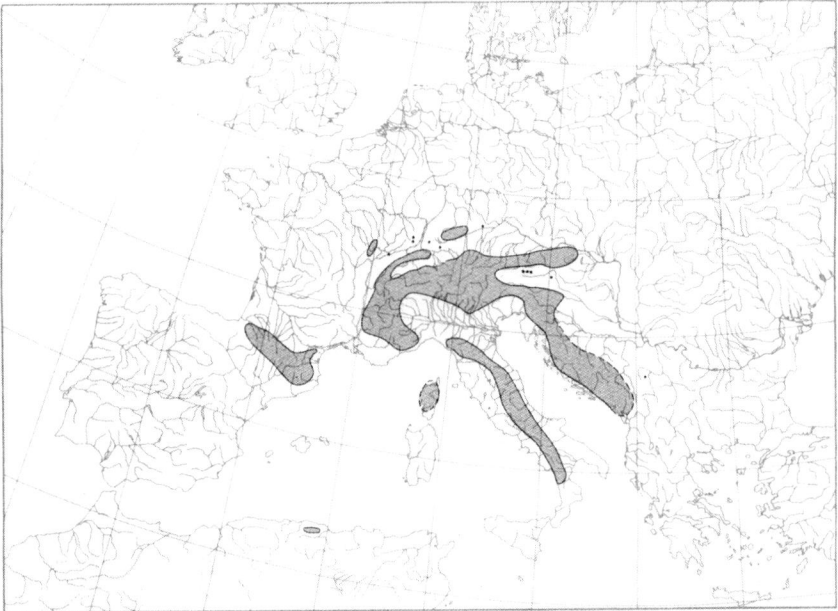

Fig. 4.21 Range of *Hieracium humile* (From Meusel and Jäger 1992)

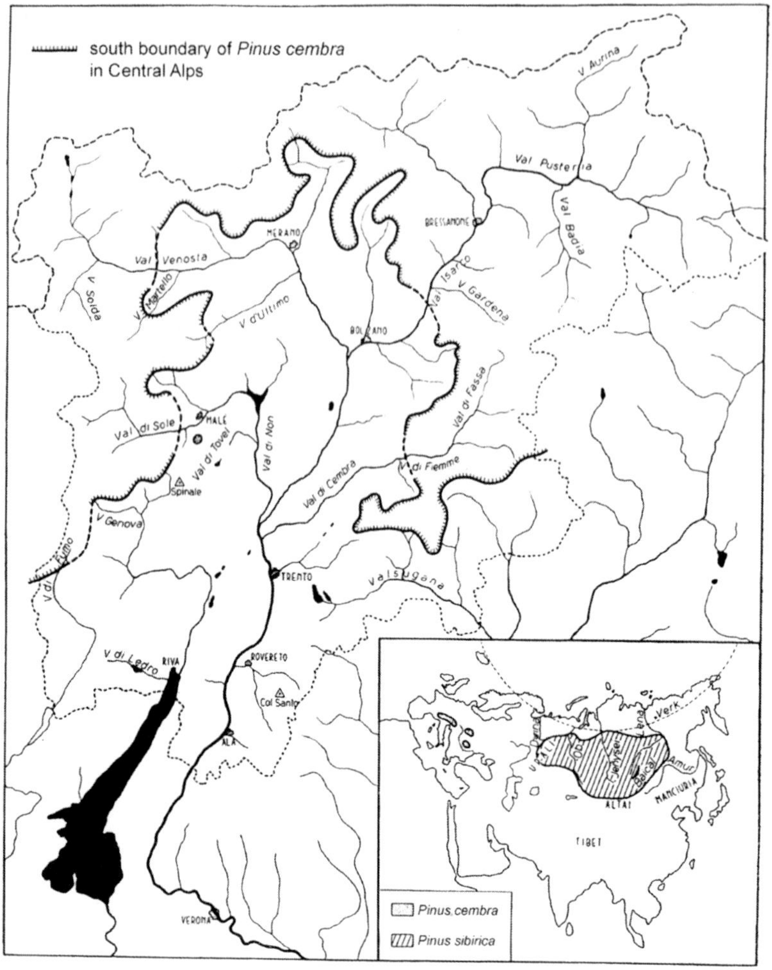

Fig. 4.22 At *lower right*: distributions of *Pinus cembra* and *Pinus cembra* ssp. *sibirica* (comple range); at *left*: the southern limit of the *Pinus cembra* range in the Trentino-Alto Adige Region (partial range) (From Marchesoni 1959)

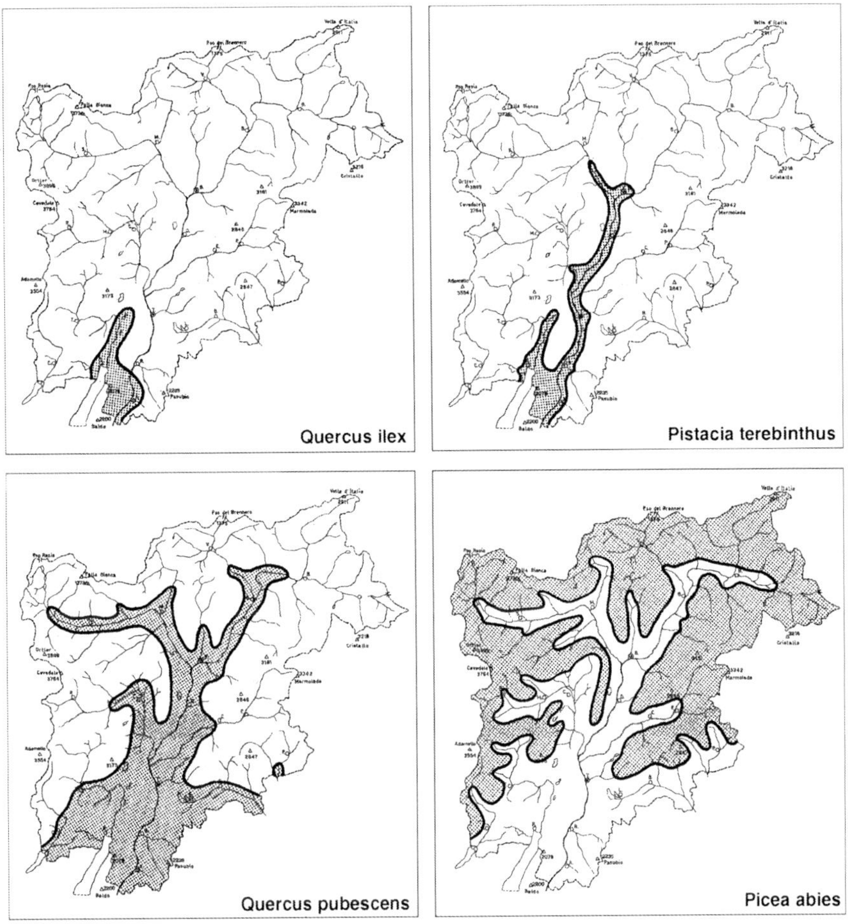

Fig. 4.23 Distribution maps of tree species in the Trentino-Alto Adige Region, partial ranges (From Gafta and Pedrotti 1998)

Fig. 4.24 Range of *Primula tyrolensis*, with numbers *1–9* showing occurrence locations (From Tomaselli 1955)

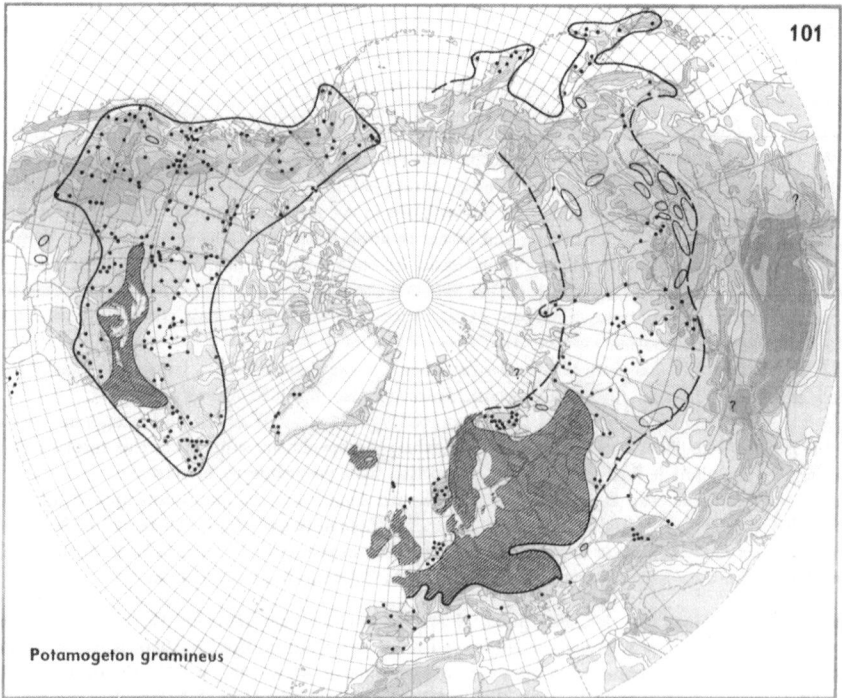

Fig. 4.25 Range of *Potamogeton gramineus* (From Hultén 1964)

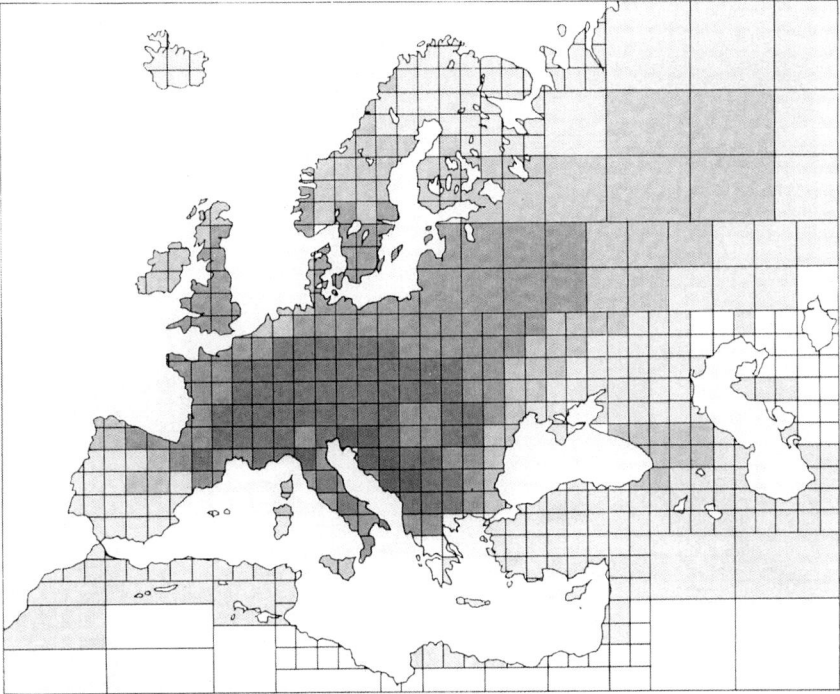

Fig. 4.26 Chorogram maps of the joint distribution in Europe of 212 nemoral species of Italian *Fagus sylvatica* forests (From Nimis and Bolognini 1990)

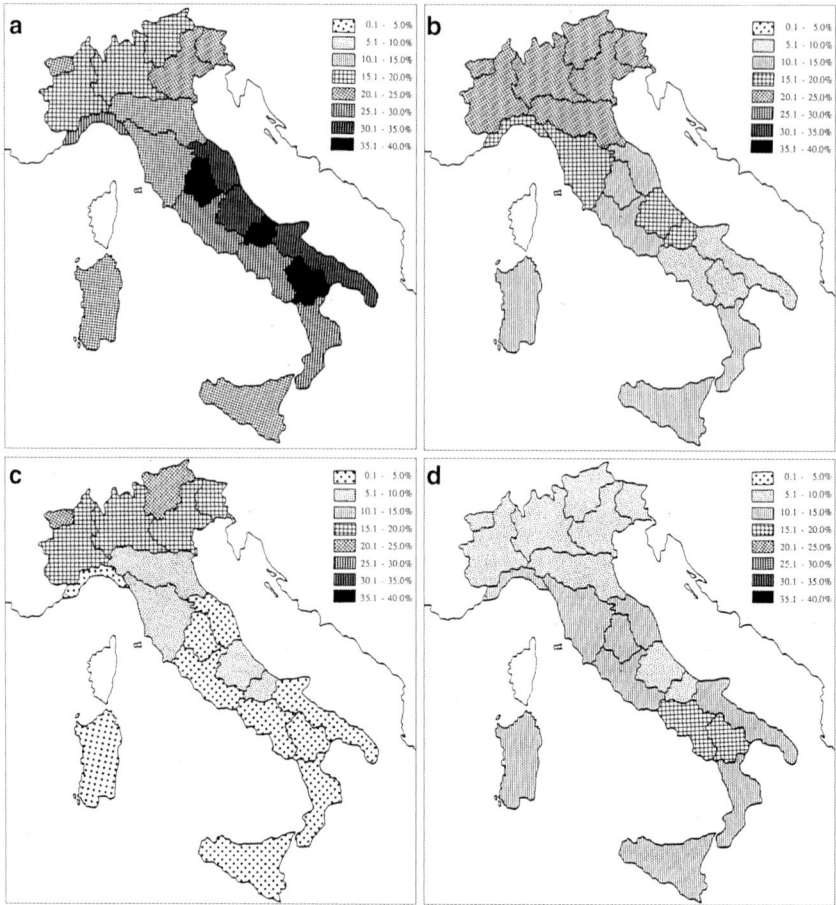

Fig. 4.27 Distribution in the Italian regions of the main chorological *Musci* elements, by frequency (classes of 5%); (**a**) temperate elements; (**b**) boreal elements; (**c**) arctic-alpine elements; (**d**) Mediterranean elements (From Cortini Pedrotti 1996)

Mapping Vegetation

5

Definition of Vegetation Mapping

A vegetation map consists of a topographic base map that shows vegetation units by appropriate colors and symbols; a vegetation map showing the concrete units that form the vegetation is based on concrete phytocoenoses.

Producing a vegetation map constitutes the last step in a cognitive process that starts with the sampling of vegetation in the field and proceeds with definition of vegetation types by recognition and classification of plant associations, or other units according to the different geobotanical schools.

Trends in Vegetation Mapping

Vegetation mapping constitutes an especially effective way to process spatial data describing the plant cover. Vegetation maps, moreover, are tools for the interpretation and description of the environment because they also contain information on soils, climate and many other ecological factors.

The main themes addressed today by vegetation mapping include such topics as: mapping still unmapped areas; new methodologies for cartographic data collection (such as imagery from satellites); mapping by Geographic Information Systems (GIS); use of ecological data on maps; mapping environments; ways of understanding biodiversity at the level of plant communities; and mapping for applied problems, such as conservation of biodiversity, planning and management of protected areas, land-use planning, agro-silvo-pastoral problems, environmental impact analysis, and mapping and evaluation of habitats.

F. Pedrotti, *Plant and Vegetation Mapping*, Geobotany Studies,
DOI 10.1007/978-3-642-30235-0_5, © Springer-Verlag Berlin Heidelberg 2013

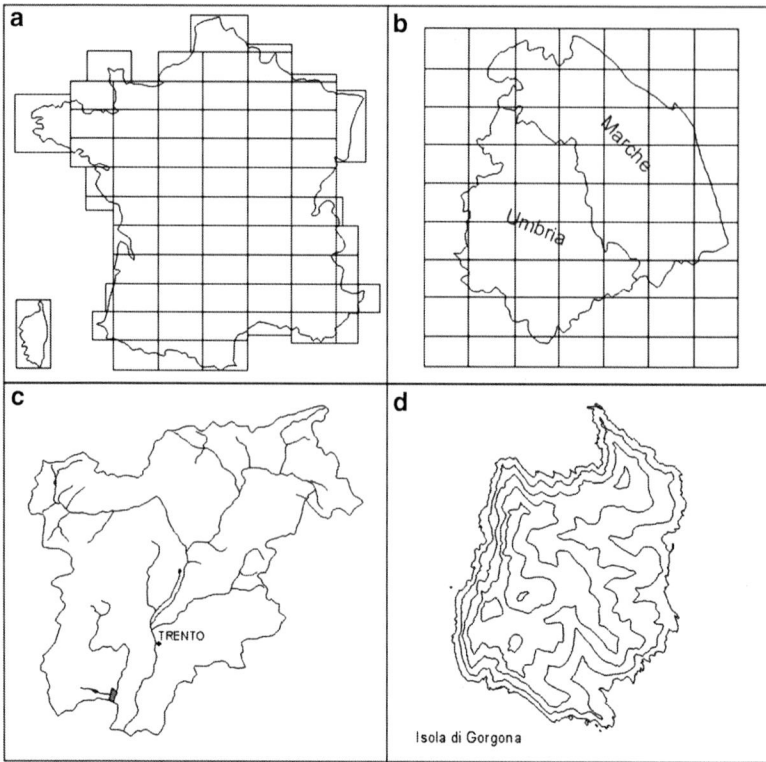

Fig. 5.1 Map format; (**a**) network employed for dividing the territory of France; (**b**) the network, at scale 1:50,000, from the Istituto Geografico Militare (IGM) in the regions Marche and Umbria; (**c**) example of a territory limited by administrative borders, the Trentino-Alto Adige Region; (**d**) example of a territory limited by geographical (physical) borders, Gorgona Island, Tuscany, central Italy (a – from Rey 1988; b – from Orsomando and Pedrotti 1992)

General Characteristics of Vegetation Maps

Format. A map has a definite format, according to the extent of the territory that it represents. In the case of monographs on a given area (a valley, an island, or such), the mapped area corresponds to that studied and thus is delimited by a fixed geographical limit. When the map is for a community, a province or a region (as fixed administrative units), the limits of the map are fixed administratively. If the territory is larger, it may be subdivided into smaller parcels, normally rectangular (Fig. 5.1), in which the vegetation is sampled completely, as for the vegetation of France at the scale 1:200,000 (Rey 1988) or the (administrative) regions Marche and Umbria at the scale 1:50,000 (Orsomando and Pedrotti 1992).

Topographic Basis. The mapping basis is provided by a topographic base map, onto which the vegetation is mapped. The quality of the topographic basis is very important, both for representing variations of the vegetation in relation to

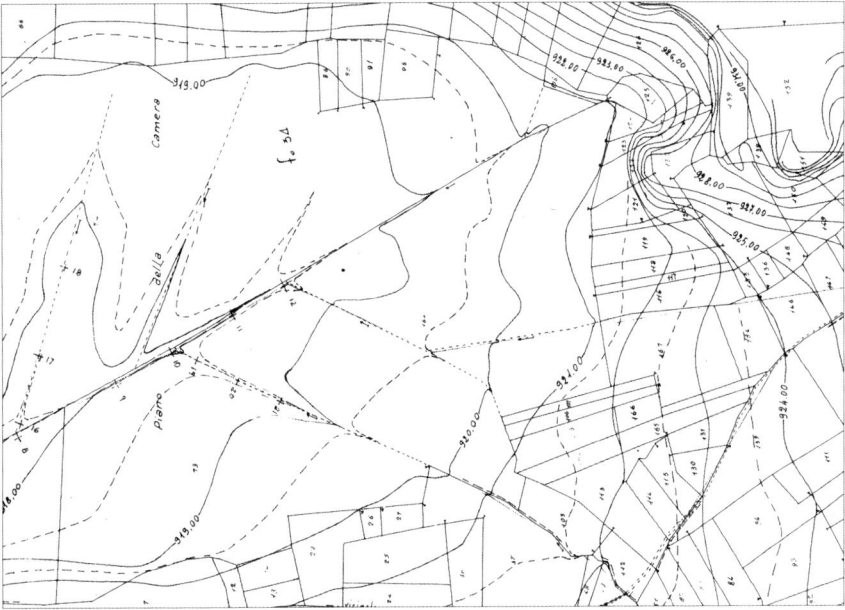

Fig. 5.2 Cadastral map, at 1:2,000, of the Piani di Montelago (upper-basin plain), in the Marche Region, central Italy; the map shows the parcels and channels, *arrows* showing the flow direction of the. This map has *contour lines*, at scale 1:2,000, re-sized at 1:5,000 for printing

morphology, orography, hydrography and other environmental characteristics, and for a greater ability to read the map. As base maps for sampling the vegetation, one can use cadastral maps, the official topographic maps of the various countries (e.g. of the IGM, Istituto Geografico Militare, in Italy) and regional technical maps. It may also be possible to obtain specially made topographic maps. Cadastral maps are always at a very fine scale (between 1:1,000 and 1:4,000) and contain many topographic elements easily recognizable as "landmarks", corresponding to which stones of particular form may be placed in the field, well fixed in the soil. Such landmarks may be small waterways, property limits, springs, roads, houses, etc. It is also possible to superimpose on cadastral maps the isometric curves (contour levels) surveyed with instruments in the field or by photogrammetry, in order to make the maps richer in cartographic elements richer and thus more easily legible and more useful in subsequent mapping (Fig. 5.2). Today all these and other operations are done much more easily by GPS, as will be described later in this chapter. By using cadastral maps it is possible to carry out more detailed vegetation surveys, always at fine scale but in areas of limited extent. The topographic maps of the IGM in Italy are at the scale 1:25,000 (*tavolette*, today called *sezioni*) (Fig. 5.45), 1:50,000 (*quadranti*, today called sheets) and 1:100,000 (sheets, no longer produced). Another possibility in Italy is provided by topographic maps, normally at the scale 1:10,000, produced in recent years by the *Regioni* and by some provinces, by means of photorestitution, using aerial photographs based on

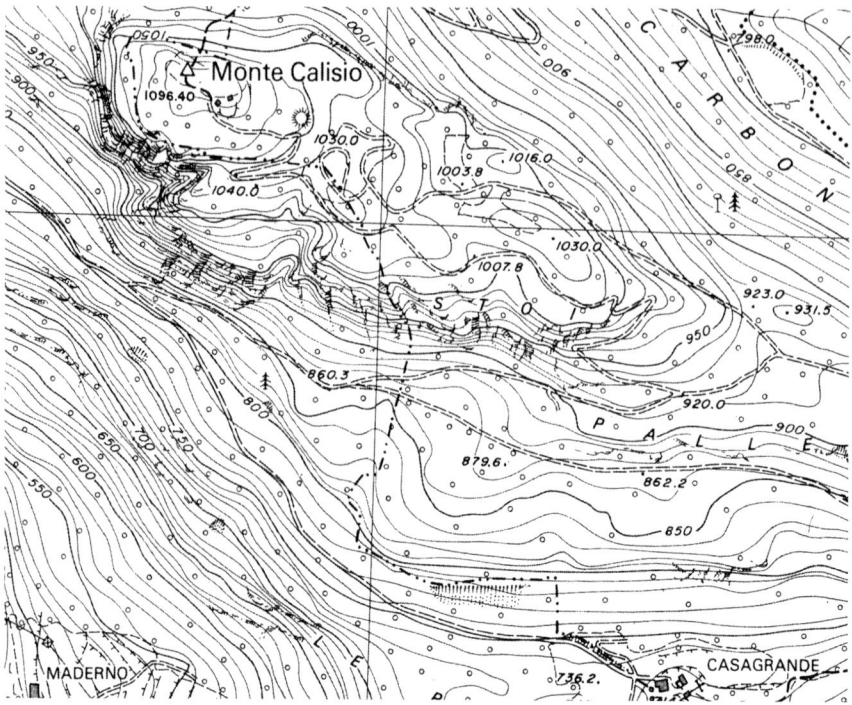

Fig. 5.3 General topograhic map of Trento Province, at scale 1:10,000, showing the area of Mt. Calisio (From *P.A.T. Trento, n. 060100*)

predetermined flight plans (*carte tecniche regionali*) (Fig. 5.3). Many territorial entities, such as the *Provincia Autonoma di Trento*, have now started producing computerized topographic maps, always at the scale 1:10,000 (Polli and Sala 2003), which are also very useful for data collection and vegetation mapping, as well as other processing.

Digital terrain model. A digital terrain model (DTM), or digital elevation model (DEM), is a three-dimensional model of data that represents the elevation of the terrain. This can be constructed from mapped contour levels, and permits three-dimensional information management. These can be used to represent the vegetation in relation to slope and exposition (aspect). Three-dimensional mapping is a simulation of topography. It can be employed also to represent the distribution of vegetation as related to altitude, exposition and slope, i.e. in relation to the geomorphology of the area studied.

Shaded relief maps. By using a digital terrain model it is often possible to derive a map that emphasizes the topographic relief by showing it with the shade that would be produced if light were coming from a particular direction. This provides a pseudo-three-dimensional view of the target territory and, in many cases, can be useful for the representation of the vegetation (Fig. 5.4).

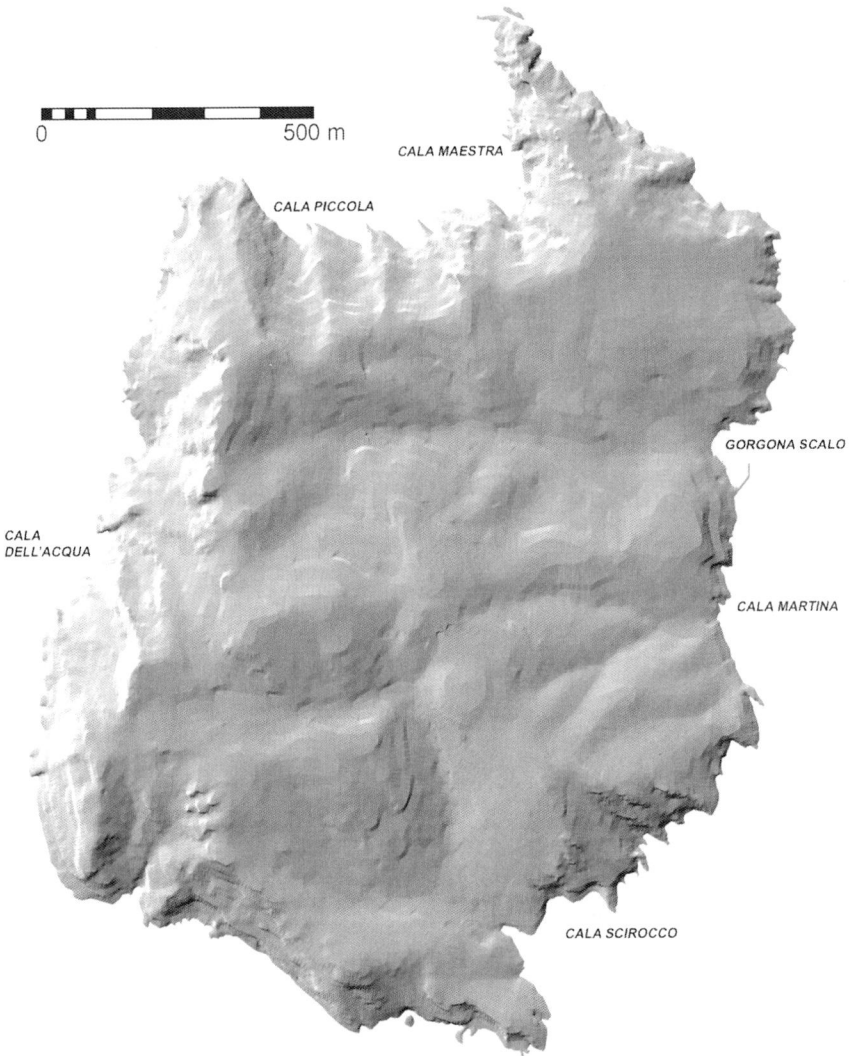

Fig. 5.4 Shaded relief maps of Gorgona Island, Tuscany, central Italy

Vegetation units. These are theoretical units that correspond to vegetation types, which may be distinct syntaxonomically, phytogeographically, dynamically, ecologically, etc., and whose definitions appear in the map legend. An example of a vegetation unit defined by phytosociolgcical nomenclature might be a *Salicetum albae*, a plant association characterized by the presence, often abundantly, of the willow *Salix alba*, which forms willow woods along water courses throughout most of Europe (Fig. 5.5).

Cartographic units. In the field vegetation units are concrete, i.e. elementary phytocoenoses or vegetation stands or individuals of an association, which can be

Fig. 5.5 A *Salicetum albae* along the Sangro River between Pescasseroli and Opi, in the Abruzzo Region of central Italy (Photo Franco Pedrotti)

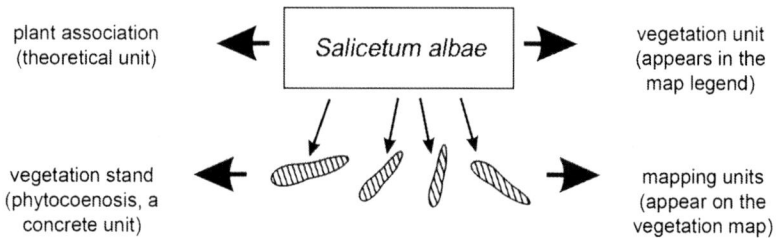

Fig. 5.6 Conceptual differences between vegetation units and cartographic units

represented cartographically; these are the *cartographic units* (Fig. 5.6). It is important to note that plant associations are a theoretical concept but exist in nature as concrete units which have first been identified and named as phytocoenoses. This was well expressed by Ivan and Doniœå (1975) when they wrote that "the plant association represents the statistical model of phytocoenoses that are similar in composition and structure". For such reasons, a plant association possesses both a concrete and an abstract character at the same time (Cristea et al. 2004).

Mosaics and interpenetrations. A mosaic is a local vegetation complex of two or more different phytocoenoses occurring in such a way as to appear like a conglomerate (according to Braun-Blanquet 1964) or a mutual interpenetration (according to other authors) of each phytocoenosis into the other(s). The types are in contact with each other and cannot be reproduced separately, in detail, on a map at the scale selected. Often it is the micro-relief of the substrate or soil patchiness that produces the multiple local phytocoenoses, forming what has come to be called a "mosaic". A mosaic can be mapped as a separate cartographic unit (*cartographic mosaic*),

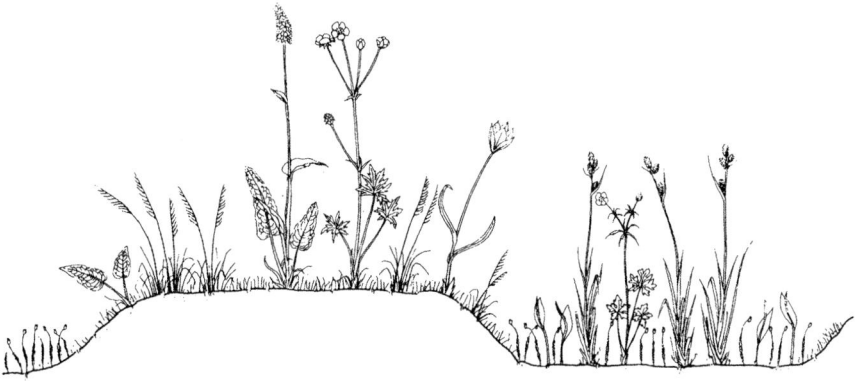

Fig. 5.7 Mosaic of *Polygono-Nardetum* (at *left*, on at dirt clod) and of *Caricetum gracilis buxbaumietosum* (at *right*, on the flat area), in the Pian Grande di Castelluccio di Norcia (plain), in the Umbria Region, central Italy. On the clod are: *Nardus stricta*, *Polygonum bistorta*, *Ranunculus acer*, *Tulipa australis* and *Aulacomnium paluste*; on the flat area are *Ophioglossum vulgatum*, *Carex buxbaumii*, *Ranunculus pedrottii* and *Polytrichum commune* ssp. *commune* (From Cortini Pedrotti et al. 1973)

such as the mosaic of *Caricetum buxbaumii* and *Polygono-Nardetum* in the Pian Grande of Castelluccio di Norcia (Figs. 5.7 and 5.8) or as the sum of the two or more associations that compose the mosaic, also showing graphically such situations by the use of the colors of the associations involved, as in the Valleys of Comacchio, where the interpenetrations of two different phytocoenoses have been mapped, by using the colors of the two phytocoenoses in alternating strips (Fig. 5.41). At broad scale, where the plant associations may be fragmented into widely spaced small phytocoenoses, the option of "vegetation mosaics" permits a synthetic but effective representation of the vegetation; this can be seen on the vegetation map of the Danube delta at scale 1:100,000, consisting entirely of vegetation mosaics that combine 59 vegetation units (Hanganu et al. 1993).

Symbols. Cartographic units of small areal extent, not mappable at the scale chosen but which must always be indicated anyway, can be shown on the map by means of conventional symbols, such as letters or small pictograms, even if this interrupts the homogeneity of the cartographic representation. Symbols are used also to indicate the presence of species of particular interest (e.g. rare species, threatened species, or species at the geographic limit of their range).

Vegetation boundaries. The cartographic units are delimited by lines that represent *vegetation boundaries*, which may be due to natural factors such as mountainous topography, geomorphology, topoclimate, underlying bedrock type, soil type, etc. (Figs. 5.9 and 5.10) or artificial factors brought about by human activity (Fig. 5.11). Boundaries due to natural factors are shown by smooth lines and delimit forms of high fractal dimensions; anthropogenic boundaries are normally more or less angular (or at least less smooth), due to land fragmentation, and delimit polygons of low fractal dimensions. Boundaries may be clear when the ecotone is

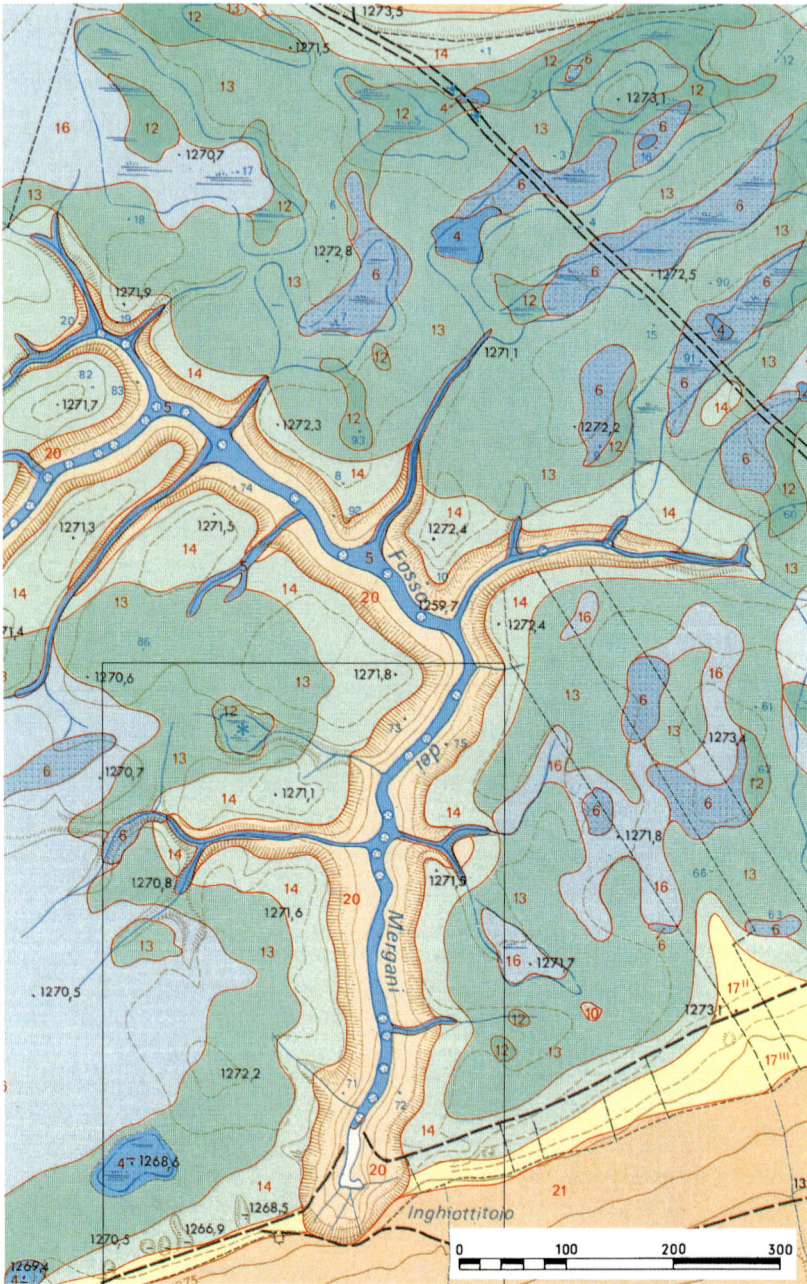

Fig. 5.8 Vegetation map of the Pian Grande (plain), at scale 1:5,000, showing an area with a natural channel and a sinkhole ("*inghiottitoio*"); the *numbers* indicate the followings associations: *4 – Caricetum gracilis, 5 – Caricetum gracilis cardaminetosum grandifoliae, 6 – Caricetum buxbaumi, 12 – Polytricho-Nardetum, 13 – Polygono-Nardetum, 14 – Filipendulo-Nardetum, 16 –* mosaic of *Caricetum buxbaumi* and *Polygono-Nardetum, 17 – Cynosuro-Trifolietum repentis, 20 – Filipendula hexapetala* community, *21 – Brachypodium januense* and *Asphodelus albus* community (From Cortini Pedrotti et al. 1973)

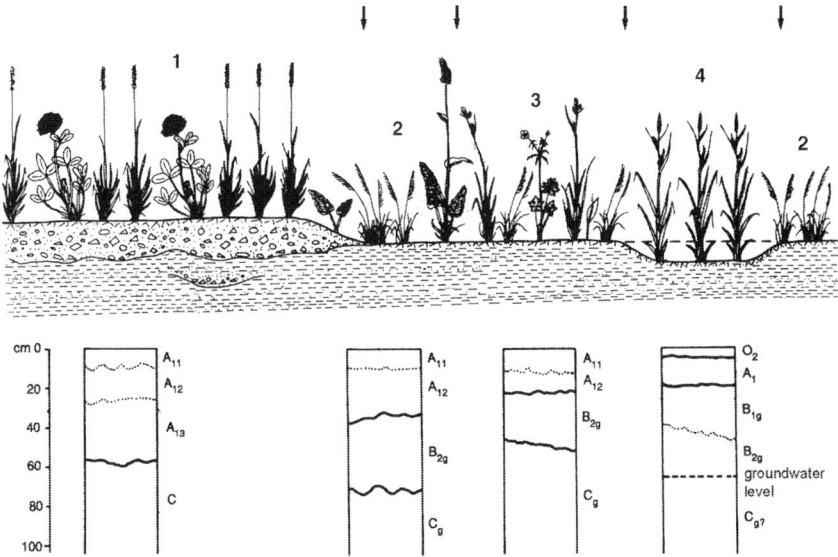

Fig. 5.9 Karst basin of the Pian Grande, showing: *1* – the association *Cynosuro-Trifolietum repentis* developed on colluvial deposits with "*terra bruna calcarea* (brown calcareous) soil"; *2* – the association *Polygono-Nardetum* on the lacustrine clay; *3* – a grouping of *Carex buxbaumii* on the lacustrine clay; and *4* – the association *Caricetum gracilis* on the dolines. The *arrows* indicate the vegetation boundaries (From Pedrotti 1997a)

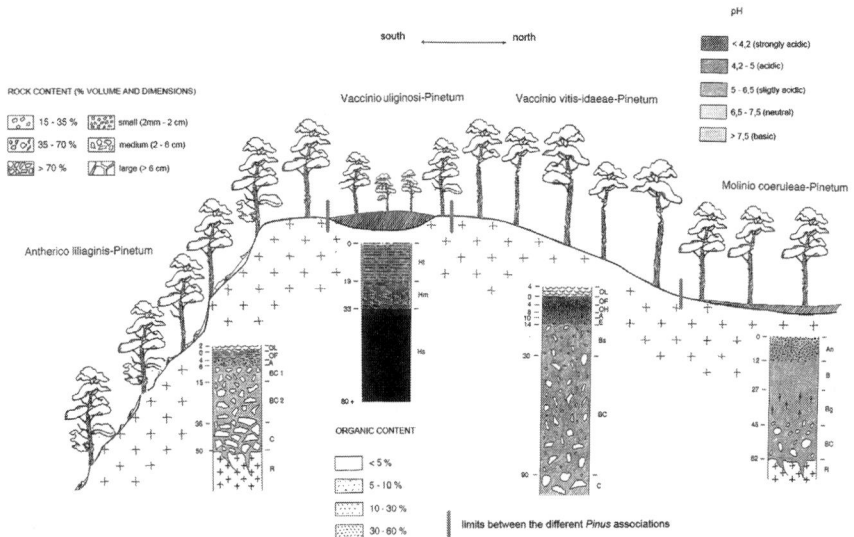

Fig. 5.10 The vegetation boundaries between the associations *Antherico liliaginis-Pinetum sylvestris*, *Vaccinio uliginosi-Pinetum sylvestris*, *Vaccinio vitis-idaeae Pinetum sylvestris* and *Molinio coeruleae-Pinetum sylvestris* are conditioned by geomorphology and pedology (From Minghetti 2003)

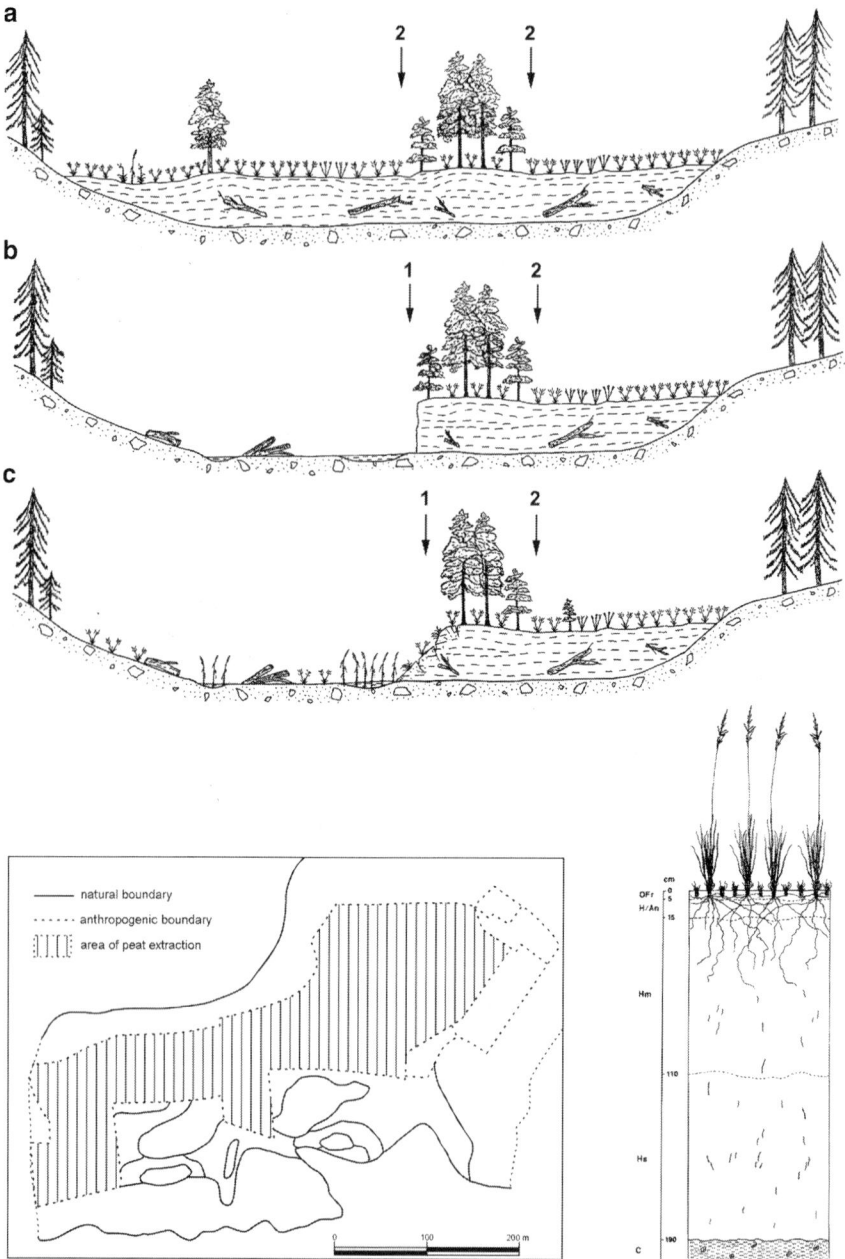

Fig. 5.11 Profiles across the peat bog "*Laghetto delle Regole*", in the Trentino-Alto Adige Region of northern Italy, before (**a**) and after (**b** and **c**) peat extraction. *Arrow no. 1* indicates the artificial vegetation limit due to peat extraction; the *arrows numbered 2* indicate natural vegetation limits (of *Molinio-Pinetum sylvestris* and *Succiso-Molinietum*). At *lower left* are natural boundaries (*curvilinear*) and artificial, anthropogenic (*straight-line*) boundaries. At *lower right* is a soil profile under the *Molinia coerulea* community (a mesic histosol with a sapric horizon, soligenous, mesotrophic, and with a surface H/An horizon). Remnants of roots are present in the peat (From Minghetti and Pedrotti 2000)

narrow, as for example in mountains, around bogs and when anthropogenic or otherwise artificial; or indistinct when the ecotone is wider, as in the case of vegetation on plains or other relatively flat topography.

Scale. The scale of the map may vary widely, depending on the mapping purpose: individual biotopes; limited areas (a valley, an island, a mountain group, or such); regions or provinces; or entire states, a continent, or the entire terrestrial globe. The themes or subjects that can be represented on the map change necessarily with changes of scale, changing the levels of study of geobotanical mapping that were listed in Chap. 1. Vegetation mapping operates with individual units based on floristics (phytosociology); the geographic scales at which it is possible to make such maps depends on the spatial variation of environmental factors, from microheterogeneous terrain (e.g. bogs, other wetlands, sandy coastlines) to macroheterogeneous terrain (e.g. plains and plateaus). In addition to what was said above and limiting ourselves only to vegetation aspects, it is good to point out that with change in scale the methods of representation of vegetation units on the map change considerably, as summarized by the following scheme:

- Maps with scale denominator less than 10,000 are defined as fine (large) scale and are of great detail, normally for small areas such as biotopes of a few tens of hectares that may include bogs, other wetlands, etc.; the vegetation of such areas can be represented cartographically in all its variability. An extraordinary example of this type of fine-scale map is the *Map of the Actual Vegetation of Ozegahara* (Japan), a large wetland complex of 8 km^2, the vegetation of which has been mapped in extreme detail at the scale 1:100 (Miyawaki and Fujiwara 1970); the map is 100 × 130 cm (Fig. 5.12).
- Maps with scale denominator between 1:10,000 and 1:25,000 are also at fine (large) scale and refer to wider areas, such as a valley or mountain group; maps at this scale can also represent the vegetation in detail, though not in as great detail as the previous case.
- Maps with scale denominator between 1:50,000 and 1:100,000 are defined as medium-scale, on which the generalization is reduced but one can still show good detail for phytocoenoses covering vast areas, such as forests; these maps permit good synthetic and immediate overviews of land vegetation units and of more local plant associations, as may be of interest in the interpretation of vegetated landscapes.
- Maps with scale denominator between 1:100,000 and 1:1,000,000 are definable as broad (small) scale, for quite large territories, in which the representation of the vegetation is possible only at the level of syntaxonomic units higher than associations (such as orders and classes) and of phytogeographic units; by this point such maps have a synthetic character and may represent the vegetation of entire countries, such as France (Ozenda and Lucas 1987), Japan (Miyawaki 1979), Spain (Rivas Martínez 1987), Romania (Ivan et al. 1993), or the Czech Republic (Neuhäuslova 2001), or of entire continents such as South America (Hueck and Seibert 1981) or a large part of Asia (Fujiwara 2008) or the circumpolar Arctic (Walker et al. 2002) or South Africa (Mucina et al. 2006).

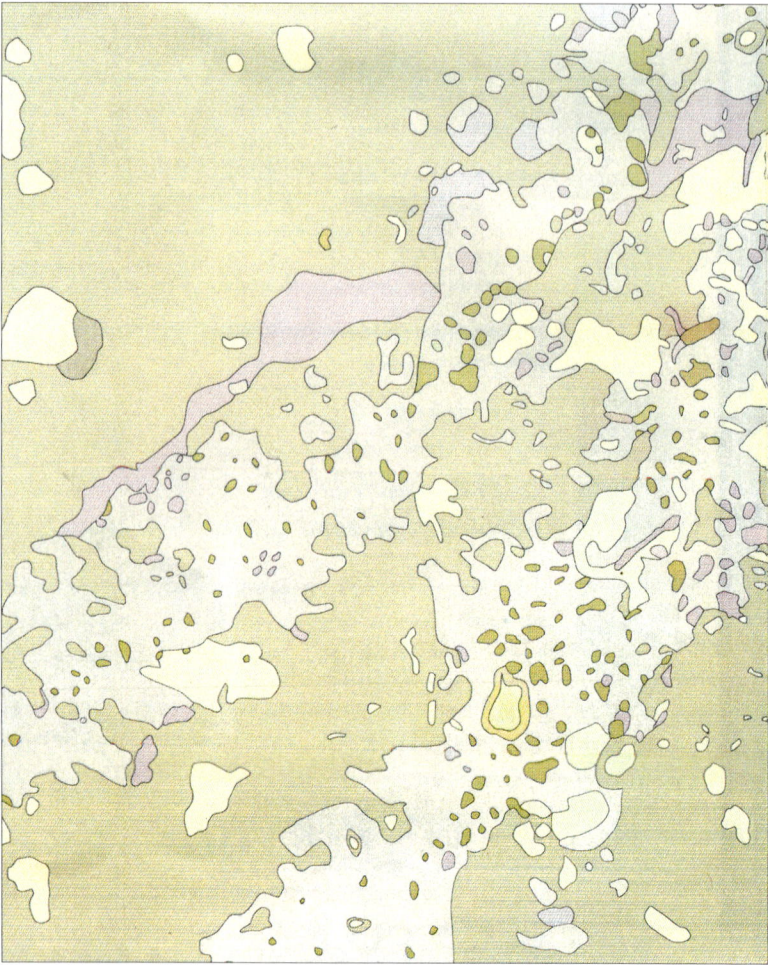

Fig. 5.12 Fragment of a vegetation map of the Ozegahara mire (Northern Honshu, Japan), at scale 1:100. The most widespread vegetation is the *Carici omiana-Spagnetum compacti* association (*sepia color*); this association is crossed by two strips of *Carici-Moliniopsietum japonicae* (juvenile phase, in *grey*), which contain many small units of the main association; others associations present are a *Carici-Moliniopsietum japonicae* (mature phase, in *yellow*) and a *Rhynchosporetum albo-yasudanae* (in *pink*) (From Miyawaki and Fujiwara 1970)

– Maps of the entire globe, finally, only permit the representation of large phyto-geographic units, such as the early maps of Brockmann-Jerosch (1919) or Rübel (1930); the world map by the latter author shows the following units, distinct physiognomically: tropical rainforests, subtropical rainforests, subtropical evergreen-sclerophyll forests, tree savannas and raingreen vegetation, grass steppe with vegetation green in winter-spring, summergreen broad-leaved forests, evergreen needle-leaved forests, and polar and alpine regions without trees.

Method of vegetation representation on the map. Cartographic representation is a language of communication designed for viewing and constitutes an unequivocal (monosemic) means of communication (Bertin 1967, 1977; Martinelli 1990). Thus the resulting map is correct if it conveys the relations between concepts previously defined (of diversity, order and proportionality) through visual relations of the same nature. For reading and understanding a map, two separate perceptions are necessary: first it is necessary to respond to the question "What do the symbols, colors, etc. mean?", namely what are the rules for reading what is contained in the map legend? There follows, then, a second question: "What are the relations among the items represented?" and how are these represented graphically?

The simplest means of representing vegetation on a map is to make a schematic design in black and white, using symbols, hatching and other types of symbolism. This method has been used mainly in the past for biotopes of small dimensions such as lakes and bogs, but also for wider areas. A second method is to make a more articulated map with the vegetation units delimited by lines (vegetation boundaries) and shown by colors. For the selection of the colors, there are no proposed or accepted rules to be respected, as have been decided, for example, on geological and soil maps.

Only Gaussen (1961b) has selected colors based on a pre-established criterion, which he, in fact, referred to the synecology of the vegetation, using warm colors for communities of xeric environments and cool colors for communities of moist environments. Among all the maps published by Henri Gaussen, the sheets Perpignan and Corsica – forming part of the *Carte de la Végétation de la France* at scale 1:200,000 – have become a classic example for the effectiveness of colors for different altitudinal belts (Dupias et al. 1965).

Data Collection for Vegetation Mapping

The process of vegetation surveying and data collection for a vegetation map first involves *identification* in the field of the different vegetation types that occur in the study area. Second, one must proceed to the *recognition* of their boundaries among the different types of phytocoenosis, each of which occupies a definite space. The third phase involves transferring these boundaries onto a map, in order to obtain the basic spatial design for the spaces occupied, which we can now call *cartographic units.*

Faliński (1990–91) listed in this way the necessary phases for collecting carto-graphic data for phytocoenoses. First, an *identification* of the phytocoenoses in the field is necessary, which must be followed by their *localization* (location in space). The next phase is the *delimitation* of the phytocoenoses under the prevailing environmental conditions where the types have developed (for example along a water course, on a slope, or in a depression). This procedure is followed separately for each phytocoenosis, and the map will be the result of the sum of the observations made for all the phytocoenoses present in the study area.

Of the operations listed, the first and second are botanical, while the third involves use of topographic survey techniques, the basis of which is in geography, geodesy, topography, photogrammetry, and other sciences.

The Botanical Problem

According to Küchler (1967), a vegetation boundary is a line that separates two different vegetation types. The botanical problem is to recognize the two adjacent types. To do this one must collect phytosociological relevés and analyze them in a table. Phytosociological tablework also involves assigning the available relevés to various associations, thus making a separation by which we can consider "theoretically" to be two or more associations (see van der Maarel and Westhoff 1964; Borza and Boæcaiu 1965; Ellenberg and Klötzli 1967; Neshataev 1971; Orloci 1972; Ferrari et al. 1972; Guinochet 1973; Ivan and Donicǎǎ 1975; Dierschke 1994; Kent and Coker 1994; Pott 1995; Chytrý et al. 1999). In the field, on the other hand, one must make a "concrete" separation that consists of identifying where (according to what line) and how (whether cleanly or diffusely) the adjacent associations come into contact.

In any case, in the past, some authors doubted the existence of vegetation discontinuities, at least in the phytosociological sense. Recently Krebs (2001), returning to the concept of the vegetation continuum (cf Whittaker 1962; McIntosh 1967; Austin and Smith 1989), posed the rhetorical question "*community boundaries*?" He responded that plant communities change gradually with environmental conditions and the species are distributed as in a continuum along such environmental gradients. In fact, numerous field studies have shown both the continuous and discontinuous character of vegetation, which depends on spatial variation in environmental factors. Whittaker (1960) reached this conclusion in his work on the Siskiyou Mountains and wrote that it is preferable to complete whatever gradient analyses may be made by adding a classification of the communities. At a speculative level the problem is certainly interesting and will not be ignored: the basic hypothesis of this book remains that of recognizing communities, in whatever ways they are different and separated by limits that are sometimes sharp and sometimes diffuse, and more difficult to recognize.

The process of "concrete separation" of two different vegetation types can be done based on the presence or absence of species in a specific characteristic combination, along a particular route or transect, and by use of differential species (identified by phytosociological tablework, already carried out). In this way it is possible to recognize and separate the "vegetation units", while the cartographic units are delimited by a line that defines the individual units that are homogeneous in terms of vegetation, i.e. that house a unique vegetation type.

When the boundary is sharp, the solution is relatively easy, as in the case of a glade with herbaceous vegetation in the middle of a forest (Fig. 5.13); in this case the boundary between the glade and forest associations is shown on aerial photographs or orthophoto maps, and does not pose problems in the field

Fig. 5.13 Artificial clearing in *Fagus sylvatica* forest, in the Abruzzo National Park, central Italy (Photo Franco Pedrotti)

or in the laboratory. When the boundary is indistinct, though, one must resort to a generalization, provided that if the ecotone is quite wide and forms a transition between two vegetation types, it may be mapped automatically. In the forest of Białowieza, for example, the contact zone between the two associations Potentillo albae-Quercetum and Tilio-Carpinetum is 16 m wide (Matuszkiewicz 1972).

Much more complex is the separation between adjacent vegetation types that are similar physiognomically and cannot be distinguished on aerial photographs. In this situation, floristic relevés must be made in the field along a transect and processed by specific numerical analysis in order to show the ecocoenotic discontinuity (Cornelius and Reynolds 1991). An example of this type of analysis (*split moving-window boundary analysis*) is a study of the vegetation boundary between the edge habitat (as ecotone) and that of the forest (Fig. 5.14) by Gafta (2002). Once identified in the field, the vegetation boundary must be transferred to the map, at the same time separating the cartographic units.

The Cartographic Problem

The main aspects involved in the cartographic survey of vegetation distributions are the choice of base map, the topographic survey of the boundaries and their subsequent transfer onto the base map.

The choice of base map depends above all on the type of investigation being conducted, but also on the cartography available, as especially in the past. Today it is very easy to have at one's disposal topographic maps that can be of help in various ways for making a vegetation map.

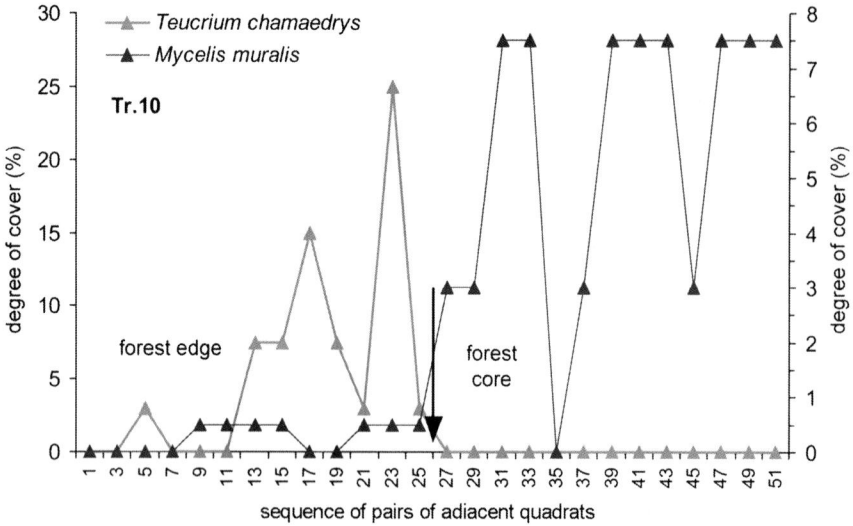

Fig. 5.14 Analysis of eco-coenotic discontinuity for studying the boundary between the *Fagus sylvatica* forest and clearing vegetation. This study was done using *Mycelis muralis* (a species from the forest) and *Teucrium chamaedrys* (from the clearing), noting their cover degree along a transect. The *arrow* shows the boundary position (From Gafta 2002)

Boundaries can be surveyed in the field by visual inspection combined with an instrument-based method (Puscaru-Soroceanu and Popova-Cucu 1966), using measuring tape, altimeter, compass, tachometer, theodolite, electrical theodolite and other suitable analog instruments. When there are many reference points on the map and in the field, these can be referred to directly, often by tape measure. If there are few reference points, as on level terrain, one must resort to construction of transects and datum-point grids fixed in the field. The number and locations of the transects depend on the relief, on the complexity of the vegetation and on map scale. On uneven topography, on which the vegetation distribution is usually strictly tied to the topography, the transects are chosen in relation to important surface features such as mountain peaks and valleys. Transects do not necessarily have to be fixed in the field but can be established by a rough route map that can then be followed on foot. Along the transect one then proceeds to survey, using an altimeter to mark the boundaries between the different vegetation types; based on the data obtained, one can then construct a vegetation profile (Fig. 5.15). One can proceed analogously in flat areas (Fig. 5.16) or areas with slight inclination (Fig. 5.17), but in this case, if necessary, it is good to take measurements with the measuring tape or by other means. The transect is marked absolutely on the terrain by means of stakes only in the case of detailed surveys at very fine (large) scale, as in the case of relevés in a bog (Fig. 5.30).

In any case, for detailed surveys in flat areas, one must set up a grid with stakes fixed in the ground and transfer this to the map by use of instruments, such as a theodolite (Fig. 5.18). Then it is possible to survey the vegetation limits, making

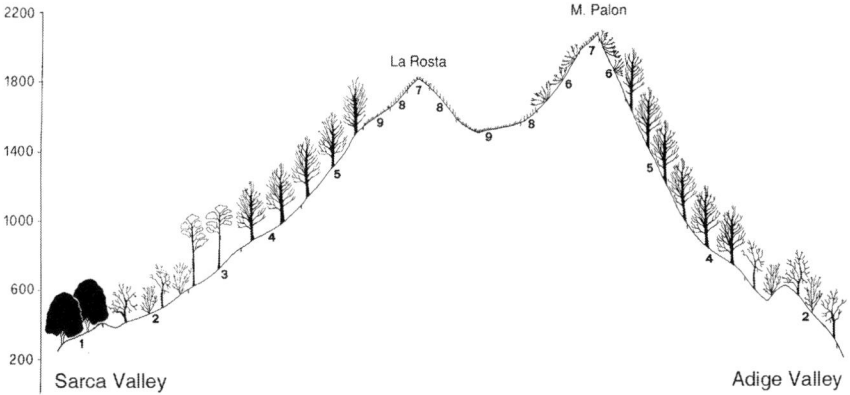

Fig. 5.15 Vegetation profile of Mt. Bondone, from the Adige Valley to the Sarca Valley, in the Trentino-Alto Adige Region of northern Italy: *1 – Celtidi australis-Quercetum ilicis*; *2 – Fraxino orni-Ostryetum carpinifoliae*; *3 – Chamaecytiso-Pinetum sylvestris*; *4 – Carici albae-Fagetum*; *5 – Cardamino pentaphylli-Fagetum*; *6 – Erico carneae-Pinetum mugo*; *7 – Seslerio-Caricetum sempervirentis*; *8 – Scorzonero aristatae-Agrostidetum tenuis*; *9 – Sieversio montanae-Nardetum* (From Pedrotti et al. 1994)

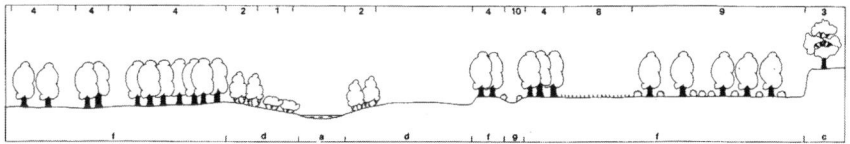

Fig. 5.16 Vegetation profile in and along the Ofanto River, San Nicola di Melfi, in the Basilicata Region of southern Italy: (**a**) the watercourse in the dry period; (**c**) paleoterrace; (**d**) recent terraces; (**f**) alluvial plain; (**g**) abandoned riverbed; *1 – Salicetum incano-purpureae*; *2 – Salicetum albae*; *3* – isolated trees of *Quercus pubescens* in fields; *4 – Populetum albae*; *8* – pastures with *Poo bulbosae-Plantaginetum serrariae* association; *9 – Populetum albae* thinned by invasion of *Prunetalia*; *10* – shrubs of the *Prunetalia* order (From Pedrotti and Gafta 1996)

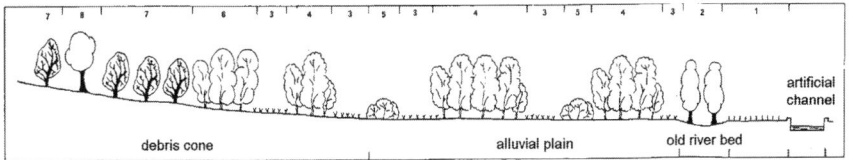

Fig. 5.17 Vegetation profile of the swamps of Roncegno, Valsugana, in the Trentino-Alto Adige Region of northern Italy: *1* – anthropogenic vegetation (*Tanaceto-Artemisietum vulgaris* and *Juncetum tenuifolii*); *2 – Salicetum albae*; *3* – clearings with associations of the orders *Magnocaricetalia* (*Peucedano palustris-Caricetum acutiformis*) and *Molinietalia* (*Scirpetum sylvatici* and *Lysimachio-Filipenduletum*); *4 – Carici acutiformis-Alnetum glutinosae*; *5 – Salicetum cinereae*; *6 – Alnetum incanae*; *7 – Robinia pseudacacia* formation; *8* – isolated trees of *Quercus robur* (From Pedrotti and Gafta 1994)

Fig. 5.18 Network of datum
points made in the field and
drawn, for vegetation survey,
on a cadastral map of a flat
sector of the Pian Grande
(plain); the scale was 1:2,000,
re-sized to 1:6,000 for
printing; the *black points*
show the position of the
datum points

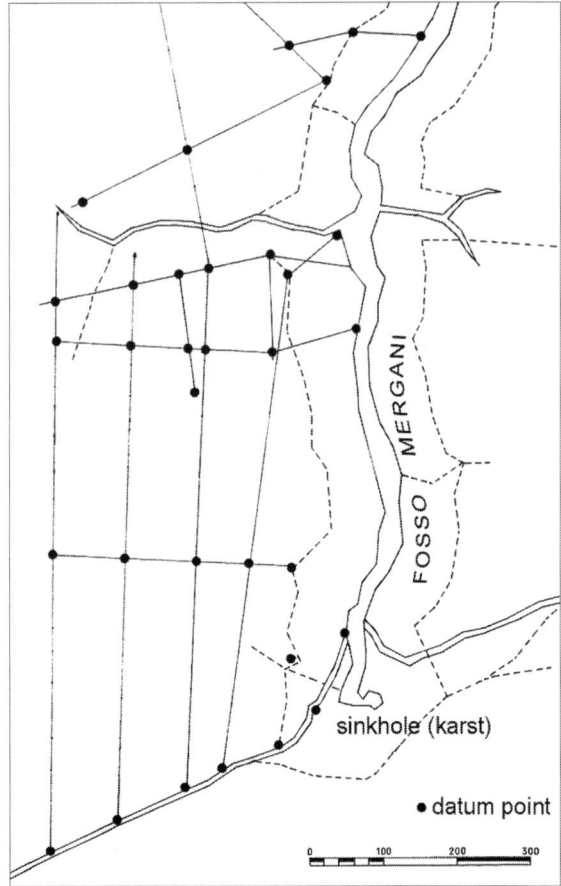

reference to the grid of fixed points for the flat part without other references and to
the topography or both (Fig. 5.19), as in the case of the *Pian Grande*, a vast, mostly
flat area furrowed by a natural channel (Fig. 5.8).

For surveys in great detail at very fine (large) scale, such as 1:100, as for
mapping facies and microphytocoenoses in an association (Fig. 5.20), one can
delimit a rectangular area of 1,000 square meters (100 × 10 m) in the field, with
stakes, and subdivide it further into smaller areas. Using both longitudinal and
transverse measurements as references, one samples the whole area and can
delimit and map the facies and microphytocoenoses and other particulars deemed
necessary (Fig. 5.21).

In any case, one has available automatic instruments for direct collection of the
sampling points in the field, such as the electronic theodolite already mentioned.

Already for some years now, the traditional methods of data collection in the
field have been superseded by *Global Positioning Systems* (GPS). GPS use permits

Fig. 5.19 Topographic map of the Pian Grande made from photorestitution, at scale 1:2,000, re-sized to 1:6,000 for printing; the map shows details of the geomorphology, like the Fosso Mergani and its tributaries, the dolines, the flat area and the mountain slopes

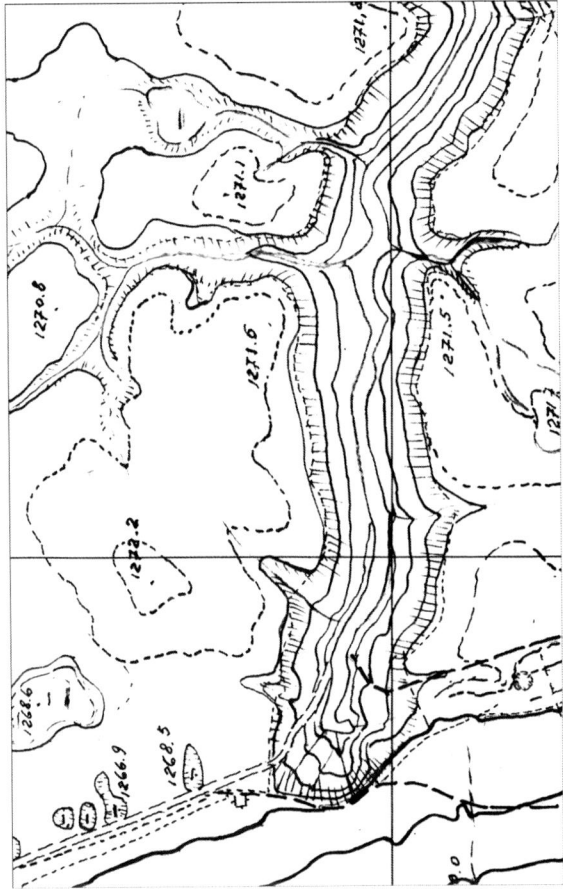

one to determine the ground position and altitude of a point with a precision that may vary from several centimeters to several meters, based on the instrument used and the targeted area of study. The system works with a network of 24 man-made earth satellites, with 21 satellites always active and 3 in reserve. These are divided into groups of four such that at each instant and at each point there are above us 5–8 satellites within range, at an elevation of about 20,000 km. The commercial GPS receivers, today at a very pleasing cost and about the size of a common cell phone, tune in automatically to the frequency of the satellites and, in a few minutes, identify the distance to at least four satellites, determining in this way the user's own geographic position on earth's surface in terms of latitude, longitude and elevation. The instruments must be exposed to the satellite signals, i.e. they do not function in closed places, such as caves, canyons and under a dense tree canopy. Two signals are generated by the satellites: the first indicates the position with a precision of about 300 m, and the second with a precision of about 50 cm. While the

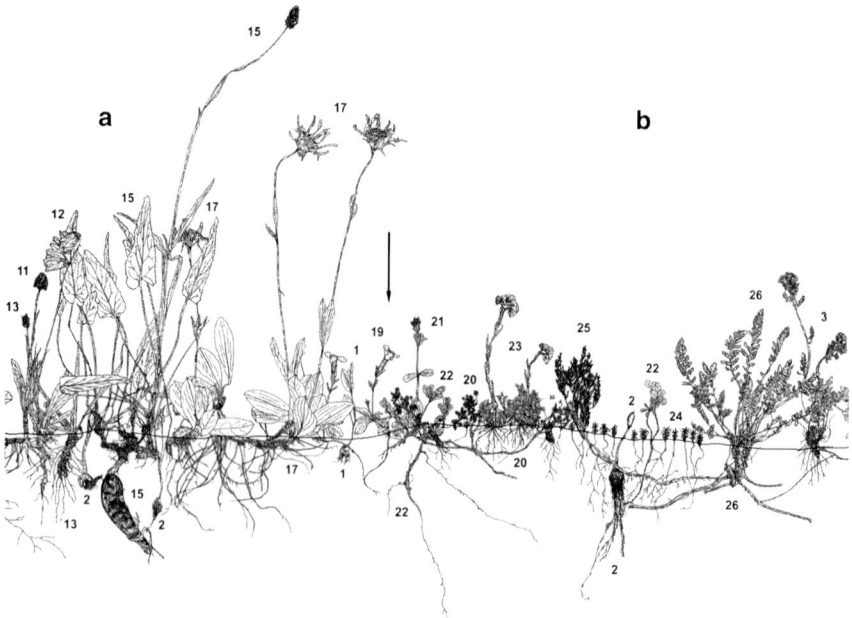

Fig. 5.20 Profile of the association *Sieversio montanae-Nardetum strictae* in the meadows of the Viotte of Monte Bondone, Trento Province, northern Italy: (**a**) typical facies; (**b**) microcoenosis of *Antennaria dioica* and *Ericaceae*; the *arrows* show the boundaries between the *Antennaria dioica* and *Ericaceae* microcoenosis and typical facies. The numbers are for the following species: *1 – Polygonum viviparum*; *2 – Crocus albiflorus*; *3 – Potentilla crantzii*; *11 – Nigritella nigra*; *12 – Campanula barbata*; *13 – Plantago atrata*; *15 – Phyteuma betonicifolium*; *17 – Arnica montana*; *19 – Gentiana verna*; *20 – Potentilla erecta*; *21 – Veronica bellidioides*; *22 – Vaccinium gaultherioides*; *23 – Antennaria dioica*; *24 – Polytrichum commune*; *25 – Calluna vulgaris*; and *26 – Pedicularis elongata* (From Doniţă et al. 2003)

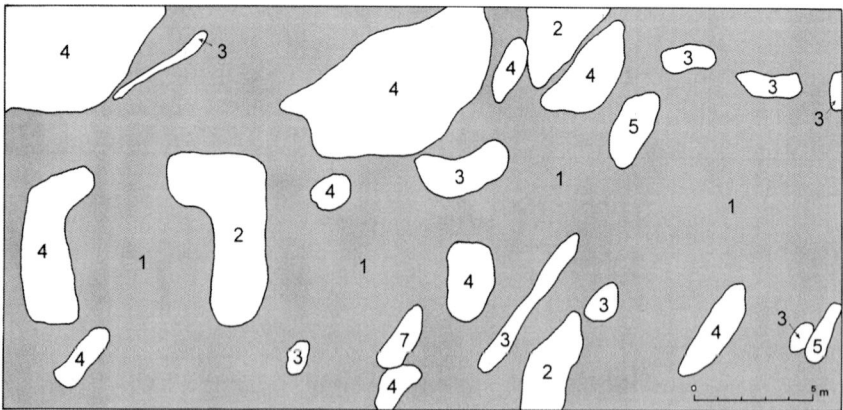

Fig. 5.21 Micro-relief of phytocoenoses of the *Sieversio montanae-Nardetum strictae*, in the Viotte del Monte Bondone: *1 –* typical facies; *2 – Ranunculus montanus* and *Potentilla aurea* facies; *3 – Deschampsia caespitosa-Molinia coerulea* facies; *4 – Antennaria dioica* and *Ericaceae* microphytocoenosis *5 – Veratrum album* facies; *6 – Trollius europaeus* facies; *7 – Ranunculus aconitifolius* facies (From Doniţă et al. 2003)

Fig. 5.22 Aerial black-white photo of the Altipiano di Piné, Trento Province (northern Italy); the original scale was 1:30,000, here re-sized to circa 1:40,000 for printing

first signal is transmitted openly, the second is transmitted only approximately, for reasons of national security.

The most sophisticated GPS receivers, in particular those mounted in vehicles, permit determination of one's position within a geographic map, permitting the user to verify his or her position in real time and a route followed.

GPS is a useful and practical instrument for a correct determination of the coordinates of places where sampling has been done, also in studies that require periodic monitoring of sites, such as permanent plots for the study of vegetation dynamics.

Also useful are aerial photographs, in black and white (Fig. 5.22), in infra-red and in color (Figs. 5.23 and 5.24). The photos are made along pre-determined flight lines, called tracks, in such a way as to cover the sampling area completely (Fig. 5.25). It is also possible to use aerial photographs from low altitude (100–800 m), thus very detailed, obtained from controlled balloons equipped for air-photo surveys (Ferrari et al. 1978; Mahito and Takeshi 1998).

Fig. 5.23 Color aerial photo of a south-exposed slope, Valcamonica, Lombardia Region, central Italian Alps

Aerial photos are examined as pairs of adjacent photographs by means of a stereoscope, by which the two photos are superimposed to provide a stereoscopic view. The photo-interpretation is done first in the laboratory, by identifying all the vegetation boundaries (polygons) recognizable on the air photos. These boundaries are then transferred to the base map by photo-restitution (i.e. the creation of a single correct image from multiple aerial photos, using a stereoplotter or, more recently, computer software). The boundaries identified are usually simply called "edges" (*fotolimiti* in Italian; see Fig. 5.26), and these delimit the types that are distinguishable on the photo (*fototipi* in Italian). These cartographic units (polygons) are

Fig. 5.24 Color aerial photo executed for detailed relevés (scale 1:2,880) of the Inghiaie biotope, Trento Province, northern Italy. On the photo we can see, from right to left: fields, moist meadows (*Selino-Molinietum*) partially still mowed (see plots) and partially abandoned and in secondary succession with *Salicetum cinereae* development; and mesophilic woodland of *Carpinion* (Photo *Provincia Autonoma di Trento 1992*)

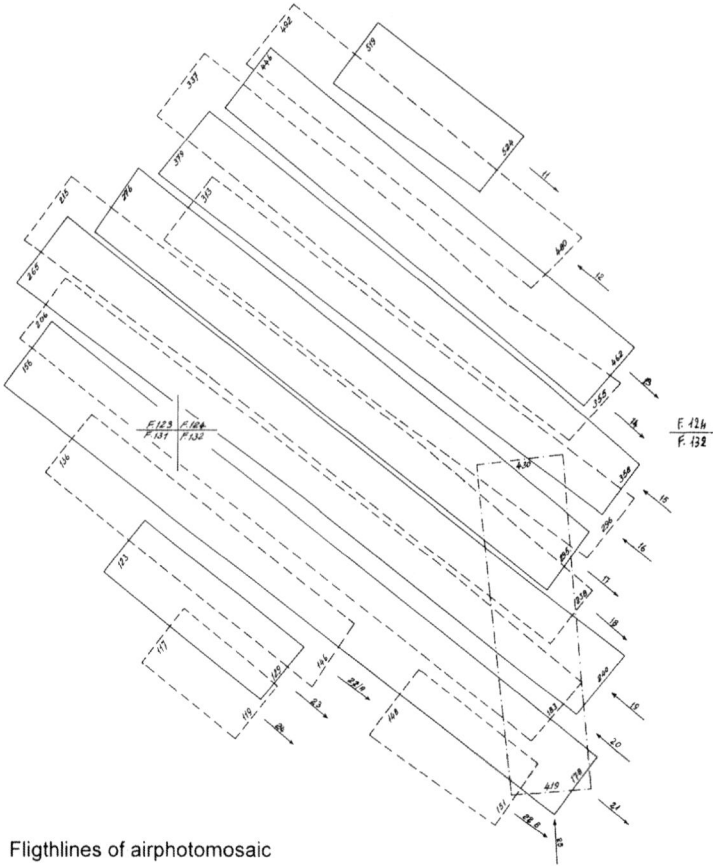

Fligthlines of airphotomosaic

Fig. 5.25 Flight lines and strips for the resumption of aerial photographs referred to 1:100,000 I.G.M. maps

recognizable on the aerial photo because they possess the same vegetation structure (Fig. 5.27). The recognition of the different vegetation units with air photos is based on differences of form, cover and different hue intensities. It is thus necessary to check the results in the field ("ground truth"), which permits complete attribution of the cartographic units identified to the vegetation units. Surveys by aerial photogrammetry are thus integrated with determinations made on the ground, in such a way as to cover as many habitat types as possible and to sample visually and by instruments, such as by altimeter. This is indispensable when one must establish the boundary between two vegetation units with similar physiognomy (thus not recognizable on the air photos) which follow each other altitudinally in mountainous areas, such as different associations of needle-leaved forest (Fig. 5.28) or prairie (Fig. 5.29). In the case of herbaceous vegetation, where a very detailed survey may be needed, air photos in black and white only help partly; more useful are infra-red

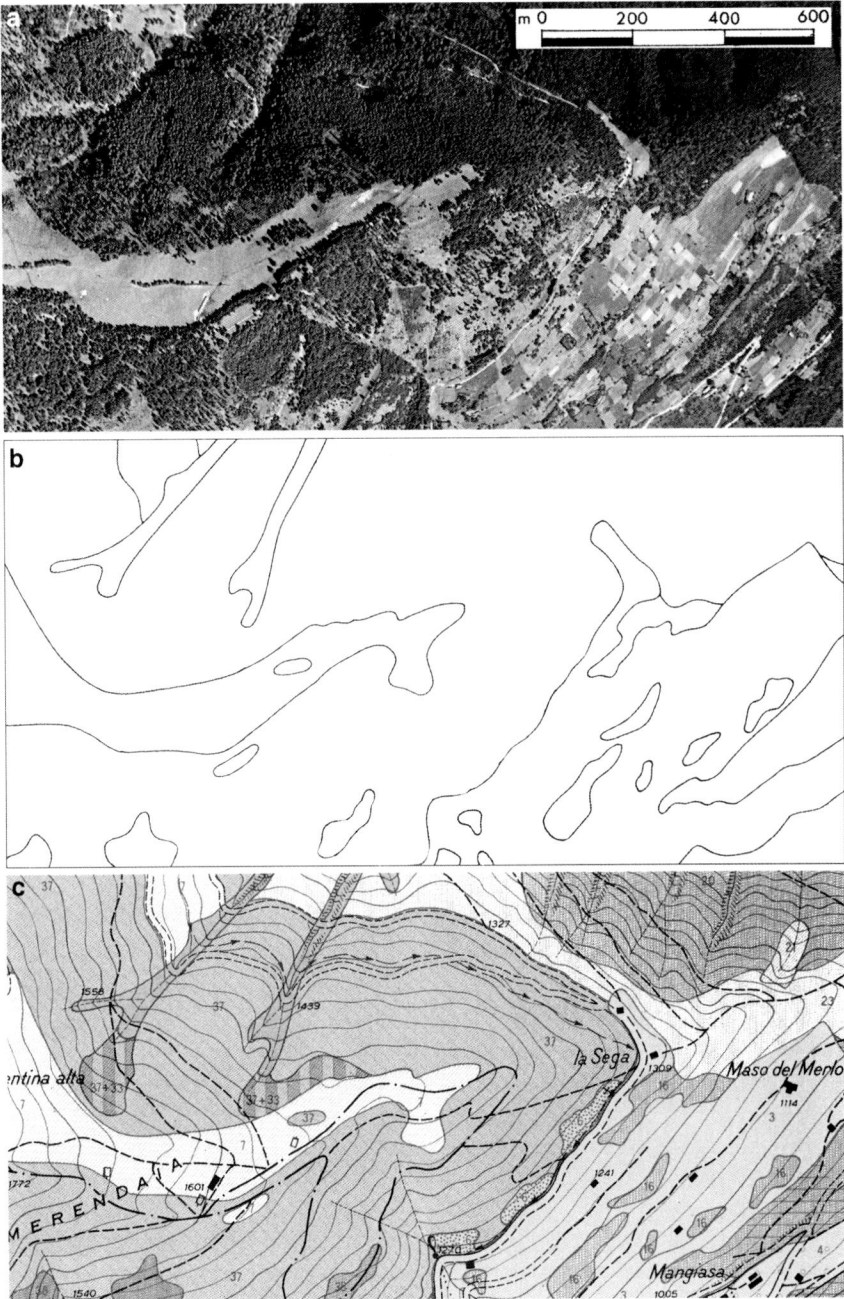

Fig. 5.26 (a) Aerial photo of the area between the villages Croviana and Presson, in the Val di Sole, Trentino-Alto Adige Region, central Italian Alps; (b) photolimits (i.e. boundaries visible on the photo); (c) vegetation map. The numbers indicate: *1* – urban settlements; *2* – crops; *3* – meadows; *4* – meadows with orchards; *5* – marshy meadows; *6* – xeric pastures; *7* – montane pastures; *15* – shrubland; *16* – deciduous broad-leaved forest; *17* – deciduous broad-leaved forest with conifer plantations; *21* – riparian forest; *23/24* – coniferous forests; *33* – hygrophilous woodland; *38* – pastures with coniferous species (From Pedrotti 1965–1968)

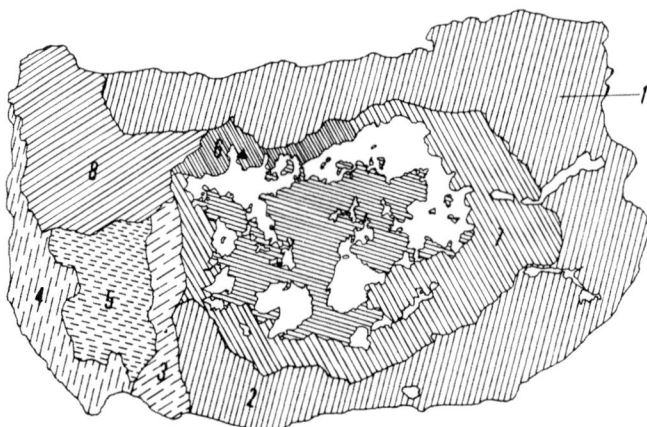

Fig. 5.27 *Phototypes* of vegetation of the Palude della Chioggiola wetland, in the Emilia-Romagna Region of the northern Apennines. The phototypes (i.e. types visible on the photo) are interpreted by comparing three aerial photos, which were made at an altitude of 200 m and with three different kinds of films. Seven phototypes have been identified; the vegetation map is reproduced in Fig. 7.27 (From Ferrari et al. 1978)

Fig. 5.28 Conifer forest on the southwestern slopes of Mt. Costalta, Piné, central Italian Alps; *1 – Homogyno-Piceetum*; *2 – Oxali-Piceetum*; *3 – Vaccinio -Pinetum* (Photo Franco Pedrotti)

and color photographs, but it is still necessary to check the result in the field, perhaps best by making transects.

For the bog Vedes in Trentino, a transect was constructed first and then a vegetation profile was surveyed from the center to the periphery (Fig. 5.30); the

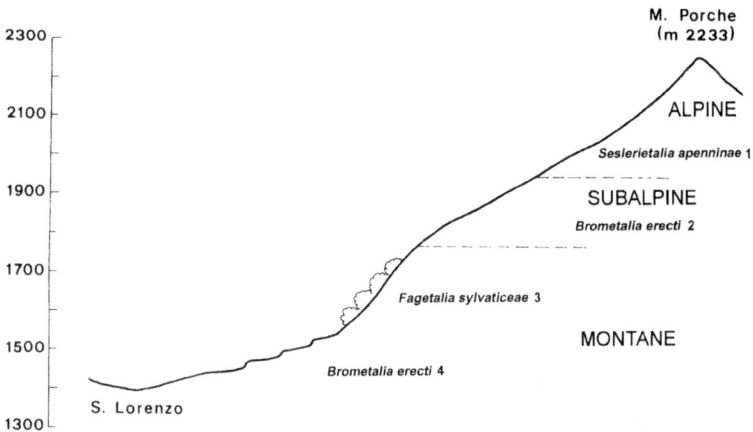

Fig. 5.29 Western slopes of the Sibillini Mountains in central Italy, where the forest vegetation was completely destroyed excepted for a few residual nuclei of *Fagus sylvatica* forest; it is impossible to identify the grassland associations in the aerial photo; *1 – Seslerietalia apenninae*; *2 – Brometalia erecti* at its upper limit of distribution; *3 – Fagetalia sylvaticae*; *4 – Brometalia erecti* pastures developed on ancient terraces after abandonment as croplands (Photo Franco Pedrotti)

cartographic survey of the vegetation was then done, successively, using air photos in combination with the transects (Fig. 5.31).

The same method, i.e. use of air photos and transects, was also employed for surveying the vegetation of the Capalbio Dunes in Tuscany (Fig. 5.32).

Also very useful is the *photomosaic*, i.e. the full set of air photos combined according to their geographic position and rectified to eliminate distortions due to the angles at which the photos were made. In the end, each photo is rendered as it would appear if viewed perfectly orthogonally (Fig. 5.33) and acquires metrics

Fig. 5.30 Vegetation profile of the Vedes mire, Trentino-Alto Adige Region, northern Italy: 0 – forests of *Picea abies* and *Larix decidua*; *1 – Caricetum rostratae* (not possible to see in the picture); *2 – Pino mugo-Sphagnetum* variety with *Betula pubescens*; *3 – Pino mugo-Sphagnetum*; *4* – cushions of *Sphagnetum magellanici*; *5* – depressions with *Rhynchosporetum albae*; *6* – depression with *Caricetum limosae*; *7* – a *Scheuchzeria palustris* community; abd *8* – a small lake in the middle of the mire (From Pedrotti 1980)

closest to those of a topographic map, to which the information may be transferred (Vianello 1998). A georeferenced photomosaic, onto which the vertex coordinates of the polygons have been transferred, is called an *orthophoto plan* (Fig. 5.34); an orthophoto plan onto which contour levels and elevation points have also been transferred, by photorestitution, plus associated toponyms, may be called an *orthophoto map* (Fig. 5.35). Orthophoto plans and orthophoto maps can be used easily, and in many cases their use can be substituted for that of aerial photographs. Given the nature of this book, for more details on the use of aerial photogrammetry in vegetation surveys, we refer the reader to specific publications such as by Howard (1970), Amadesi (1993), Vianello (1998), Bezoari et al. (2002) and others.

For very dense sampling, one must resort to various methods simultaneously, which can be applied to different parts of the same map, according to the type of vegetation present and the difficulty of interpretation.

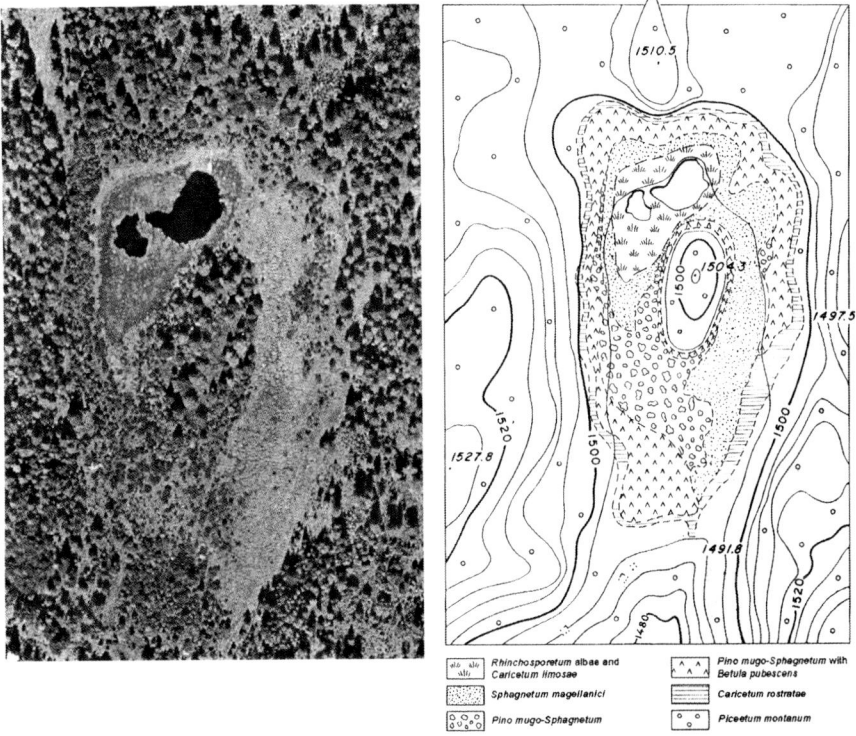

Rhinchosporetum albae and Caricetum limosae	Pino mugo-Sphagnetum with Betula pubescens
Sphagnetum magellanici	Caricetum rostratae
Pino mugo-Sphagnetum	Piceetum montanum

Fig. 5.31 At the *left*, an aerial photo used for the vegetation survey of the Vedes mire; at the *right*, the map of the mire vegetation (From Pedrotti 1980)

Generalization

In each phase of the survey and during the definitive drawing up of the map, it is always indispensable to proceed to a *generalization* that involves both the botanical aspect (Isachenko 1962) and the cartographic aspect at the scale of the mapping (Töpfer 1979). The geographic generalization can be stressed more or less according to the detail to be attained and the scale at which one is working, and thus must submit to the criteria of the generalization, i.e. those used for the topography, hydrography, hypsometry, etc.

As for the botanical aspect, one must give up on representing all the details collected in the field, for two main reasons: one of a graphic nature, eliminating all those details that it is too difficult to show on the map; and one of a scientific nature, in that the map content is not just a "photograph" of the vegetation in the field but presents a homogeneous scientific interpretation of it at all points in the area surveyed; the degree of detail of the representation must always be the same over the whole area studied.

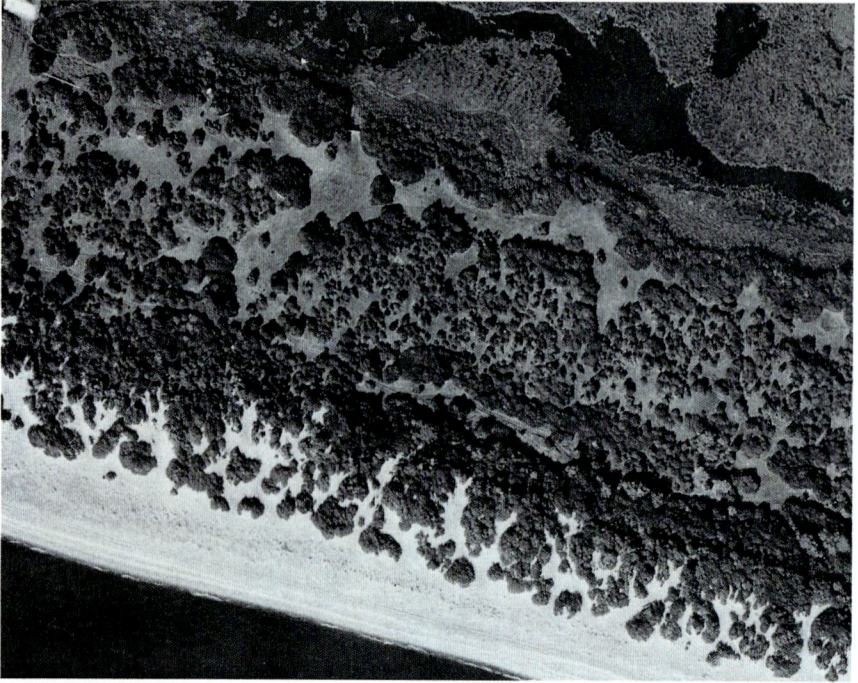

Fig. 5.32 Aerial photo of Capalbio dune, Tuscany Region, central Italy, and vegetation map. The *colors* show the following associations: *Juniperetum macrocarpae-phoeniceae* (*light green:*) with *Phillirea angustifolia* patches (in *medium green*); *Oleo-Lentiscetum* (*dark green*) with *Juniperus macrocarpa* patches (in *darker green*); *Helianthemetalia guttati* (*orange*); *Crucianelletum* (*red*); *Agropyretum mediterraneum* (*red*); *Juncetum maritimi* (*green*) (From Pedrotti et al. 1975)

Fig. 5.33 *Photomosaic* of the Tronto Valley, area of Sperlonga, Marche Region, Adriatic central Italy, 1:50,000

Fig. 5.34 *Orthophoto plan* of the area of Castelluccio di Norcia and Pian Grande, at scale 1:25,000 (Table 325/b); here it is possible to see: meadows of flat areas, crops, the village of Castelluccio, and the eroded slopes of Mt. Vettore (*Servizio Urbanistica Regione Marche 1987*)

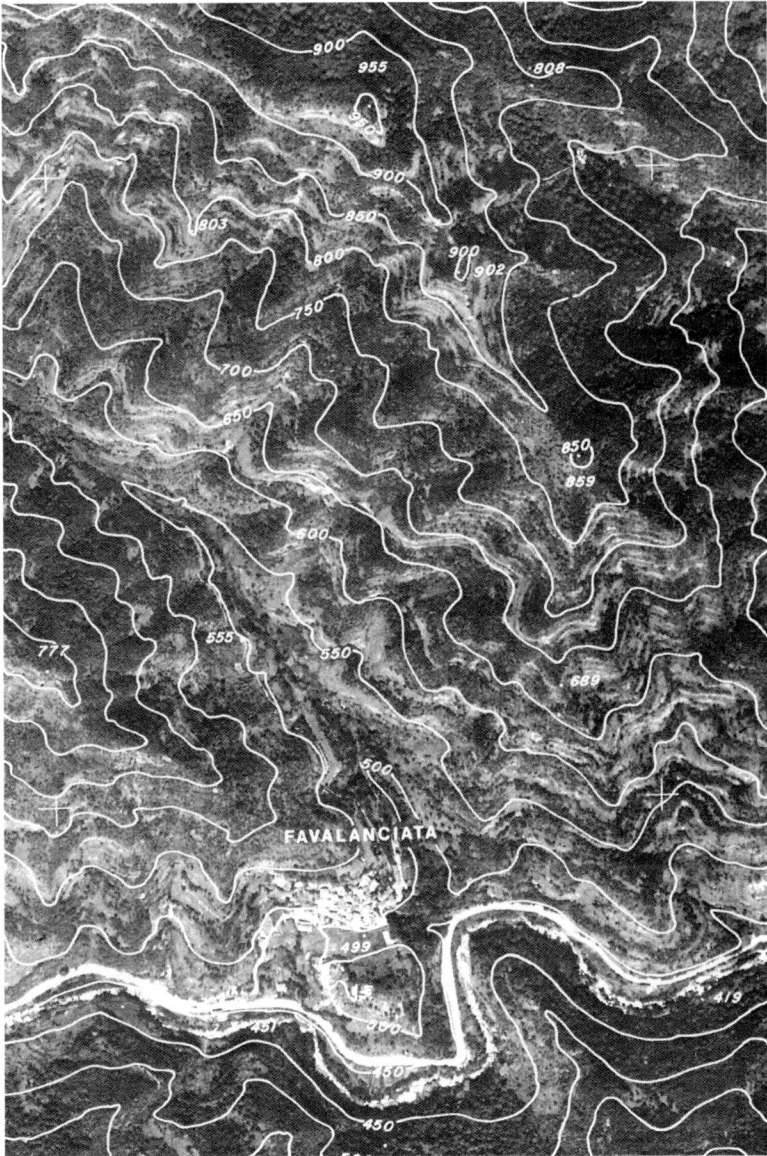

Fig. 5.35 *Orthophoto map* of the Tronto Valley, Marche Region, Adriatic central Italy, at scale 1:10,000 (*Servizio Urbanistica Regione Marche 1979*)

Faliński (1990–91, 1999) presents a very detailed list of cases suggesting all the possible generalizations that can be made about syntaxonomy, dynamic states, phases of succession and degeneration forms of plant communities.

As for the cartographic aspect (vegetation boundaries), the generalization consists above all in elimination of some of the limiting vertices so as to make them more linear, compatibly with the level of accuracy defined beforehand.

Computer Cartography

For all practical purposes, computerized cartography began in the late 1960s at Harvard University with the development, by H. T. Fisher, of the computer program SYMAP (synagraphic mapping) for use in city planning. SYMAP contained algorithms for the production of contour, choropleth, and proximal (nearest-neighbor) maps, and was run, as were all programs in those days, from punch-cards, on a large mainframe computer, with mapped results printed on a high-speed line printer.

The term 'synagraphic' refers to the fact that the quantitative content of the map was calculated and the spatial arranging (mapping) were done at the same time, in the same program execution. In other words, the result could be a predictive map, with results obtained from models or other computational algorithms built into SYMAP through a special subroutine.

The first such model-driven computer map involving vegetation, and perhaps the first such world computer map in any science, was the SYMAP-based "Miami Model" world map of annual net primary productivity (NPP) predicted from a relation between field-estimated annual NPP and annual averages of temperature and precipitation, presented in Miami in November 1971 (Lieth and Box 1972; see Fig. 5.36). This was followed by maps of gross primary productivity (Lieth and Box 1977) and autotrophic respiration, and all were quantified by computerized "volumetry" to provide the first systematic quantitative estimates of the main biotic components of the global carbon budget (Box 1978). Other maps, including one of annual photosynthetic efficiency, were produced by mathematical overlaying and other proto-GIS techniques (Box 1979a, b, c).

Computer maps of predicted world distributions of plant growth-form types, based on climatic envelope models, were produced by Box (1981). By the mid-1980s, results of ecological simulation models were also being presented geograph-ically, as world or other computer maps (e.g. Box 1988). By the 1990s this included results from dynamic simulation models, including the climatic potential upper limit for phytomass accumulation (and thus carbon sequestration) on the world's land areas (Fig. 5.37).

True vegetation maps, however, with types shown by different colors, had to wait until color software and printers became generally available, beginning in the latter 1980s. The first modern, colored world computer maps of vegetation types, for predicted natural vegetation (biomes), were by Prentice et al. (1992) and Box (1995; see Fig. 10.3 in Chap. 10).

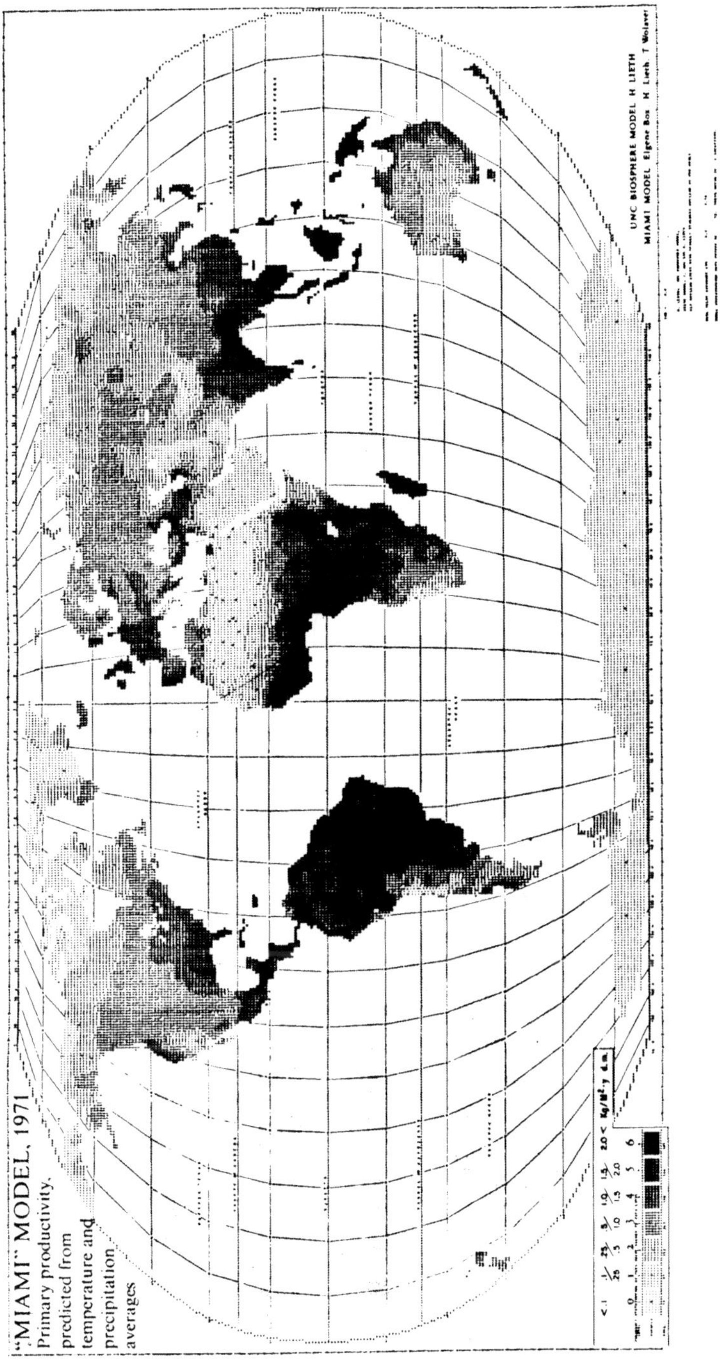

Fig. 5.36 The "Miami Model": net primary productivity predicted from average annual temperature and precipitation (From Lieth and Box 1972; cf. Lieth 1975)

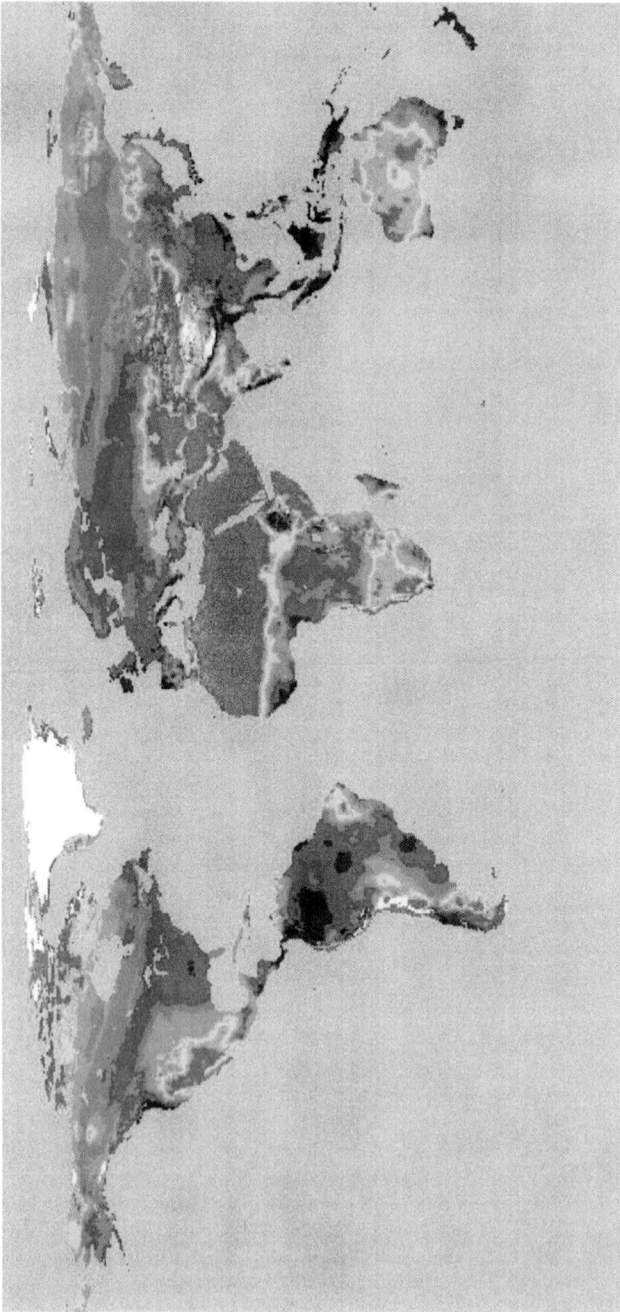

Fig. 5.37 Potential standing phytomass of the land areas in equilibrium with current climate. The potential maximum amount of plant biomass (phytomass) that can accumulate at a site, assuming that nutrients and other factors are not limiting, is determined by climate, through its effects on the

Mapping with Satellite Imagery

Photos from space were made as early as 1946, from sub-orbital V-2 rockets. Mapping with satellite imagery, however, really began with the first Landsat satellite in 1972, the first real-time imagery from 1977, and the widely used Thematic Mapper sensor, with seven data bands, on the Landsat 4 satellite launched in 1982. The first global coverage by satellite sensors came with the first polar-orbiting satellites in the early 1980s, especially the so-called Advanced Very High-Resolution Radiometer (AVHRR) on the NOAA satellites. By combining visible and near-infra-red bands in a ratio called the Normalized Difference Vegetation Index (NDVI), it was possible to enhance the capture of reflected green wavelengths and monitor vegetation over large areas (cf Tucker 1978, 1979; Tucker et al. 1985). It was soon recognized that annually integrated NDVI values correlated well with annual net primary productivity (NPP), though not with instantaneous measures such as standing biomass (except at low levels of vegetation density, essentially without trees). These relationships were validated globally by Box et al. (1989), and criteria for "geographic validation" of geographic models were suggested by Box and Meentemeyer (1991).

Satellite data represent the second worldwide data basis for modelling the physical environment, after climatic data. Climate-based models predict potentials, but satellites see what is actually on the ground and, as such, represent a useful complement to climate-based models. As a counterpart to the climate-based NPP map of Fig. 5.36, a satellite-based world NPP map was produced later from the NPP-NDVI relationship (Box and Bai 1993; see Fig. 5.38). The increasing availability of satellite data, however, meant that it was and still is very easy to produce "pretty pictures", and many applications of satellite data for monitoring and mapping are not rigorously validated. Also, satellite data still often cannot provide the desired detail on vegetation maps, even at local scale, as illustrated below with the example of the Foresta Umbra on the Monte Gargano massif (Fig. 5.39).

More sophisticated techniques for making vegetation maps with satellite imagery are summarized by various authors, including Lillesand et al. (1999), McCoy

---

Fig. 5.37 (Continued) rates of production (photosynthesis) and maintenance costs (respiration) of the vegetation. These climate-driven processes were simulated using climate data, for each pixel on the map, permitting the vegetation to "grow" until it reached equilibrium with the local climate. Equilibration occurs when the respiratory losses by the increasing phytomass balances the potential photosynthesis of the foliage, which increases rapidly early in stand development. The resulting potential maximum phytomass, shown on the map, represents an upper limit to the amount of carbon that could ever possibly be sequestered in terrestrial vegetation – if the land were abandoned and allowed to grow up to its maximum potential. Phytomass levels on the map (in kilograms of dry phytomass), and the landscape they generally represent, are as follow: *brown* (deserts with no vegetation 0 kg); *light brown* (semi-desert, tundra, 0–1 kg): *medium green* (woodlands) 10–30 kg; *yellow* (grasslands, shrublands, 1–5 kg); *lightest green* (scrub/open woodlands, 5–10 kg); *dark green* (closed forests) 30–70 kg; *medium blue* (tall rainforests) 70–80 kg; *dark blue* (tallest rainforests) >80 kg. Polar and high-mountain icecaps are shown in *white* (From Miyawaki and Box 2006)

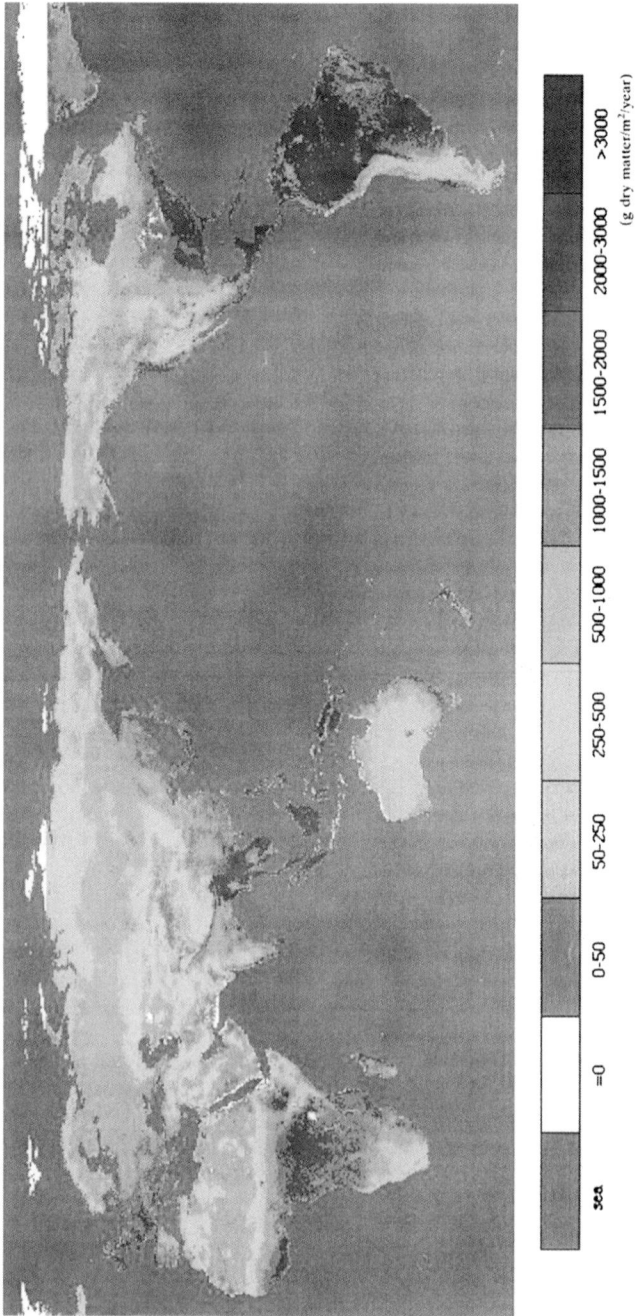

Fig. 5.38 Terrestrial actual net primary productivity of the actual vegetation cover of the world's land area, estimated from NOAA AVHRR Satellite data (1985–87 averages) (From Box and Bai 1993)

(2005), Kamagata et al. (2006), and Xie et al. (2008). Remote sensing involves study of the electromagnetic radiation emitted or reflected by the terrestrial surface, energy received from outside, usually from the sun. The images that can be used in remote sensing may be classified by three fundamental parameters: the sensor (photocamera, scanner or radar), the platform (airplane, helicopter or satellite) and the spectral bands employed (frequency and wavelength). For the systems of data acquisition one can distinguish analog images (photographs), in which the objects show variations that are continuous and progressive in color or hue; and digital images (discrete images). The first provide high spatial detail but cover only a limited part of the electromagnetic spectrum and of space, and require physical support, namely photographic film. The second operates over a much wider portion of the electromagnetic spectrum, and the images are shot by rectangular pixels of identical dimensions.

Observation of the earth with satellite technology offers the possibility to widen the spectrum studied and to obtain data from the same area at different times, because the satellite overflies the same area periodically, for example flying over a landscape on the same path every 16 days at the same time of day (~10:00 a.m. at our latitudes).

Interpretation of the images is based on the various responses to the different wavelengths. Regional studies still often use imagery from the family of Landsat satellites, which now have 7 bands in the electromagnetic spectrum, with radiometric representation at 8 bits. Many other satellites also send earth imagery continuously, among them SPOT (4 channels) with a resolution of between 10 m and 1 m in panchromatic, and Ikonos (5 channels, 11 bits) with a resolution of between 20 and 4 m in multispectral.

Raw images are first geo-referenced and ortho-rectified, and then are passed on for interpretation by various techniques that permit one to distinguish peculiarities of the matter observed. The transformation is possible with the help of both hardware and software components. Figure 5.38 shows a Landsat image (L5, Thematic Mapper, 7 channels, 183 × 172 km) from bands 2, 3 and 4, to which are assigned, respectively, the colors blue, green and red (technique for color composites in RGB mode). With this combination of bands, deciduous forest vegetation appears in bright red, while evergreen sclerophyll vegetation appears as dull red. Areas strongly affected by human activity (quarries and mines, recent urban expansion, etc.) show a white response, and agricultural areas appear as green.

To enhance certain characteristics of the area to be studied, one simple and commonly used operation for processing the spectral bands is analysis by principal components (PCA). Spectral enhancement represents an improvement of digital images for understanding the information coming from a few [cor]related bands and obtaining a smaller number of new bands that can be interpreted more easily.

Unsupervised "classification" groups pixels into classes according to statistical aggregation techniques; in order that a cluster represent a single class, the distribution of pixels must be unique for each class.

Supervised "classification" is based on choices made through direct knowledge ("training data") of the features to be investigated in some sampling area.

Fig. 5.39 Promontory of Gargano, Apulia Region, Adriatic southern Italy, as obtained from Landsat[TM] color composite imagery; conventional colors bands 2-3-4 from 26-VI-2001. In this picture it is possible to see deciduous forests (*light red*) and evergreen mediterranean forests (*dark red*); the *square red* patches are crops areas (*Processed by Sergio Ruggieri, Camerino*)

Homogeneous areas of pre-selected classes (thematic features) are identified, and each pixel on the full image is assigned to the class to which its spectral signature is most similar, based on an automatic classification algorithm. The images generated represent a thematic map of the classes selected, to which it is then possible to associate colors or symbols for better cartographic representation (Figs. 5.40 and 5.41). An estimate of the accuracy of the "classification" (assignment of pixels to classes) is made at the end of the process, comparing in a matrix the training data with the "classified" map produced and obtaining an error matrix. The training data for each class must be relatively homogeneous, i.e. all pixels must represent that class, and they must be numerous enough to represent and permit understanding all the variability in the class.

In the management of protected natural areas and of the natural environment in general, remote sensing permits inventory and control of the state of the environment, as in monitoring temporal changes on the ground (*change detection*) by comparing images taken at different times; evaluating changes in the forests, in cultivated and urban areas; and the monitoring of invasive alien species (Ustin et al. 2002). In forest management, remote sensing is used to estimate forest cover (extent, areal percentage, etc.), survey the health of the forests, estimate burned areas, analyze vegetation phenology and distribution, etc.

Up to now, however, satellite imagery still cannot recognize and delimit all the vegetation units occurring in a given territory, so verification on the ground remains always mandatory.

In this regard we mention the case of the Foresta Umbra ("shady forest") on the top of the Monte Gargano, a large rocky promontory on the Adriatic coast of middle

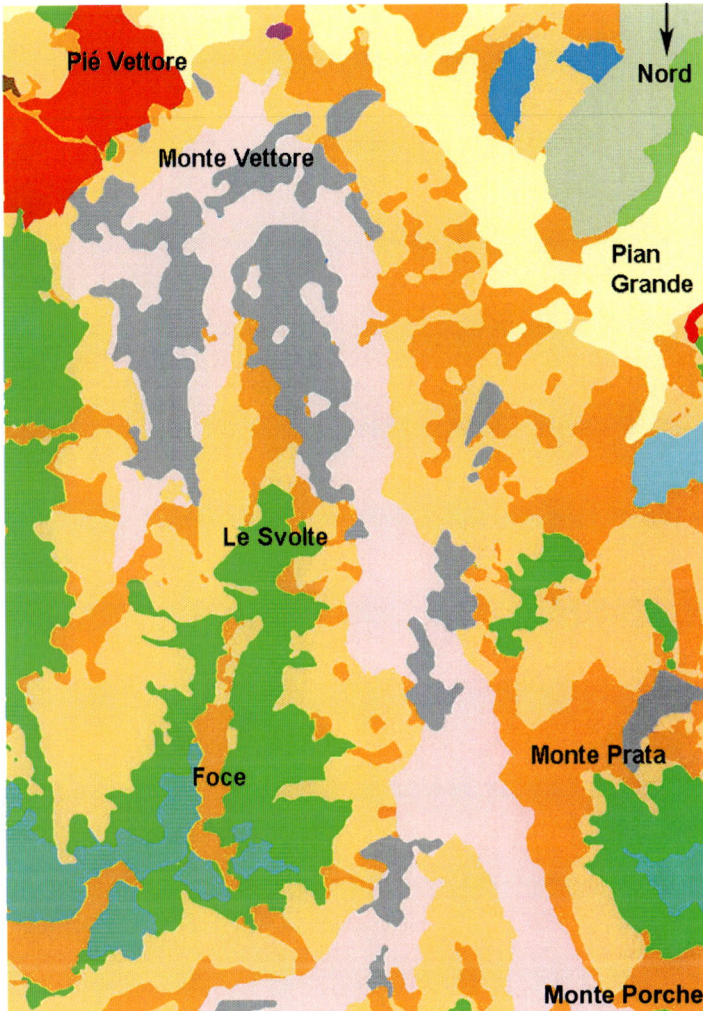

Fig. 5.40 Vegetation map of the Sibillini Mountains, area of the Lago di Pilato valley, Marche Region, central Italy, at scale 1:50,000; see also Fig. 5.41

Italy. This is a vast forest formed by tall beech (*Fagus sylvatica*) and hornbeam (*Carpinus betulus*) trees, with the two associations Aremonio-Fagetum and Doronico-Carpinetum, which constitute the primary vegetation. The map of the Foresta Umbra obtained from field sampling is shown at left in Figs. 5.42 and 5.43 and that obtained from satellite imagery on the right. The delimitation of the deciduous forest is very good on the map made from satellite data, albeit with the caution that it was not possible to subdivide the two associations of beech and hornbeam. Missing completely, though, is the vegetation of the quite numerous dolines (Taxo-Fagetum), of small glades (secondary meadows), and of somewhat

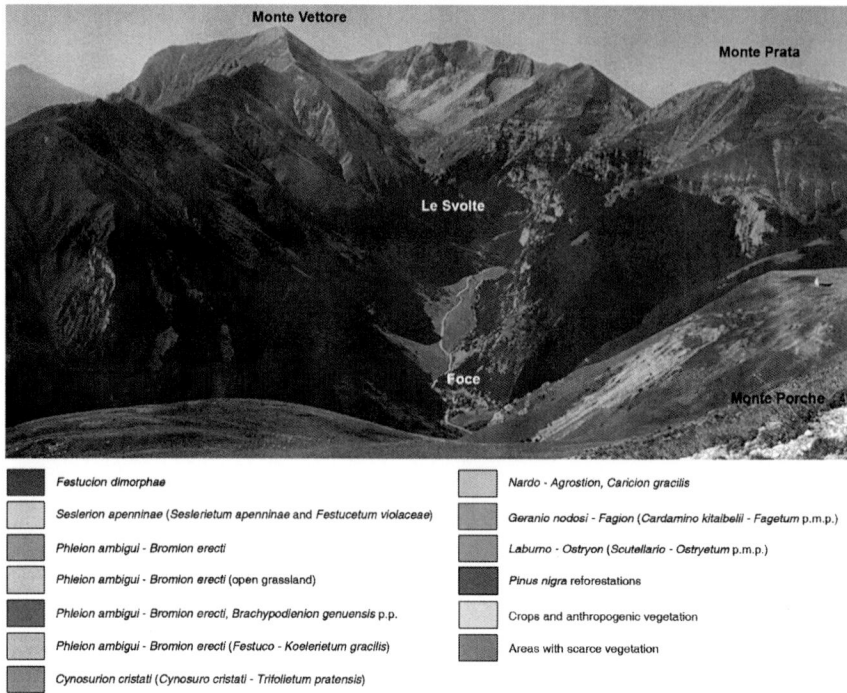

Festucion dimorphae

Seslerion apenninae (Seslerietum apenninae and Festucetum violaceae)

Phleion ambigui - Bromion erecti

Phleion ambigui - Bromion erecti (open grassland)

Phleion ambigui - Bromion erecti, Brachypodienion genuensis p.p.

Phleion ambigui - Bromion erecti (Festuco - Koelerietum gracilis)

Cynosurion cristati (Cynosuro cristati - Trifolietum pratensis)

Nardo - Agrostion, Caricion gracilis

Geranio nodosi - Fagion (Cardamino kitaibelii - Fagetum p.m.p.)

Laburno - Ostryon (Scutellario - Ostryetum p.m.p.)

Pinus nigra reforestations

Crops and anthropogenic vegetation

Areas with scarce vegetation

Fig. 5.41 Panorama of the area shown in Fig. 5.40, with the village Foce (945 m), Le Svolte (1,050 m), Mt. Vettore (2,456 m), Mt. Argentella (2,200 m) and Mt. Porche (2,233 m); Marche Region, central Italy (Photo by Jessica Mazzarelli)

rocky slopes (*Quercus ilex*). From this one can deduce that the map obtained by remote sensing is very useful as a reference basis for a subsequent map made by traditional methods. The same considerations are valid for the vegetation map of Monti Sibillini (central Italy); by comparing the vegetation map and the landscape section shown in Fig. 5.40, one can see a good correspondence in the distribution of *Fagus sylvatica* and also in the vegetation of the clearing in the valley bottom (near Foce). Still this map presents two problems, one of accuracy of the vegetation boundaries, and another in the number of associations represented (not all could be discerned by the remote sensing) and the number of cartographic units. Seen strictly botanically, satellite imagery can perform the same functions as aerial photography for identifying and delimiting vegetation types. This is of great value for furnishing a general view of the distribution of vegetation and for the solution of many applied problems.

A new methodology uses spectral feature analysis of data from the Airborne Visible/Infrared Imaging Spectrometer (AVIRIS). From this is was possible to obtain very detailed maps of forest vegetation (Fig. 5.44) as well as non-forest vegetation (Kokaly et al. 2003); such maps can be used for problems of environmental monitoring but may also aid in the production of subsequent vegetation maps in certainly much more detail.

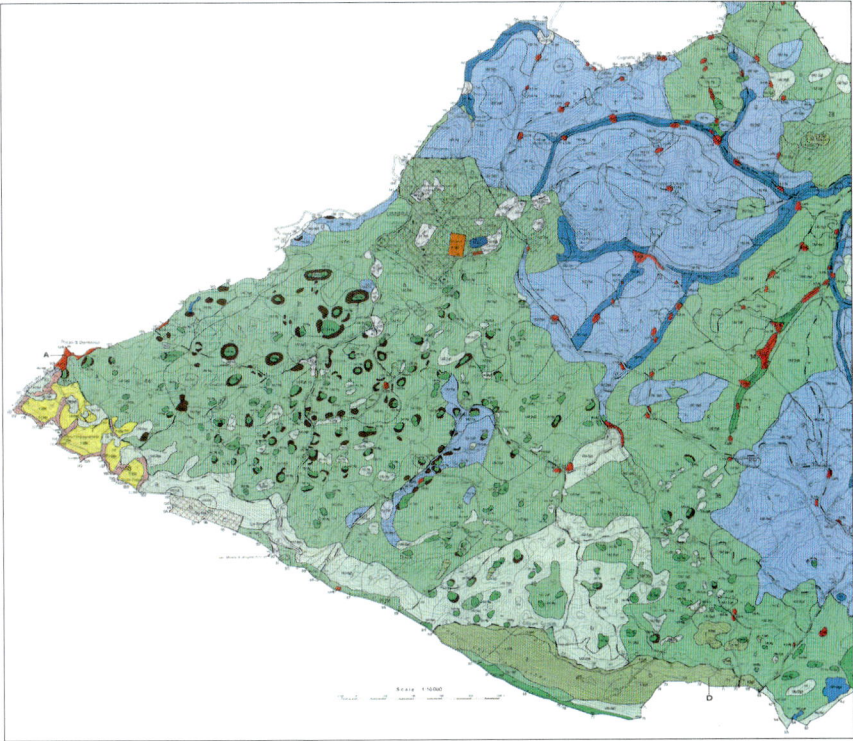

Fig. 5.42 Vegetation map of the Foresta Umbra, Apulia Region, Adriatic southern Italy, at scale 1:10,000 but with the size reduced to permit comparison with the map in Fig. 5.43 (From Pedrotti and Faliński 2002)

Confidence Limits in Vegetation Mapping

Cartographic surveys of vegetation are based, first of all, on the premise that the data surveyed are to be placed on a map, with implications for scale, classification of vegetation units, level of detail to be attained, etc.

The confidence one can have in the methods of vegetation mapping depends on the initial theoretical premises and on correct application of these methods, in both stages of the work, i.e. in the field and in the laboratory. Interesting, in this regard, is the cartographic experience obtained in an area of 6 km^2 in the Białowieza National Park (Poland), where a map of the actual vegetation was made by six research groups working independently (Faliński 1994). Five groups used the classical phytosociological method, while the sixth made use of an integrated synusial method. According to a comparative analysis of the six maps produced (Greco et al. 1994), the methodology normally employed in phytosociological mapping provided a high level of confidence (60 %): the authors of five of the six were in fact substantially in agreement, at least at the physiognomic level. The level of

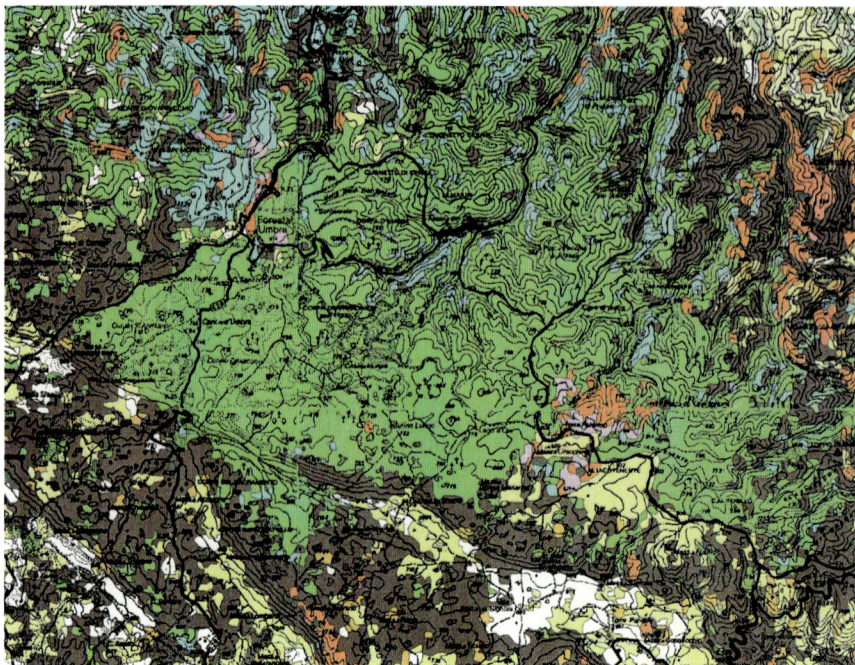

Fig. 5.43 Vegetation map of the Foresta Umbra obtained from the satellite image from Fig. 5.42; the *green color* shows deciduous forests (*Aremonio-Fagetum* and *Doronico-Carpinetum*), *brown* shows *Quercus pubescens* forests, and *yellow* shows clearings with herbaceaous vegetation (*Processed by Sergio Ruggieri, Camerino*)

agreement diminished rapidly, though, falling to 44 % if all the syntaxa reported were considered, many of which were only facies or variants at low levels in the hierarchy, identifiable only at a scale finer than that utilized (1:10,000). Agreement depends also on the initial selection of the research groups, since some did not map subassociations and facies of the forest association and others gave up on mapping the widespread but very small meadow and wetland associations.

Confidence in the methods used for phytosociological mapping of the vegetation was much higher in areas with higher degrees of naturalness, such as the various sectors inside the Białowieza National Park, while the areas with greater uncertainty of representation were those characterized by substitute communities. In this last case there were frequent errors in localization of the mapped types, especially in an area outside the Park and subject to cutting by humans (Pedrotti and Venanzoni 1994a).

On the whole, the classical phytosociological method of mapping vegetation units appears to be quite adequate for representing reality; the few discrepancies among the authors appear only in less relevant aspects. Interpreting one or another of the five comparable maps, one can see that no information has been lost. For the sixth map, it is necessary finally to admit the possibility that one research group out of six did produce discrepant results that could not be generalized, as these results were related

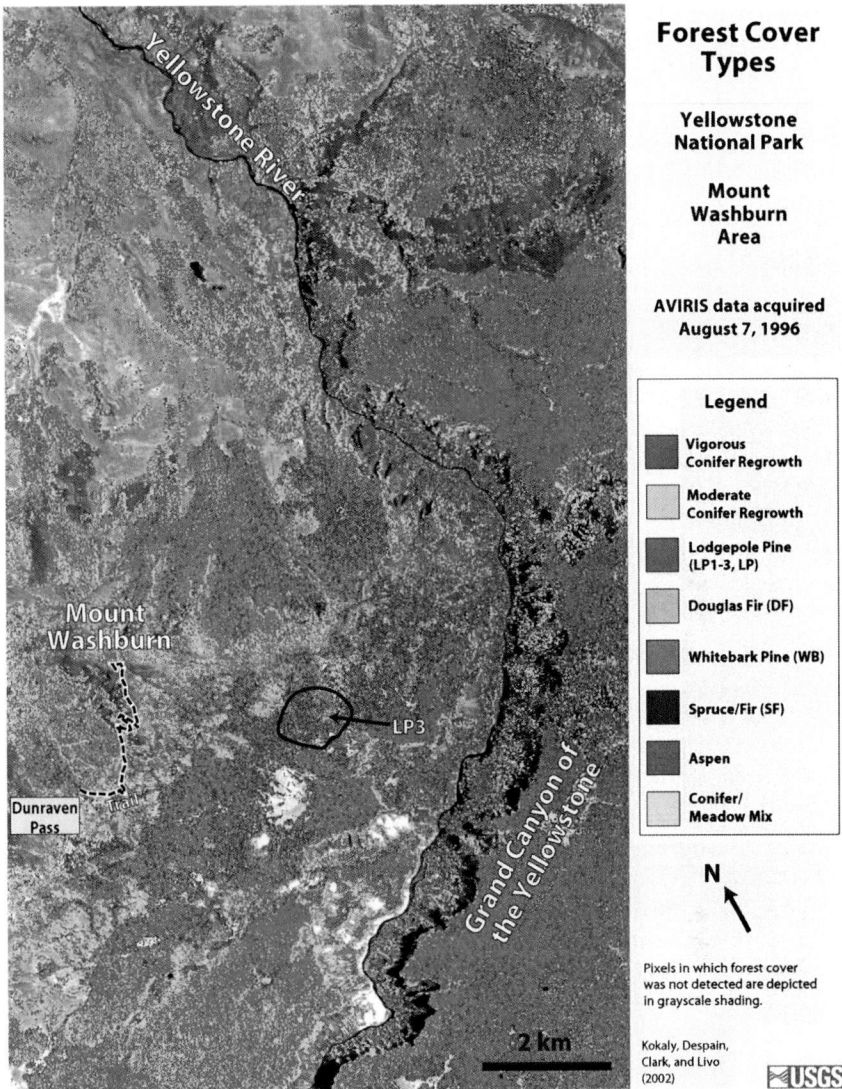

Forest Cover Types

Yellowstone National Park

Mount Washburn Area

AVIRIS data acquired August 7, 1996

Legend

- Vigorous Conifer Regrowth
- Moderate Conifer Regrowth
- Lodgepole Pine (LP1-3, LP)
- Douglas Fir (DF)
- Whitebark Pine (WB)
- Spruce/Fir (SF)
- Aspen
- Conifer/ Meadow Mix

N

Pixels in which forest cover was not detected are depicted in grayscale shading.

Kokaly, Despain, Clark, and Livo (2002)

Fig. 5.44 Map of forest cover types of the Mount Washburn area of Yellowstone National Park (Wyoming, USA), derived from AVIRIS data and the USGS Tetracorder export system (From Kokaly et al. 2003)

to the initial choice of premises for making the map. For the accuracy of the boundaries surveyed and, consequently, how the cartographic units are represented, it seems that some authors considered this much more important than others.

In conclusion, one must emphasize that geobotanical mapping is a type of cartography, both thematic and analytical, that can be done only at the end of a cognitive process of interpreting the vegetation as a biological/ecological fact. The cartographic representation of natural data does not make the claim to "transform reality into certainty of the representation", as Gambino (1991) would have wanted; in fact, from a theoretical viewpoint, the "certainty of representation" (Farinelli 1989) is simply and only a function of the author's depth of knowledge of the vegetation. It depends on various circumstances, in particular the accuracy of the botanical data collection in the field and the use of more precise survey techniques (see also Chap. 7).

Editing Vegetation Maps

After the vegetation survey in the field and the making of a field sketch map (Fig. 5.45), it is necessary to proceed in the laboratory to revise the work done so far and to rework it according to two directives. One is graphical and related to the method for representing the cartographic units surveyed, revising attentively the complex of curves that constitute the vegetation boundaries; the second is related to the content, selecting the associations (or other vegetation units) that will be represented on the map and that will constitute the definitive legend. To the map surveyed in the field, which is normally a map of actual vegetation, it is possible to add complementary maps, such as for the geographic location of the territory mapped, hypsometry, geology, climatology, pedology and potential vegetation (Fig. 5.46). In order to avoid errors in reading the map, when the number of colors is large and thus can easily be confused, one can add an index number (or letter) in each mapped unit, which must appear also in the map legend.

The map will normally be printed at a scale equal to or broader than that used for the field survey; if not, the precision in form of the mapped units diminishes proportionally with the enlargement.

Cartography with Geographic Information Systems

Vegetation maps can today be digitalized, providing access to all the advantages offered by information systems; and there are many advantages, such as precision, speed of production, and workspace required.

Of the various possibilities, the most useful is the geographic information systems (GIS), so called because its fundamental characteristic is the certainty that all data are geo-referenced. That means that all elements (points, lines, polygons, and pixels) are identifiable by coordinates according to a single, specific reference system (UTMG, Gauss-Boaga, etc.). The GIS is an instrument for acquiring, storing, studying, transforming and visualizing georeferenced data with an aim to satisfy a wide range of application needs. The set of data manipulated represents a model of the real world that can be used to study environmental phenomena, to plan actions to be taken, and to optimize decision-making processes. The data in a GIS are constituted by two components: a spatial component that describes the position (coordinates) of an object according to a defined reference system, and a descriptive component, carrying alphanumeric information called attributes.

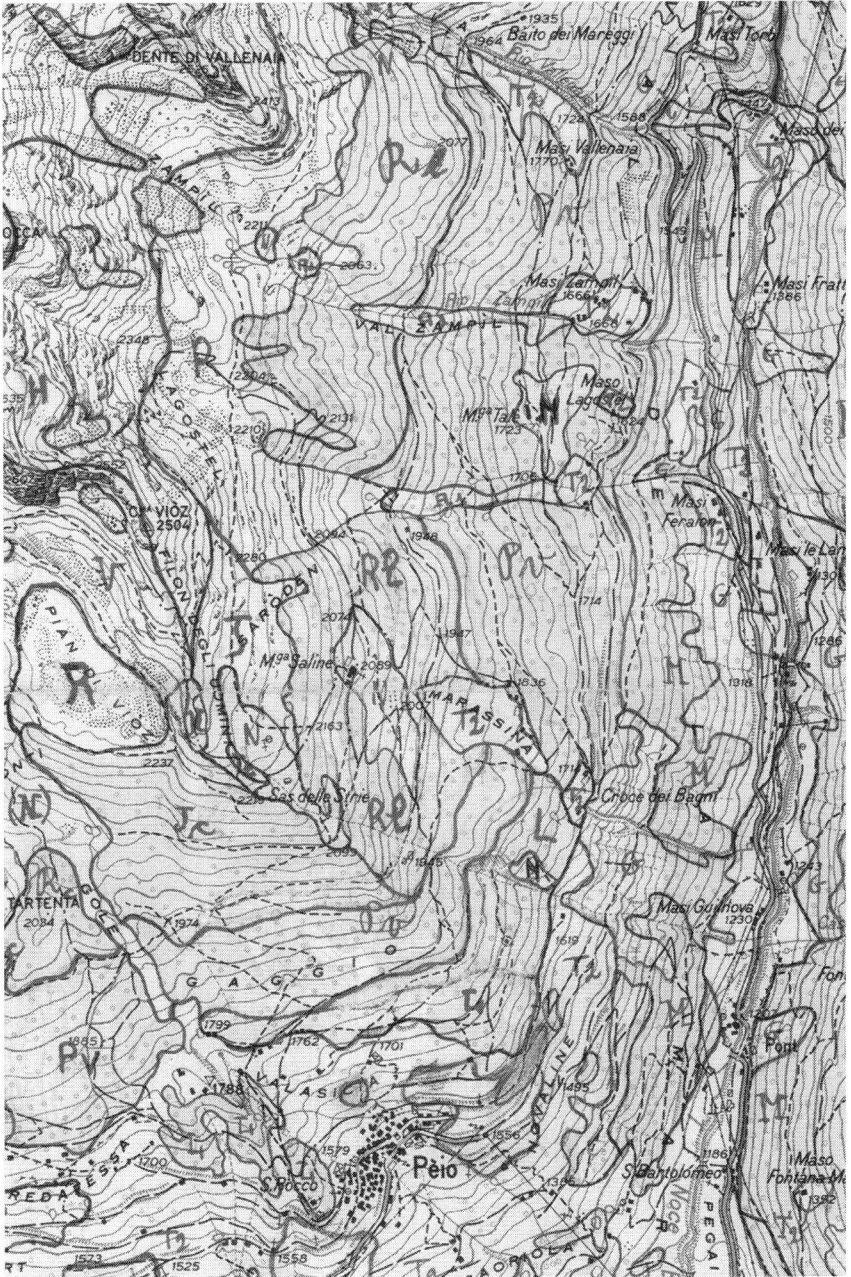

Fig. 5.45 Field sketch (draft) of one I.G.M. map (scale 1:25,000) employed for vegetation survey in the Stelvio National Park, northern Italy

authors
map title

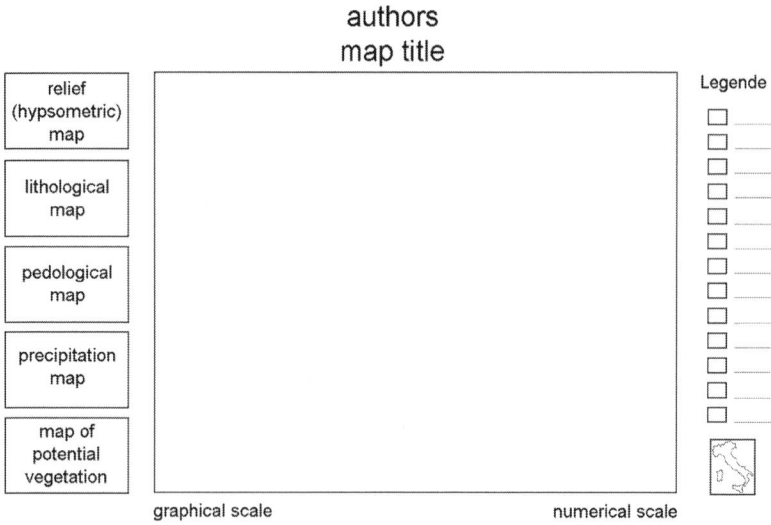

Fig. 5.46 Model table format for the vegetation map

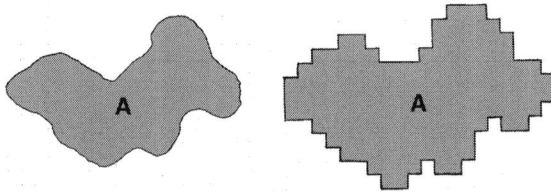

Fig. 5.47 The same patch A (base unit) displayed in vector format (at *left*) and in raster format (at *right*); the shapes of the patches are the same, but the areas and perimeters do not coincide

The reproduction of territory may be done according to two models, rasters of pixels and vectors connecting points. Rasters represent a regular grid in which each cell, called a pixel, expresses an alphanumeric value; vectors, on the other hand, describe the objects by points, lines, polygons and nodes through the coordinates of one or more significant points. Vector data are more faithful to reality, even though more data are necessary for their use (Fig. 5.47). The data model has an influence on the possible operations, so one uses data in the format more consistent with the operations to be performed, converting raster data into vector data and vice-versa. Vector information is organized in themes, each bearing some content relative to some topic (road networks, hydrography, urban, etc.); each theme is supported, in the GIS, by an information stratum called a *layer*. A geographic map may be produced by the superimposition or other processing of various layers of information. Data management is done by using archiving systems denoted DBMS (*Data-Base Management Systems*), which may be combined in a single system or separated into a geometric data-base and an interconnected alphanumeric structure. Data manipulation is done at various levels, with the possibility of analysis through logical functions, aggregation or extraction of elements, arithmetic functions such as summation and difference, overlay functions, combination of two or more data layers, or geometric functions; it is also possible to create *buffers* around particular elements.

Computerized mapping of vegetation means creation and management of a bank of data relative to topology (relative position), spatial relationships (distances) and other attributes (degree of cover, phytomass, phytodiversity, etc.) that characterize the cartographic units. A "computerized" map is thus also a bank of data that are easily accessible, and this offers great possibilities for modeling the distribution of vegetation types in relation to climate, geomorphology, anthropogenic factors, and other factors that vary spatially. The various layers (topography, geology, vegetative cover, etc.), when geo-referenced, can interact and be integrated permitting creation of models involving information present in each thematic layer.

For applications, there are two basic categories of GIS. *Land Information Systems* (LIS) are management systems dedicated primarily to services such as cadastres, technological networks, etc. *Decision Support Systems* (IDSS) are systems designed to support decision-making groups on projects tied to land transformation.

The potential of GIS, represented above all by their extreme speed of processing large quantities of data over wide areas, makes them a very efficient tool in the field of geobotanical cartography.

These systems can be used also for construction of thematic maps made by overlaying cartographies made in different years (vegetation dynamics).

Finally one might note that it is also possible to computerize "classical" vegetation maps produced on paper in the past. These are valuable documents, rich in information, and have great value; they should be maintained as useful temporal references. With computerization, these maps become available for successive modification and processing.

Subjectivity and Positional Errors in Vegetation Mapping

A vegetation map cannot claim to be an exact replica of the complex, real vegetation features.

This concerns two points, the scientific content of the map and the vegetation boundaries.

In addition, a vegetation map can not represent all the features of the vegetation, but only the part we have decided *a priori* to consider.

In making vegetation maps, there are always some errors during the field survey and related to instrument and software performance during aerial photography, rectification of the photographs, georeferencing and digitising. Human subjectivity, of course, always represents another possible source of error.

Aspinall and Pearson (1995) recognised three components of potential human errors in thematic maps, involving class identity, heterogeneity within a polygon and boundary location. Delineating polygons during photointerpretation involves a generalisation that depends on the accepted level of vegetation heterogeneity within polygons but also on visual accuracy and the experience of the interpreter. In addition, assigning polygons to vegetation units involves abstraction.

In order to record the position and extent of vegetation units, a 'hard' boundary is drawn around an area that is relatively but sufficiently homogeneous in terms of floristic composition, namely the vegetation patch. The boundaries drawn, however, do not necessarily reflect the existence of sharp transitions between vegetation patches (Green and Hartley 2000). Generally, natural and semi-natural landscapes feature ecotones, for which the use of fuzzy or 'soft' boundaries would be more appropriate. The problem lies not only in drawing the boundary but also in the complexity of its shape, which in turn reflects how much vegetation detail is to be represented on the map. Therefore, it may be practically useful and theoretically important to quantify the uncertainty in boundary location.

The error associated with the position of between-patch boundaries can be illustrated through the so-called epsilon bands that are overlaid over the limits of the polygons. Epsilon bands enclose a zone that has a specific probability of including the 'true' location of the boundary (Dunn et al. 1990; Goodchild 1993). To estimate how large a statistically defined epsilon band should be, one must have some knowledge on the magnitude and distribution of deviations associated with the boundaries.

An empirical approach to quantifying positional boundary errors requires comparison of several surveys conducted by different people, as in the case of the Białowieza forest, noted before. In that case, six surveyors identified the vegetation boundaries with the help of aerial photographs and field verification.

Today, this comparison can be done in greater detail with maps produced by GIS, which may involve photointerpretations done by different people and overlays of the resulting layers. Undoubtedly, some boundaries will vary widely (Drummond 1990; Walsby 1995) and consequently, will form many "sliver polygons" of classes that differ in the two layers (Chrisman and Lester 1991). The interpreted vegetation limits can be considered to be independent estimates for the position of the true boundary (Green and Hartley 2000). The magnitude of the deviations can be measured in a GIS at a number of randomly located points as the width of the sliver polygon. If measured many times, these widths would produce a probability distribution function for the positional deviations of the overlay. By using intensive re-sampling procedures (e.g., bootstrapping), one can then estimate the standard error of the mean deviation and a reasonable confidence interval. In this way, appropriate epsilon bands of boundary uncertainty can be appraised and transposed as buffer zones of different widths that encompass the 'true' position of the vegetation boundary at a given level of alpha probability.

In practice, epsilon bands are especially useful for the representation of 'soft' ('fuzzy') boundaries that are often revealed only at fine scales i.e., where large ecotones occur between (semi)-natural vegetation patches. Even 'hard' boundaries, however, can become polygons (transitional vegetation units) at finer scales, as for forestfringe vegetation (*Trifolio-Geranietea*).

Types of Vegetation Maps

6

There are many types of vegetation maps, for various reasons. First of all, vegetation associations possess particular intrinsic characteristics based on their floristic composition, structure, synecology, syndynamics and synchorology. Each of these may be represented on a map, yielding maps of quite different meaning. Second, maps may vary according to scale and definition of the vegetation units. Third, a vegetation map depends on the theoretical conceptions of the different geobotanical schools and thus on the vegetation interpretation and classification resulting from these different approaches. Finally, maps may have "mixed" characteristics, i.e. may involve two or more thematic bases, such as actual vegetation and altitudinal belts.

For such reasons, the cartographic typologies proposed by various authors differ considerably, because each is inspired by different criteria, emphasizing some map characteristics more than others. So, no classification is presented here but only a list of the main types of vegetation map. The list was compiled keeping in mind the evolution of geobotanical thought since the middle of the 1800s, when the first vegetation maps were made. Listed first are those maps that can be defined as "fundamental" because they refer to the classification of vegetation and so represent the starting point for the production of other maps, such as maps of dynamics or phytoecology. In treating the different map types, some description is also given of the theories that inspired the various schools of geobotany.

The following types of vegetation map are considered here:

Physiognomic maps

Phytosociological maps

Phytosociological maps of actual vegetation (syntaxonomic units)

Integrated phytosociological maps (synphytosociological, involving vegetation series or sigmeta)

Phytosociological maps of potential vegetation (syntaxonomical climax units)

Phytogeocoenological maps (phytogeocoenoses)

Maps of coeno-associations

Maps of vegetation "bands" (cingoli, ceintures, Gürtel)

Maps of vegetation dynamics

F. Pedrotti, *Plant and Vegetation Mapping*, Geobotany Studies,
DOI 10.1007/978-3-642-30235-0_6, © Springer-Verlag Berlin Heidelberg 2013

Synchorological maps
Phytoecological maps
Maps of the state of vegetation conservation (naturalness, synanthropization, etc.)

Physiognomic Maps

Physiognomic maps show the basic physical structure of the vegetation (forest, shrubland, grassland, etc.), as based on the main growth forms (trees, shrubs, grasses, etc.) of the dominant or co-dominant species in the vegetation formation. The result is that the vegetation formations are defined rather generically, such as deciduous forests, conifer forests, formations of evergreen sclerophylls, etc. Moreover, the definitions and classifications differ with various authors, who have been occupied with general description of the vegetation, from Grisebach (1872) and Brockmann-Jerosch (1930), until Dansereau (1951, 1957), Fosberg (1961), Ellenberg and Müller-Dombois (1967), Tomaselli (1970a) and others. Recently, Rivas-Martínez (1996) distinguishes 22 climate-based types of forest for the whole world.

The first vegetation maps were physiognomic, among them the "Phytogeographic Map of the Bernina Massif" (Switzerland) at the scale 1:50,000 by Rübel (1912b), which provided a prelude to the phytosociological maps. Also, the *Carta Geobotanica della Repubblica Dominicana* (Geobotanical map of the Dominican Republic) at scale 1:1,100,000 by Ciferri (1936) is a physiognomic map that distinguished vegetation formations on an ecological-physiognomic basis, such as hyperxerophytic and subxerophytic forest, moist savanna, etc.

These were rather general maps that constitute first attempts to portray the vegetation of a particular territory; normally these maps are at broad scale and show vast territories. Still it is possible, though, to enrich the content of these maps, as was done for example in Australia by Beard (1979), who classified the vegetation formations in three categories, expressed by an index formed from three letters (formula of Beard-Webb) that indicate: the growth form in the most important vegetation layer (tall, medium or short trees, shrubs, dwarf shrubs, bunch and cushion grasses, other herbaceous plants, mosses and lichens); the dominant genera; and the degree of ground cover. These maps were made at scales of 1:250,000 and 1:1,000,000, with a synthesis map for all of Western Australia at the scale 1:3,000,000 (Fig. 6.1).

The *Map of the Vegetation of China* (1:4,000,000) by the Institute of Botany, Academia Sinica (Hou 1979, 1983), distinguishes the following vegetation formations: needle-leaved forest, broad-leaved forest, shrub formations, desert, steppe, savanna, grassland, and wetlands (Fig. 6.2). These units have been subdivided based on phytoclimatic regions and on dominance by one or a few species. The presence of species of particular phytogeographical significance, such as *Metasequoia glyptostroboides*, is indicated by particular symbols.

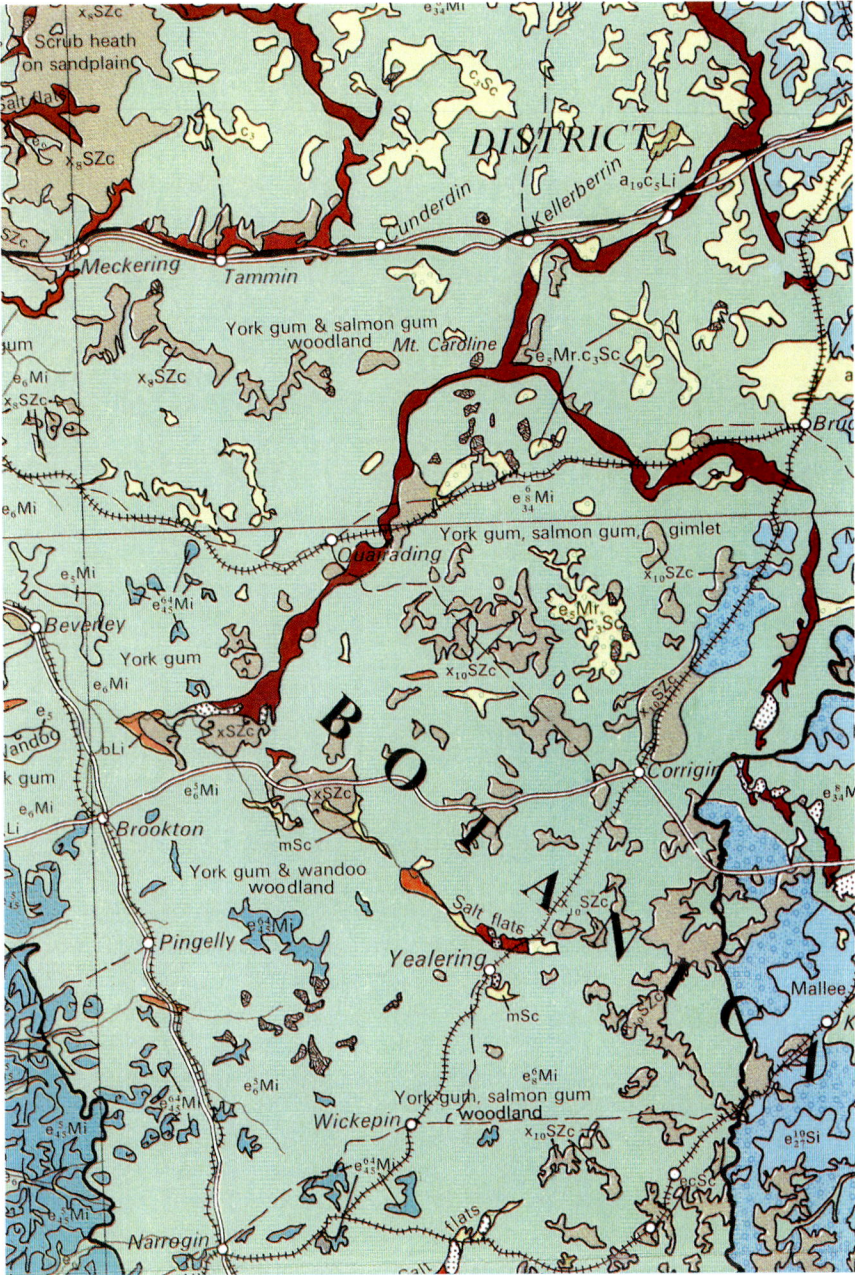

Fig. 6.1 Vegetation map of Western Australia, Swan sheet, at scale 1:1,000,000. The colors represent the following vegetation types: *green-blue* = woodland of *Eucalyptus calophylla*, *E. wandoo*, *E. astringens*, *E. accedens*; *light green* = woodland of *Eucalyptus marginata* with other species of *Eucalyptus*; *light red* = low woodland of *Banksia attenuata* and *B. menziesii*; *yellow* = shrubland of *Acacia* sp. and *Casuarina* sp.; *blue* = shrubland of *Eucalyptus tetragona* (tall-shrubby *Eucalyptus* called "mallee"); *sepia* = shrubland of *Proteaceae-Myrtaceae*; and *brown* = succulent steppe of *Arthrocnemum* sp. (From Beard 1981)

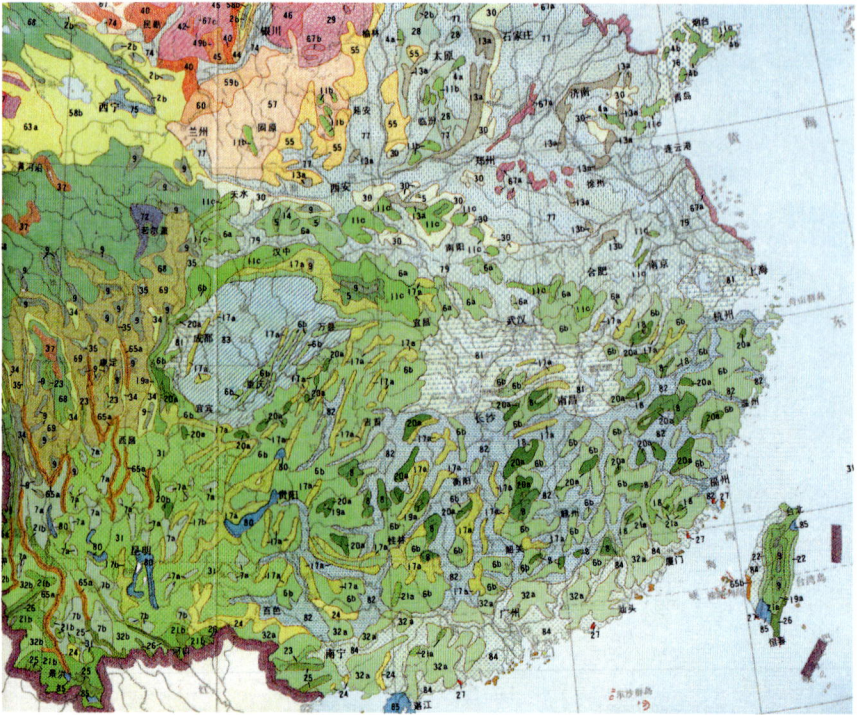

Fig. 6.2 Vegetation map of China, at scale 1:14,000,000, south-eastern part; this is a physiognomic map with 85 formations (From Hou 1983)

Phytosociological Maps

In phytosociological mapping by the Zürich-Montpellier School (Braun-Blanquet 1964), it is possible to identify various levels of integration and thus of cartographic representation, according to the units to be mapped. The units are plant associations and series of vegetation or sigmeta (or sigma-associations), which belong to levels II and III in the list of map types from Chap. 1. It is also possible to distinguish subtypes of phytosociological map, in particular phytosociological maps of actual vegetation (i.e. maps of syntaxonomic units), integrated phytosociological maps (i.e. maps of vegetation series or sigmeta or sigma-associations), and phytosociological maps of potential vegetation (i.e. maps of climax syntaxonomical units).

Geo-synphytosociological maps, i.e. maps of geosigmeta, are no longer vegetation maps in the strict sense, but rather are maps of grand complexes of vegetation (see Chap. 8).

Phytosociological Maps of Actual Vegetation

Phytosociological maps of actual (or "real") vegetation represent the vegetation that is observed in the field at the moment of survey. Such maps show the spatial distribution of the vegetation units belonging to various syntaxa of the hierarchical phytosociological system, i.e. associations, subassociations, variants, facies, alliances, orders and classes; these are classical phytosociological maps. The plant associations are represented with various colors according to the phytosociological system; for example, all the associations belonging to a particular alliance or order may be represented by the same color but with different tones. Up until now there still do not exist any codifications or guidelines for the use of colors in phytosociological maps (Fig. 6.3).

In environments such as high mountains or undrained bogs, scarcely influenced by man, phytosociological maps of actual vegetation show the primary spatial differentiation of the vegetation in natural environments. Where original (or "primary") phytocoenoses have been disturbed by human activities, such as deforested areas and occupied today by meadows or agricultural fields, phytosociological maps instead show the secondary differentiation of vegetation. These constitute an inventory of the plant resources of a particular area and of biodiversity at a phytocoenotic level.

One of the first modern phytosociological maps was that of Braun-Blanquet (1937–1943), showing an area near Montpellier in Languedoc (1:20,000). The vegetation there was represented by associations, subassociations, stages, facies and variants, with colors, symbols and abbreviations. The legend of the map is very detailed and includes short descriptions of the associations, combined into alliances.

From phytosociological maps of actual vegetation it is possible to deduce numerous botanical and environmental indicators. From one map of the actual vegetation of the Białowieza forest, Faliński (1999) obtained six derivative maps showing the vertical structure of phytocoenoses, species richness, dynamic stage (based on springtime presence of various biological forms), main environmental conditions, soil moistness, and light levels.

Indices can also be added to phytosociological maps, to identify the constituent cartographic units. In such a way one can extract objective information on the frequency, fragmentation, form and spatial distribution of communities in the area, and on phytocoenotic diversity.

Geographic information systems (GIS) are particularly convenient for this kind of subsequent processing, and the many possibilities to derive information from vegetation maps by automated methods are far from being completely determined.

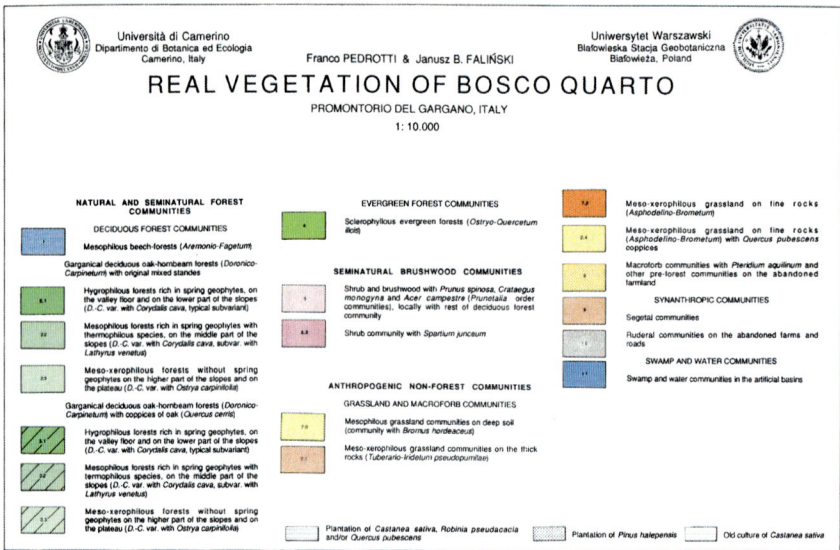

Fig. 6.3 Vegetation map of the "Bosco Quarto" forest, Apulia Region, southern Italy (From Faliński and Pedrotti 1990)

Phytosociological Maps of Actual Vegetation with Zonal Units

Zonal units are the various bioclimatic zones and altitudinal belts seen from a global perspective, and mapping with these, in combination with associations or other more "concrete" vegetation units, can produce a very interesting kind of map. Only a single example is known so far, namely the *Vegetaţia* (Map of the Vegetation of Romania) at scale 1:1,000,000 by Doniţă and Roman (1979). This map (Fig. 6.4) has a mixed character because in it the vegetation associations (actual vegetation) are regrouped according to bioclimatic zones and altitudinal belts of vegetation (see Chap. 10). Zones correspond to units of broad-scale potential. In this map the vegetation associations were combined into three groups and mapped accordingly: (1) Zonal units based on altitude, i.e. according to altitudinal belts; (2) Zonal units based on latitude, i.e. according to vegetation zones; and (3) Intrazonal and azonal units.

Integrated Phytosociological Maps

Integrated phytosociological maps, or synphytosociological maps, represent vegetation series, also called sigmeta, sigma-associations or synassociations according to the concepts of Tüxen (1979), Rivas-Martínez (1985) and Géhu (1987, 1991a). A vegetation series is the quantified spatial sum of the vegetation associations that compose it and refers to an ecologically homogeneous area, called a *tesela* in Spanish (meaning an ecotope or "patch"). The term *sigmetum* (from the letters $\sum$) makes reference to the summation of the associations that compose the series (Fig. 6.5); these plant associations would all develop into the same type of potential vegetation, because the area is ecologically homogeneous.

Subsequently, Rivas-Martínez (2005a, b) has distinguished something called a *perma-sigmetum*, that is a sigmetum formed of permanent communities in unusual or extreme environments, such as the polar regions, extreme deserts, the summits of high mountains, rock walls, etc.

Compared with the dynamic series of vegetation by Gaussen (1954) and Ozenda (1964), the various stages that compose the series are indicated here by the plant associations actually appearing.

Each vegetation series is composed of several plant associations, from the initial pioneer association to intermediate associations and finally to the final associations, which constitute the more mature vegetation of the area and which represent the climax vegetation ("head" association of the series).

Integrated phytosociological maps show the ecological determinism (climatic, geomorphologic, etc.) of the sigmeta and the dynamic link that unites the associations of the series; these associations represent the spatio-temporal differentiation of the vegetation.

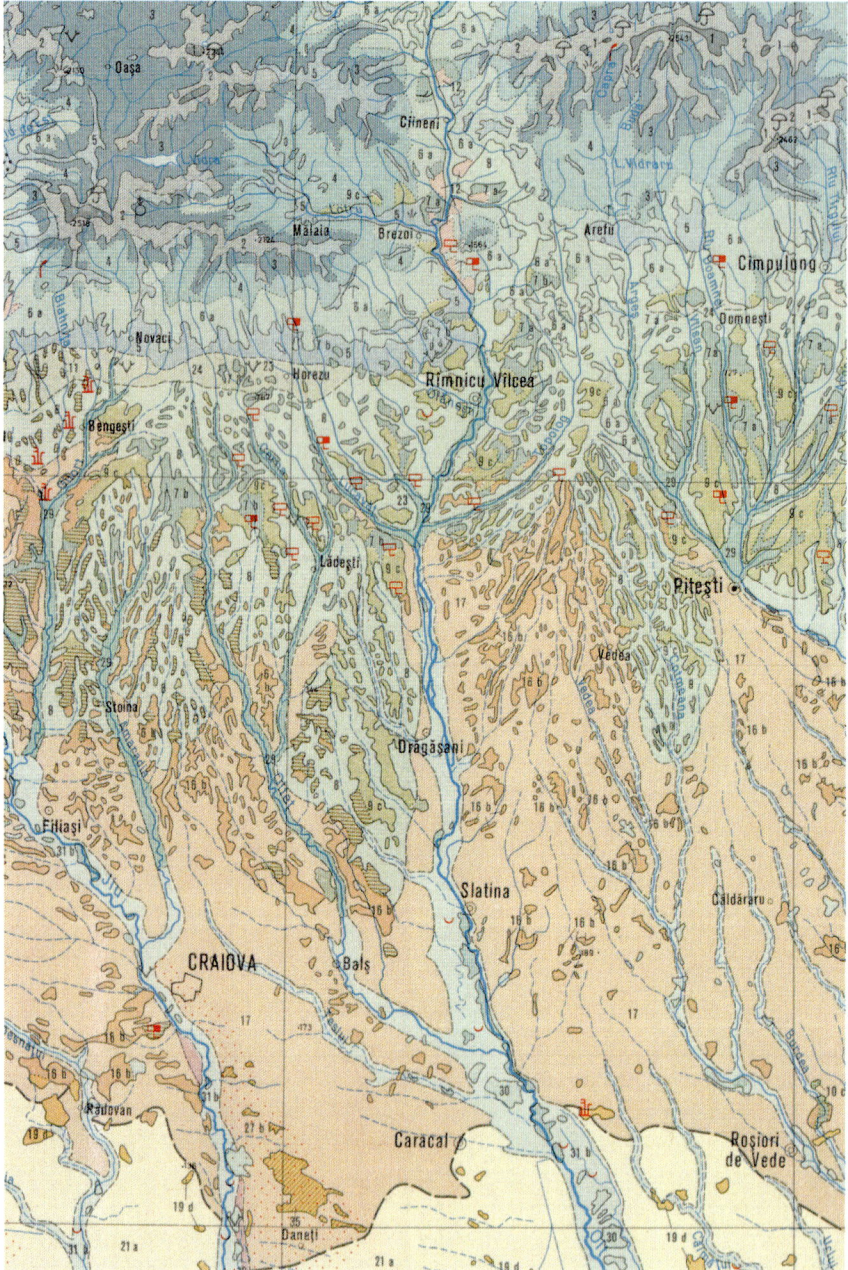

Fig. 6.4 Vegetation map of Romania, at scale 1:1,000,000, showing the area between the Carpathian Mountains (*in the north*) and the Olt river plain (*in the south*). The vegetation units are grouped as: zonal altitudinal units (belts), zonal latitudinal units, intrazonal units, and azonal units. The *green-blue* colors indicate associations of the alpine, subalpine and montane belts; *yellow-orange* indicates associations of the foothill (colline) belt and lowlands. On this map it is possible to see the relationship between the actual vegetation (generally small fragmented areas) and potential vegetation (large areas) (From Doniță and Roman 1979)

Fig. 6.5 Vegetation series of *Pinus sylvestris* of inner valleys of the central Italian Alps [*Astragalo-Pineto* sigmetum], Agumes, Val Venosta, Bolzano Province. The series is composed by: *A* – dry meadows of *Festuca valesiaca* and *Carex supina* (*Festuco-Caricetum supinae*); *B* – scrub of *Berberis vulgaris, Rosa montana, R. coriifolia, R. glauca,* etc. (*Berberido-Rosetum*); and *C* – forest of *Pinus sylvestris* (*Astragalo-Pinetum*) (Photo Franco Pedrotti)

Cartographically, this representation is done by using a suitable set of colors. On integrated phytosociological maps a given series is represented by a single color, which will have lighter tones for initial associations in the series and will proceed to darker tones for intermediate and then final associations (Fig. 6.6). An integrated phytosociological map on which only the climax associations are represented corresponds to a map of the potential vegetation, such as the *Mapa de Series de Vegetación de España* (Map of the Vegetation Series of Spain) at scale 1:400,000 by Rivas-Martínez (1987) and the *Mapa de Series de Vegetación de Navarra* (Map of the Vegetation Series of Navarra) at scale 1:200,000 by Loidi and Bascones (1995).

Such maps are also called "synphytosociological" maps, since they disclose diversity at a level higher than that of associations, namely the sum of the associations that constitute a vegetation series or sigmetum.

An integrated phytosociological map provides simultaneously a representation of the actual vegetation, through articulation of the different plant associations, and of their potential. In fact the associations are combined not according to the phytosociological system but by their dynamics; for such reasons, one can understand immediately, on an integrated phytosociological map, to which climax associations the various plant groupings in the series are tending or whether these already represent the climax vegetation of the area.

Integrated phytosociological maps contain, then, more information than the classical phytosociological maps of actual natural vegetation or those of potential vegetation.

CALCAREOUS SUBSTRATE

Series of *Fagus sylvatica*

Forest (*Polysticho-Fagetum*)
Mantle (*Prunetalia*)
Meadow (*Campanulo glomeratae-Cynosuretum*)
Meadow/Pasture (*Brizo mediae-Brometum erecti*)
Pasture (*Seslerio nitidae-Brometum erecti*)

Series of *Ostrya carpinifolia*

Forest (*Scutellario-Ostryetum*)
Mantle (*Spartio-Cytisetum sessilifolii*)
Meadow/Pasture (*Brizo mediae-Brometum erecti*)
Pasture (*Asperulo purpureae-Brometum erecti*)

MARINE-SAND SUBSTRATE

Series of *Quercus cerris*

Forest (*Aceri obtusati-Quercetum cerris*)
Mantle (*Junipero communis-Pyracanthetum coccineae*)
Meadow/Pasture (*Centaureo bracteatae-Brometum erecti*)
Pasture (*Coronillo minimae-Astragaletum*)

SAND SUBSTRATE

Series of *Fagus sylvatica*

Forest (*Carici sylvaticae-Fagetum*)
Mantle (*Prunetalia*)
Meadow/Pasture (*Achilleo collinae-Cynosuretum*)
Heath (*Calluno-Genistion*)

Serie of *Quercus cerris*

Forest (*Aceri obtusati-Quercetum cerris* subass. *pyretosum*)
Mantle (*Berberidion*)
Meadow/Pasture (*Centaureo bracteatae-Brometum erecti*)

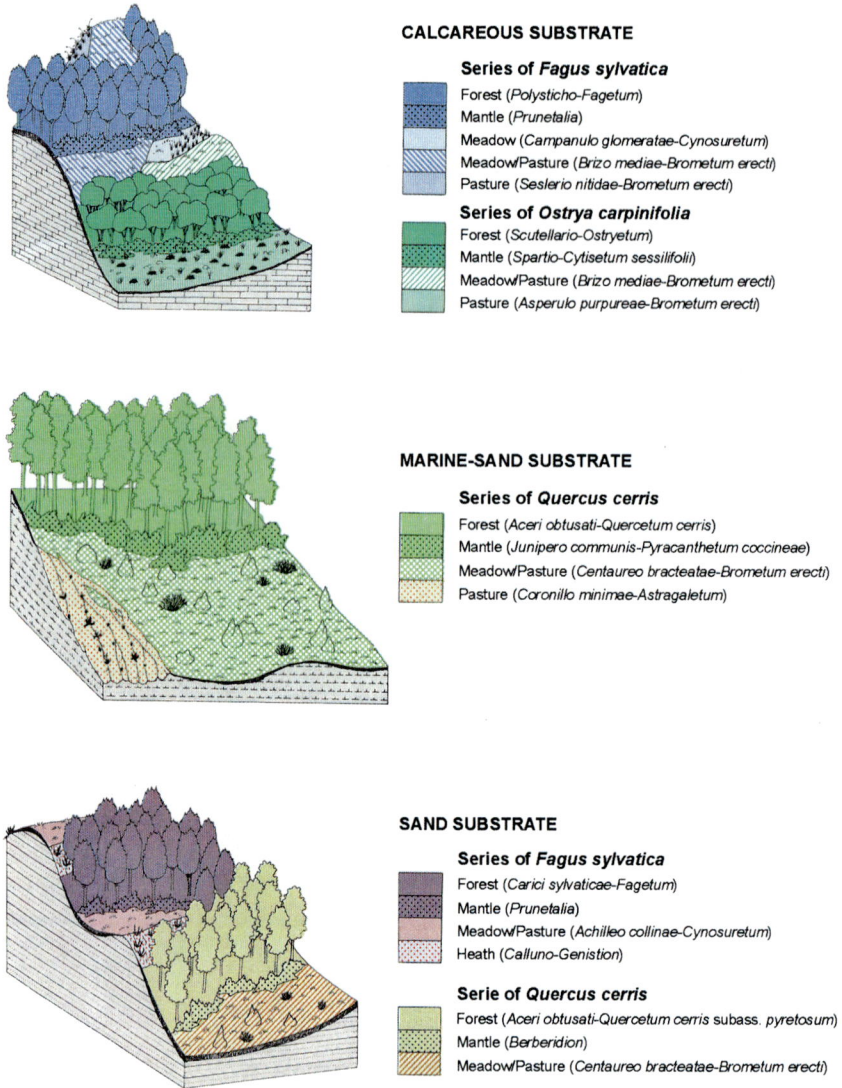

Fig. 6.6 Schematic representation of five vegetation series at Cagli, Marche Region, central Italy (From Biondi et al. 1990)

An example of an integrated phytosociological map at scale 1:50,000 is the *Carta della Vegetazione del Foglio Foligno* (Map of the Vegetation of the [official] Foligno sheet) on which 10 series of vegetation are shown, each of which is mapped with all the associations that compose it (Figs. 6.7, 6.8, and 6.9). Another map, at a

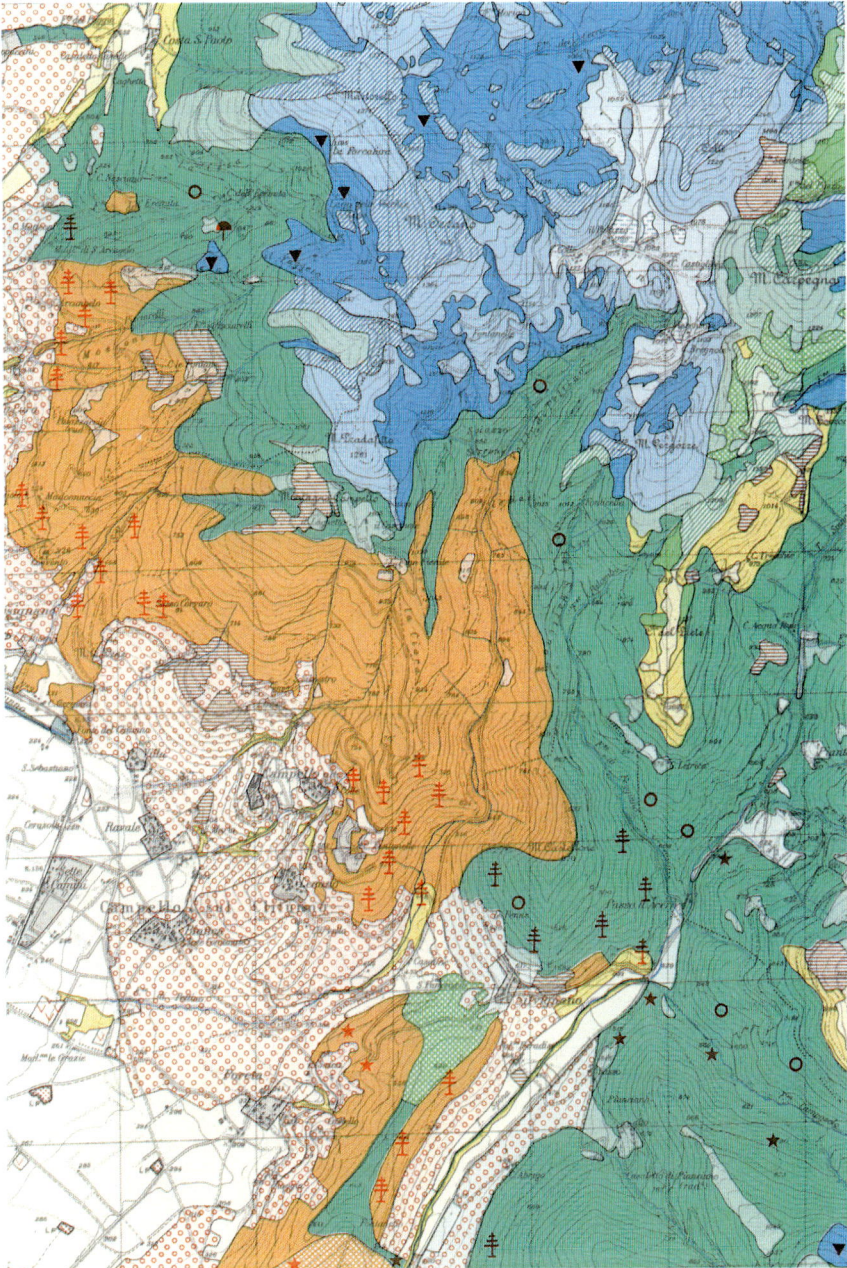

Fig. 6.7 Vegetation map (Foligno sheet), at scale 1:50,000, of the Campello sul Clitunno, Umbria Region, central Italy (From Orsomando 1993)

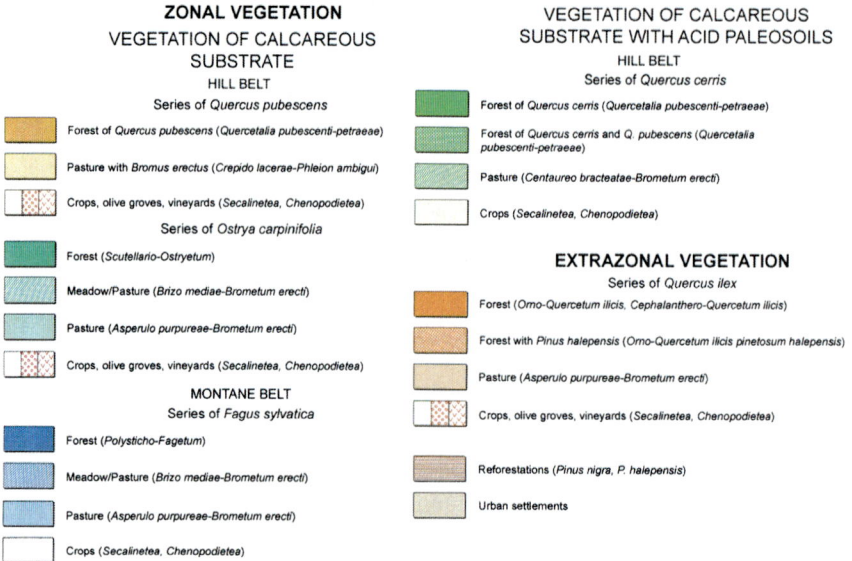

ZONAL VEGETATION
VEGETATION OF CALCAREOUS SUBSTRATE
HILL BELT
Series of *Quercus pubescens*

Forest of *Quercus pubescens* (*Quercetalia pubescenti-petraeae*)

Pasture with *Bromus erectus* (*Crepido lacerae-Phleion ambigui*)

Crops, olive groves, vineyards (*Secalinetea, Chenopodietea*)

Series of *Ostrya carpinifolia*

Forest (*Scutellario-Ostryetum*)

Meadow/Pasture (*Brizo mediae-Brometum erecti*)

Pasture (*Asperulo purpureae-Brometum erecti*)

Crops, olive groves, vineyards (*Secalinetea, Chenopodietea*)

MONTANE BELT
Series of *Fagus sylvatica*

Forest (*Polysticho-Fagetum*)

Meadow/Pasture (*Brizo mediae-Brometum erecti*)

Pasture (*Asperulo purpureae-Brometum erecti*)

Crops (*Secalinetea, Chenopodietea*)

VEGETATION OF CALCAREOUS SUBSTRATE WITH ACID PALEOSOILS
HILL BELT
Series of *Quercus cerris*

Forest of *Quercus cerris* (*Quercetalia pubescenti-petraeae*)

Forest of *Quercus cerris* and *Q. pubescens* (*Quercetalia pubescenti-petraeae*)

Pasture (*Centaureo bracteatae-Brometum erecti*)

Crops (*Secalinetea, Chenopodietea*)

EXTRAZONAL VEGETATION
Series of *Quercus ilex*

Forest (*Orno-Quercetum ilicis, Cephalanthero-Quercetum ilicis*)

Forest with *Pinus halepensis* (*Orno-Quercetum ilicis pinetosum halepensis*)

Pasture (*Asperulo purpureae-Brometum erecti*)

Crops, olive groves, vineyards (*Secalinetea, Chenopodietea*)

Reforestations (*Pinus nigra, P. halepensis*)

Urban settlements

Fig. 6.8 Legend of the previous map (Fig. 6.7)

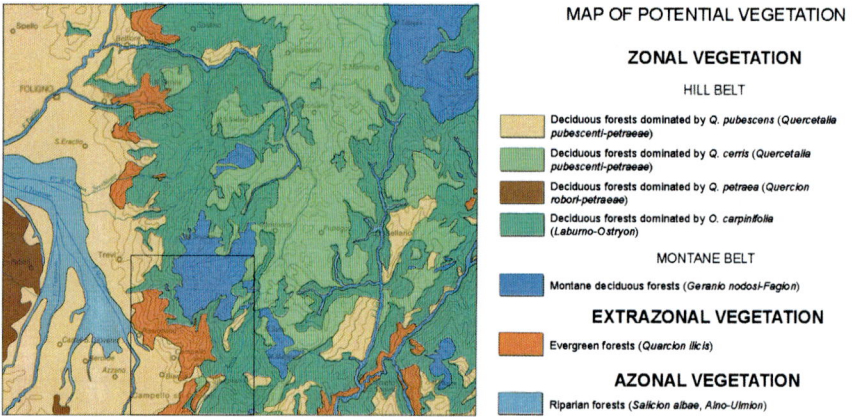

MAP OF POTENTIAL VEGETATION

ZONAL VEGETATION

HILL BELT

Deciduous forests dominated by *Q. pubescens* (*Quercetalia pubescenti-petraeae*)

Deciduous forests dominated by *Q. cerris* (*Quercetalia pubescenti-petraeae*)

Deciduous forests dominated by *Q. petraea* (*Quercion robori-petraeae*)

Deciduous forests dominated by *O. carpinifolia* (*Laburno-Ostryon*)

MONTANE BELT

Montane deciduous forests (*Geranio nodosi-Fagion*)

EXTRAZONAL VEGETATION

Evergreen forests (*Quercion ilicis*)

AZONAL VEGETATION

Riparian forests (*Salicion albae, Alno-Ulmion*)

Fig. 6.9 Map of potential vegetation (Foligno sheet), at scale 1:250,000; the *rectangle* corresponds to the surface from Fig. 6.7 (From Orsomando 1993)

finer scale (*circa* 1:10,000) is the *Carta Fitosociologica Integrata del Lago di Loppio* (Integrated Phytosociological Map of Loppio Lake, in the central Italian Alps) (Fig. 6.10); this lake is a small biotope with two surrounding vegetation series, namely a series of *Salix alba* formed by six associations and developing on basic clays (Fig. 6.11) and a series of *Salix cinerea* formed by 4 associations and developing on acidic peat (Fig. 6.12).

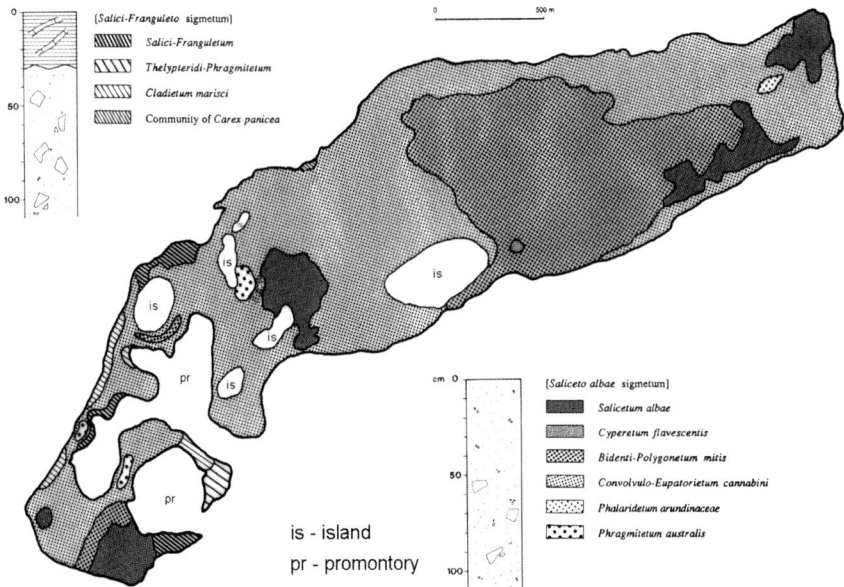

Fig. 6.10 Integrated phytosociological map of the bottom of Loppio Lake, Trentino-Alto Adige Region, northern Italy. A *Salici-Franguleto* sigmetum is developed on peat and a *Saliceto albae* sigmetum on lacustrine clay (From Gafta and Pedrotti 1994)

It is good to note that the definition of "integrated phytosociological cartography" is intended to refer to a cartography that integrates both space and time (Rivas Martínez 1985). A cartographic study of vegetation that involves various kinds of data (climate, soil, vegetation, etc.) can also be called a "complex" or "integrated" cartography if we wish, but not in the sense of Rivas-Martínez.

Phytosociological Maps of Potential Vegetation

These maps refer to the *potential natural vegetation* of Tüxen (1956), eventually redefined by van der Maarel and Westhoff (1973) as "the vegetation that would finally develop (terminal community) if all human influences on the site and its immediate surroundings were to stop at once, and if the terminal stage were to be reached at once" [grammar corrected]. More recently, the classical definition of potential natural vegetation has been amplified by Kowarik (1987), placing more emphasis on the influence of irreversible anthropogenic changes. Leuschner (1997) goes further and introduces the temporal dimension, thereby proposing the concept of potential vegetation adapted to a certain habitat. The potential vegetation would develop strictly in relation to successional changes that take place in the soil. Finally, Chytrý (1998) proposed the concept of *potential replacement vegetation*, defined as "the hypothetical vegetation which is in balance with climatic and soil

Fig. 6.11 Series of *Salix alba* [*Saliceto albae* sigmetum] of Loppio Lake: C.F – *Cyperetum flavescentis*, B.-P.M. – *Bidenti-Polygonetum mitis*, C.-E. – *Convolvulo-Eupatorietum cannabini*, PH.AR. – *Phalaridetum arundinaceae*, PHRAG.A – *Phragmitetum australis* (From Pedrotti 1988c, 1997b)

factors currently affecting a given habitat, with environmental factors influencing the habitat from outside, such as air pollution, and with an abstract anthropogenic influence (management) of a given type, frequency and intensity". For such reasons, there is a possible series of types of potential replacement vegetation for each habitat, corresponding to different human influences.

A map of natural replacement vegetation is especially useful at fine scale (>1:25,000), in as much as it permits evaluating the degree of remoteness of the actual vegetation from its state of final equilibrium, which represents an important datum (or benchmark, sensu Box 1995) for land management.

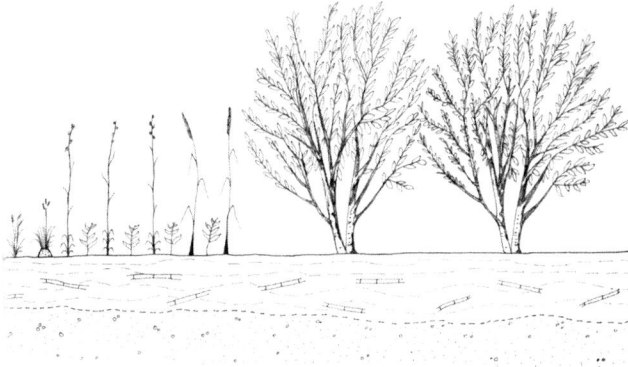

Fig. 6.12 Series of *Salix cinerea* [*Saliceto cinereae* sigmetum] of Loppio Lake, P – *Carex panicea* community, C – *Caricetum elatae,* CL.MAR. – *Cladietum marisci,* TH.-PH. – *Thelypteridi-Phragmitetum* (From Pedrotti 1988c, 1997b)

Maps of potential vegetation in the classical sense represent, then, types of vegetation more complex than a given habitat (site) can support, namely the *climax,* a concept originally defined by Clements (1912, 1928) and successively refined by other authors for various vegetation situations in various parts of the world. If one accepts the concept of natural replacement vegetation, the potential vegetation may not always correspond to the climax but rather to similar or related types, such as paraclimax, pseudoclimax, or preclimax.

Except for the last virgin forests still present on the planet and a few other areas, such as high mountains, the vegetation cover has been subjected everywhere to intense actions by man, which have modified it more or less profoundly or eliminated it completely. In all cases where man has affected the vegetation, one

Fig. 6.13 Zonal vegetation of the montane belt of the Monti della Laga, Abruzzo Region of central Italy: *Fagus sylvatica* forest (*Solidagini virgaureae-Fagetum sylvaticae*) (Photo by Rossella Vallozzi)

speaks of potential vegetation as a theoretical model of reference. Where man's influence has been less or non-existent, the actual vegetation coincides with the vegetation climax.

Maps of potential vegetation, then, show the primary differentiation of the vegetation due essentially to the combination of the many abiotic and biotic factors.

According to the environmental context in which it develops, the potential vegetation may appear zonally (in distinct bioclimatic vegetation zones), azonally or intrazonally (dependent on specific edaphic or hydric conditions), or extrazonally (dependent on microclimate) (Ivan 1979). *Zonal* vegetation is that appearing in the "vegetation zones" that follow latitude and geographic position of land masses or in the "vegetation belts" that follow altitude in uplands (Fig. 6.13). *Azonal* vegetation is not tied to particular zones or belts of vegetation but may appear in all or at least many of the vegetation zones or belts, such as the vegetation of wet environments: lakes, wetlands, water courses, etc. (Fig. 6.14). *Extrazonal* vegetation is developed outside its own zone, at sites with particular microclimatic conditions, for example stands of evergreen *Quercus ilex* are zonal along the Adriatic coast of Italy but are considered extrazonal on microclimatically favored sites in interior central and northern Italy (Fig. 6.15). *Intrazonal* vegetation is provided by associations that are inserted into a zone or belt on sites with particular conditions, such as the vegetation of some pine stands in valleys in the Alps that occur on rocky sites within the altitudinal belt of deciduous forests (Fig. 6.16).

Maps of potential vegetation can be made at quite different scales, according to the type of research, but normally the scale is broad; maps of potential vegetation are often linked to maps of actual vegetation, in order to supplement the

Fig. 6.14 Azonal riparian forest (*Carici remotae-Fraxinetum oxycarpae*) along the Saccione river, Torre Fantine, Apulia Region, southern Italy (Photo by Franco Pedrotti)

Fig. 6.15 Extrazonal vegetation: *Quercus ilex* forest (*Cephalanthero-Quercetum ilicis*), on southern slopes between Foligno and Spoleto, Umbria Region, central Italy (Photo by Franco Pedrotti)

Fig. 6.16 Intrazonal vegetation: *Pinus sylvestris* slopes (*Molinio litoralis-Pinetum sylvestris*), Trentino-Alto Adige Region, northern Italy (Photo by Paolo Minghetti)

information available about the actual study area (Figs. 6.7 and 6.9). Maps of potential vegetation may also be made for the territory of entire countries or an entire continent; examples of such maps include those for Germany at scale 1:200,000 (Trautmann 1973), Poland at scale 1:2,000,000 (Matuszkiewicz 1984), Spain at scale 1:400,000 (Rivas Martínez 1987), Romania at scale 1:2,000,000 (Ivan et al. 1993), the Czech Republic at scale 1:500,000 (Neuhäuslova 2001), plus Greece and other countries of the eastern Mediterranean area (Quézel and Barbero 1985). Today most European countries possess maps of potential vegetation. Among regional maps there is the *Carta della Vegetazione Naturale Potenziale della Sicilia* (Map of the Potential Natural Vegetation of Sicily) at scale 1:500,000, one of the first maps of this type in Italy; this map shows the zonal vegetation and, by symbols, some examples of azonal vegetation such as riparian forests of *Platanus orientalis* (Gentile 1968). The *Carta della Vegetazione Naturale Potenziale dell'Umbria* (Map of the Potential Natural Vegetation of Umbria) at scale 1:200,000 represents zonal, extrazonal and azonal vegetation, as well as some permanent communities in moist or rocky environments. The production of such maps is based on the preceding maps of phytoclimate and actual vegetation at the same scale (Fig. 6.17) (Orsomando et al. 1998, 1999).

Recently, a large map of the potential vegetation of all of Europe, from the Ural Mountains to the Straits of Gibraltar, on nine sheets, has been developed at the scale 1:2,500,000 (Bohn et al. 2000a, b, 2003). This map constitutes a broad synthesis of the vegetation of Europe as it exists up to today. On this map are represented about 700 types of vegetation, grouped into the three categories of zonal, extrazonal and azonal vegetation; each mapped unit represents an association or group of several

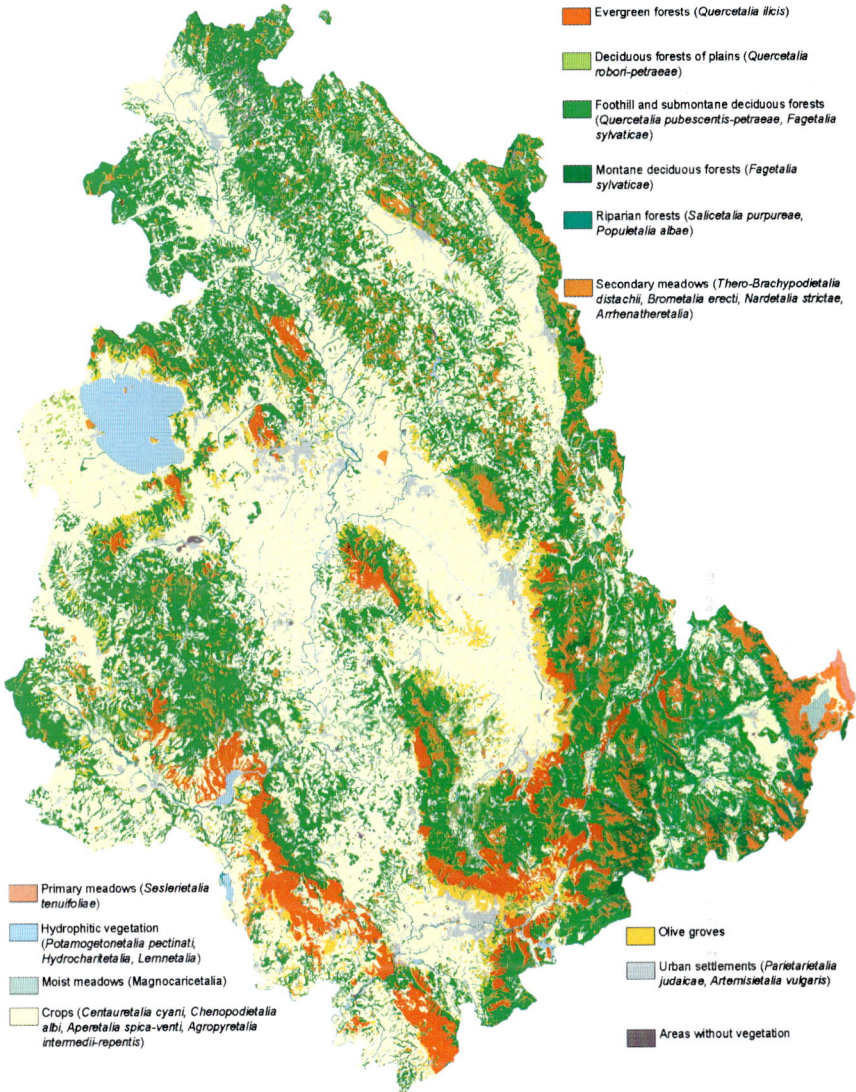

Fig. 6.17 Geobotanical map of actual natural vegetation of the Umbria Region, central Italy, at scale 1:100,000, here re-sized to circa 1:700,000 for printing; the vegetation is fragmented into little strips on plains and hills, or totally absent; on the mountains the forest maintains its continuity (From Orsomando et al. 1998)

related associations (Fig. 6.18). A more detailed portion of the European Vegetation Map, for south-central Italy (from the Marche to Calabria), is shown in Fig. 6.19.

Maps of potential vegetation can also be made without reference to phytosociological types. This is the case in the *Map of the Potential Natural Vegetation of the Conterminous United States of America* by Küchler (1964), on which the units

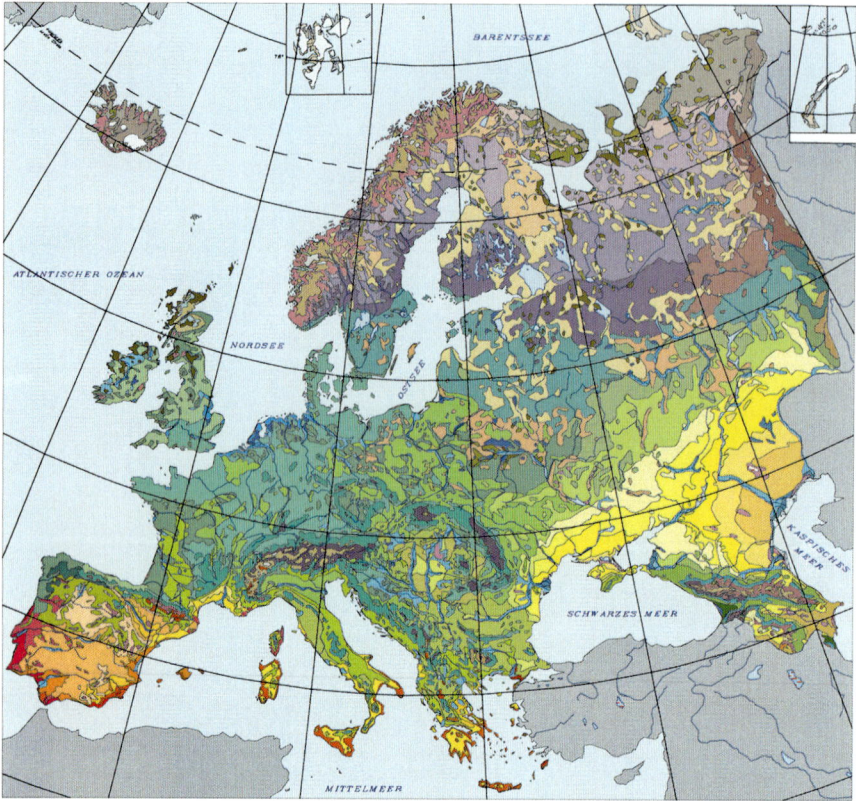

Fig. 6.18 Vegetation map (potential vegetation) of Europe (From Bohn et al. 2000a, b; *Federal Agency for Nature Conservation (BfN) 2011)*

represent vegetation formations grouped into higher categories by geography (e.g. eastern forests, western forests) and physiognomy (e.g. needle-leaved forests, broad-leaved forests) (Fig. 6.20). Much the same applies to the maps of potential vegetation by Russian authors, which are based on phytogeocoenoses, such as the *Geobotanical Map of the USSR* by Lavrenko and sochava (1954, 1956) (Fig. 6.21).

Although the definition of the units that compose a map of potential vegetation may be made using various criteria, it must be said that, in substance, the maps produced have the same significance.

Maps of both potential and actual vegetation find use in landscape ecology. In fact, the cartographic representation of potential vegetation can be used as an alternative to neutral models of the landscape, which have been used as references for comparison with spatial patterns observed and illustrated on maps of actual vegetation (Ricotta et al. 2002). In such a way, a neutral model becomes more realistic in that it takes into consideration the ecological limitations due to geomorphological, pedological and climatic factors (not found in probabilistic models) that usually are considered as determinants of potential vegetation.

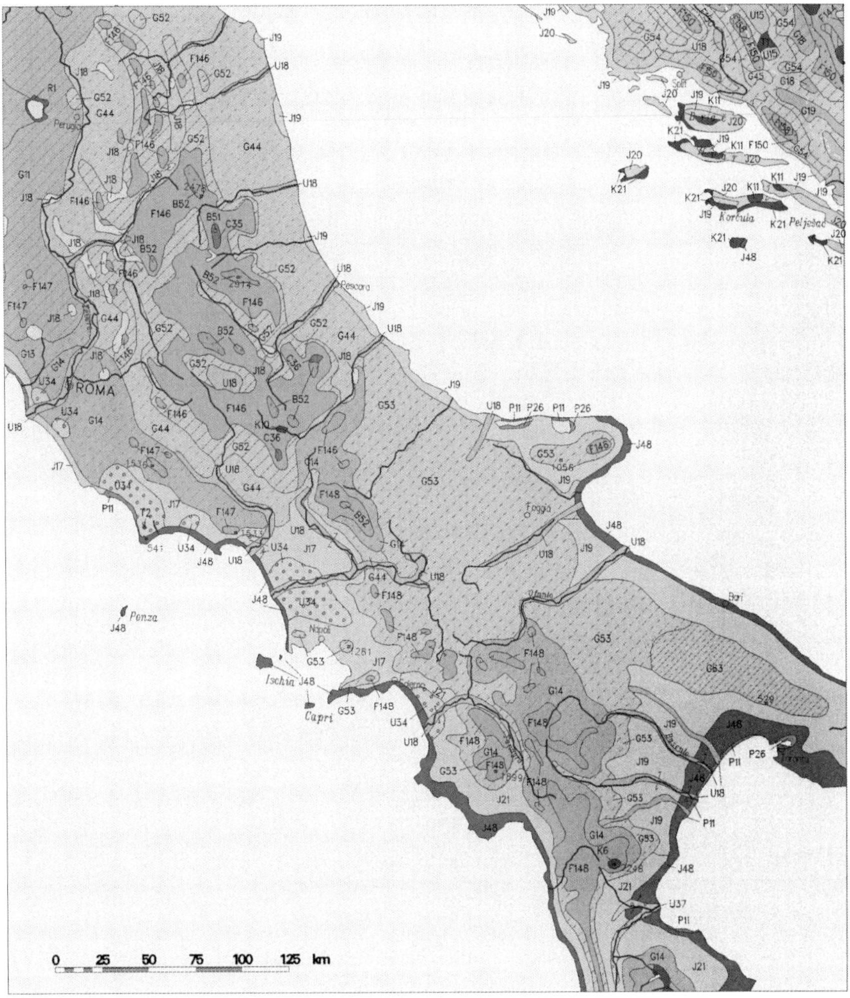

© Bundesamt für Naturschutz, Bonn 2002

Fig. 6.19 Vegetation map of Europe, sheet 8, south-central Italy and Dalmatia. Legend for the Italian portion: *B51* – primary Apennine meadows on silicate rocks (*Bellardiochloa variegata, Nardus stricta*), *B52* – primary Apennine meadows on carbonate rocks (*Sesleria tenuifolia, Carex kitaibeliana, Festuca macrathera*), *C35* – Apennine subalpine heaths (*Vaccinium myrtillus, Hypericum richeri*), *C36* – Apennine pinewood (*Pinus mugo*) with *Silene pusilla* and *Juniperus communis* ssp. *alpina* on carbonatic rocks, *F146* – central-Apennine forests of *Fagus sylvatica* (partially with *Abies alba*), with *Geranium nodosum* and sometimes *Trochiscanthes nodiflora, F147* – relict submontane *Fagus sylvatica* forests with *Ilex aquifolium, Daphne laureola, Cyclamen hederifolium, F148* – *Fagus sylvatica* forests (partially with *Abies alba*) from the southern Apennines with *Geranium versicolor, Campanula trichocalycina, G11* – mixed forests from northern Apennine with *Quercus cerris, Q. petraea* (sometimes with *Q. pubescens*), *Physopermum cornubiense, Anemome trifolia* ssp. *albida, G14* – *Quercus cerris* southern Apennine forests (sometimes with *Q. frainetto*), with *Melittis melissophyllum* and *Lathyrus* sp., *G44* – central Apennine forests of *Quercus pubescens*, with *Viola alba* ssp. *dehnhardtii, G52* – mixed central

Maps of Phytogeocoenoses

These maps represent the vegetation according to the conceptions of the Russian School (Sukachev and Zonn 1961) and are based on the concept of phytogeo-coenoses, which are defined by the dominant species, the structure and differentiation in layers, and dynamic aspects. The *Geobotanical Map of the USSR* at scale 1:4,000,000 of Lavrenko and Sochava (1956) is based on a very complex legend that contains 109 types of phytogeocoenosis, with supplementary symbols to indicate species of particular interest as well as other notes (Fig. 6.21).

Such maps have also been made at very fine scales, such as the map of the steppe in one part of the Ukraine at scale 1:2,500 (Fig. 6.22); 26 phytocoenoses are represented on this map, indicated by one or more dominant species (Didukh et al. 1984).

Maps of Coeno-Associations

In a very synthetic way, one can say that a coeno-association is an abstract unit defined by an original combination of synusiae that recur statistically in sampled phytocoenoses (Gillet 1986, 1988); the phytocoenosis, then, is a spatio-temporal organization formed by different ecological and structural units corresponding to synusiae (called "associations" by Gillet) and identified by their layers (trees, shrubs, herbs and mosses) and by different factors such as soil, light, etc. (e.g. for epigeic, epilithic and epiphytic cryptogam communities) and mosaics of

Fig. 6.19 (Continued) Apennine deciduous forests of *Fraxinus ornus, Ostrya carpinifolia, Quercus pubescens*) with *Acer obtusatum, Scutellaria columnae, Melittis melissophyllum, G53* – Adriatic southern-Apennine forests of *Quercus virgiliana, Q. pubescens* (sometimes with *Q. congesta*), with *Carpinus orientalis, Fraxinus ornus, Anemone apennina, Cyclamen hederifolium, G63* – Apulian meso-supramediterranean forests of *Quercus trojana, J17* – Ligurian mesomediterranean and Tyrrhenian forests of *Quercus ilex* with *Viburnum tinus, J18* – relictual forests of *Quercus ilex* of the internal central Apennines with *Cephalanthera longifolia, Melica uniflora, Anemome apennina, Melittis melissophyllum*, on carbonate rocks, *J19* – meso-mediterranean Adriatic forests of *Quercus ilex* with *Fraxinus ornus, Ostrya carpinifolia, J21* – southern Italian forests of *Quercus ilex* with *Teucrium siculum, Festuca drymeia, J48* – thermo-mediterranean forests of *Ceratonia siliqua, Olea europaea* ssp. *oleaster, Pistacia lentiscus*, sometimes *Chamaerops humilis, K6* – *Pinus heldreichii* forests, *K10* – Apennine pinewoods (*Pinus nigra* ssp. *nigra*) with *Genista sericea* and *Laserpitium garganicum, P11* – dune vegetation (*Cakile maritima, Sporobolus pungens, Calystegia soldanella, Anthemis maritima, Ammophila arenaria, Echinophora spinosa, Crucianella maritima*) and scrub of *Juniperus oxycedrus* ssp. *macrocarpa* and *J. phoenicea, P26* – halophyitic vegetation (*Salicornia* sp., *Arthrocnemum* sp., *Puccinellia* sp., *Juncus* sp., *Limonium* sp., etc.), *R1* – freshwater reeds (*Phragmites australis, Typha angustifolia, Schoenoplectus lacustris*, etc.), *U18* – Apennine alluvial forests of *Fraxinus angustifolia* s.l., *Quercus robur, Ulmus minor, Salix alba, Populus alba* and *P. nigra, U34* – Tyrrhenian alluvial forests of *Fraxinus angustifolia* s.l., *Quercus robur* (south of Rome also with *Quercus frainetto*), with *Carex pendula* and *C. remota, U37* – Mediterranean alluvial formations with *Nerium oleander* and *Vitex agnus-castus*, sometimes with *Platanus orientalis* (From Bohn et al. 2000a, b)

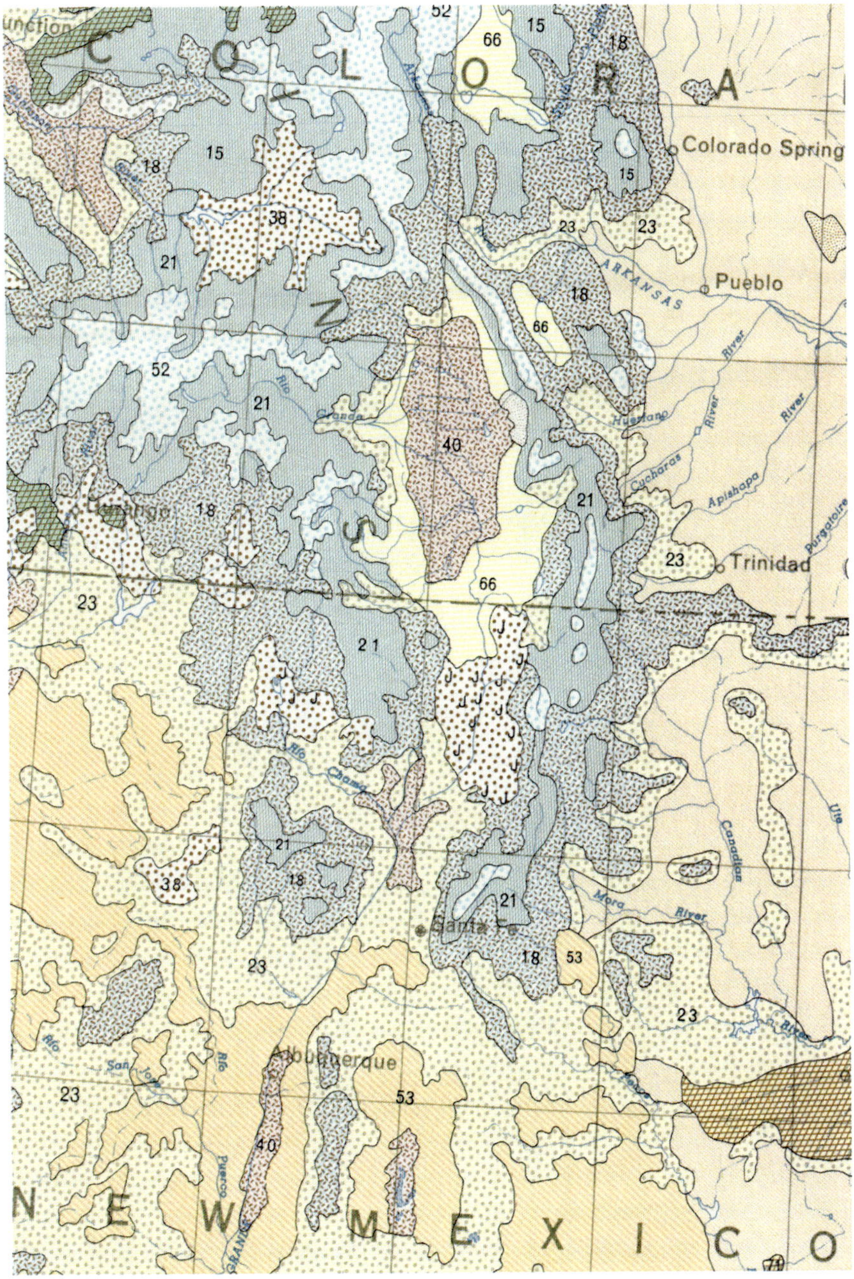

Fig. 6.20 Map of the potential natural vegetation of the conterminous United States, at scale 1:3,168,000: *15* – Western forests of *Abies lasiocarpa* and *Picea engelmannii, 18* – Forest of *Pinus ponderosa* and *Pseudotsugsa menziesii, 21* – Southwestern forest of *Abies lasiocarpa* var. *arizonica* and *Picea engelmannii, 23* – Woodland of *Juniperus monosperma* and *Pinus edulis,*

herbaceous communities. Coeno-associations are named with a nomenclature derived from classical phytosociology, for example a Carici elongatae-Alnocoenetum glutinosae, which is composed of 12 synusiae or "associations", as shown on a map of the coeno-associations of a part of the Białowieza forest, at scale 1:10,000 (Julve and Gillet 1994).

Maps of Vegetation Series

These are maps of vegetation series according to the conceptions of Gaussen (1954) and Ozenda (1964, 1982); according to these authors a series is constituted by "the climax, all species groupings that lead progressively to this climax, and those others that are derived from degradation of the vegetation". On a map, each series is represented by a particular color and the various stages in that series are represented by various tones of the same color, always darker as one approaches the climax forest. The stages in the series are indicated by the physiognomy of the vegetation, for example garrigue, macchia or forest. Criteria for selecting the colors for such maps have already been suggested above.

Figure 6.23 shows a vegetation map of the western Alps at scale 1:500,000; it is the last map produced by the Grenoble school and constitutes a synthesis of the maps collected earlier in the journal *Documents de cartographie écologique* (Ozenda and Borel 2006). This map shows 43 vegetation series and subseries grouped into the altitudinal belts: thermo-mediterranean, meso-mediterranean, western supra-mediterranean, eastern supra-mediterranean, central-European colline, montane, subalpine, alpine and nival.

In Italy some maps of vegetation series have also been made, namely for Monte Amiata (Arrigoni and Nardi 1975), for the *Alpi Apuane* (Tuscany) and for the northern Apennines from Cisa al Gottero and Cinque Terre (Ferrarini 1972, 1988).

Also the first maps of the Marche and of Umbria, at scale 1:50,000, made by the Botanical Institute of Camerino, can be included in this category; an example is the Fabriano sheet, on which the various stages in the series are indicated with phytosociological nomenclature (Pedrotti 1976). In subsequent years other maps were made, always in the Marche and Umbria, which can be considered integrated phytosociological maps (see, for example, the Foligno sheet, in Fig. 6.7).

Fig. 6.20 (Continued) 37 – Scrub of *Cercocarpus ledifolius* and *Quercus gambelii, 38* – Great Basin sagebrush (*Artemisia tridentata*), *40* – Scrub of *Atriplex confertifolia* and *Sarbobatus vermiculatus, 52* – Alpine meadows and barren areas (*Agrostis* sp., *Carex* sp., *Festuca* sp.), *53* – Grassland of *Bouteloua gracilis* and *Hilaria jamesii, 55* – Sagebrush steppe of *Artemisia tridentata* and *Agropyron spicatum, 65* – Grassland of *Bouteloua gracilis* and *Buchloe dactyloides, 66* – Grassland of *Agropyron smithii* and *Stipa comata, 70* – Prairie of *Artemisia filifolia* and *Andropogon scoparium* (From Küchler 1964)

Fig. 6.21 Map of the vegetation of the former USSR, east-central Siberia: *3* – mountain tundra of eastern Siberia, *18* – conifer forests of *Pinus sibirica, 23* – mixed conifer forests with *Larix sibirica* (23s) and *L. dahurica* (23d), *24* – *Larix-Pinus* forests of the Trans-Baikal-Amur region, *30* – *Larix* forests of central Siberia, with *L. sibirica* (30s) and *L. dahurica* (30d), *36cr* – mountain forests of *Larix dahurica* in eastern Siberia, *41* – *Betula* section *albae/Populus tremula* forests, *80* – *Koeleria gracilis/ Festuca sulcata* steppes of the Altai-Trans-Baikal region, *81* – *Cleistogenes squarrosa-Stipa capillata* and *Aneurolepidium pseudoagropyrum-Stipa capillata* steppes (From Lavrenko and Sochava 1954)

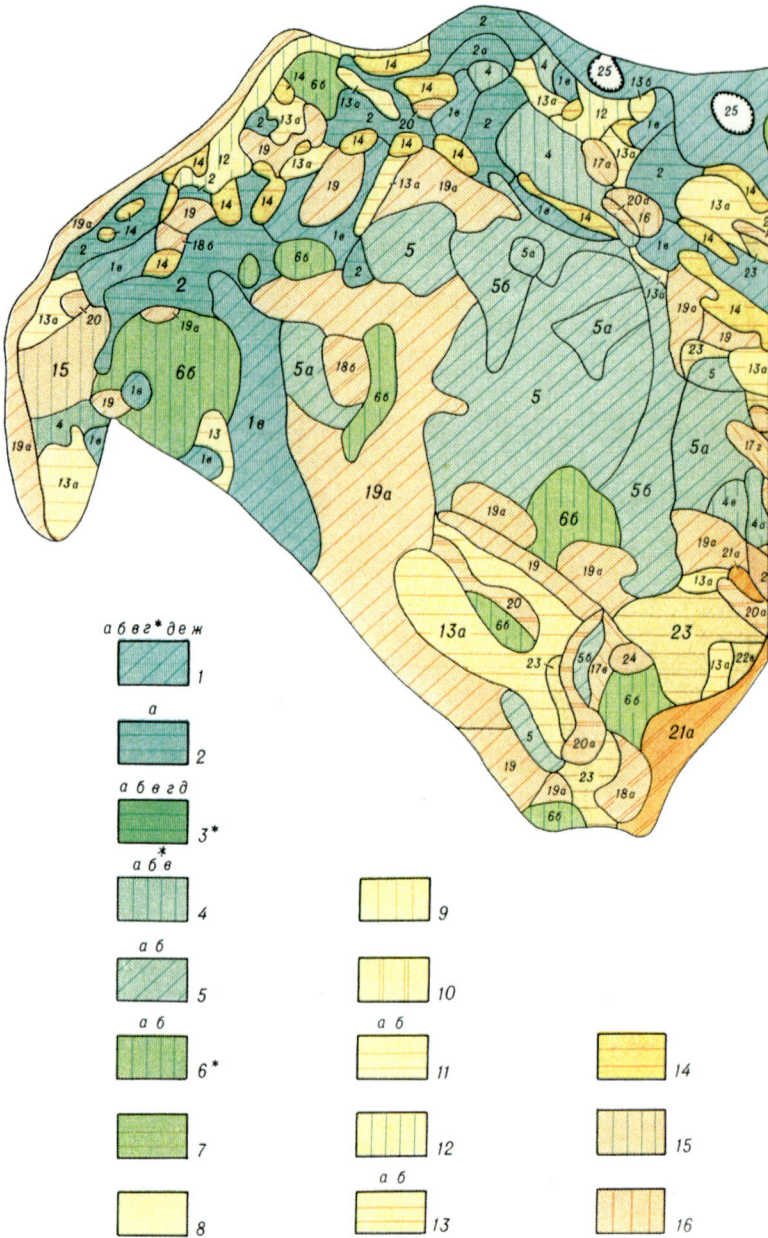

Fig. 6.22 Map of meadow phytocoenoses of Podolia (Ukraine), at scale 1:2,500; the numbers *1–16* indicate phytocoenoses with dominant or co-dominant species, for example *1 Festuca pratensis + Agrostis tenuis + Lotus corniculatus*; *2 Briza media + Festuca pratensis*; *3 Agrostis tenuis + Festuca pratensis*; *4 Festuca rubra + Briza media*; *5 Festuca rubra + Agrostis tenuis*; *6 Agrostis vinealis + Briza media*; *7 Agrostis vinealis + Carex humilis*, etc. (From Didukh et al. 1984)

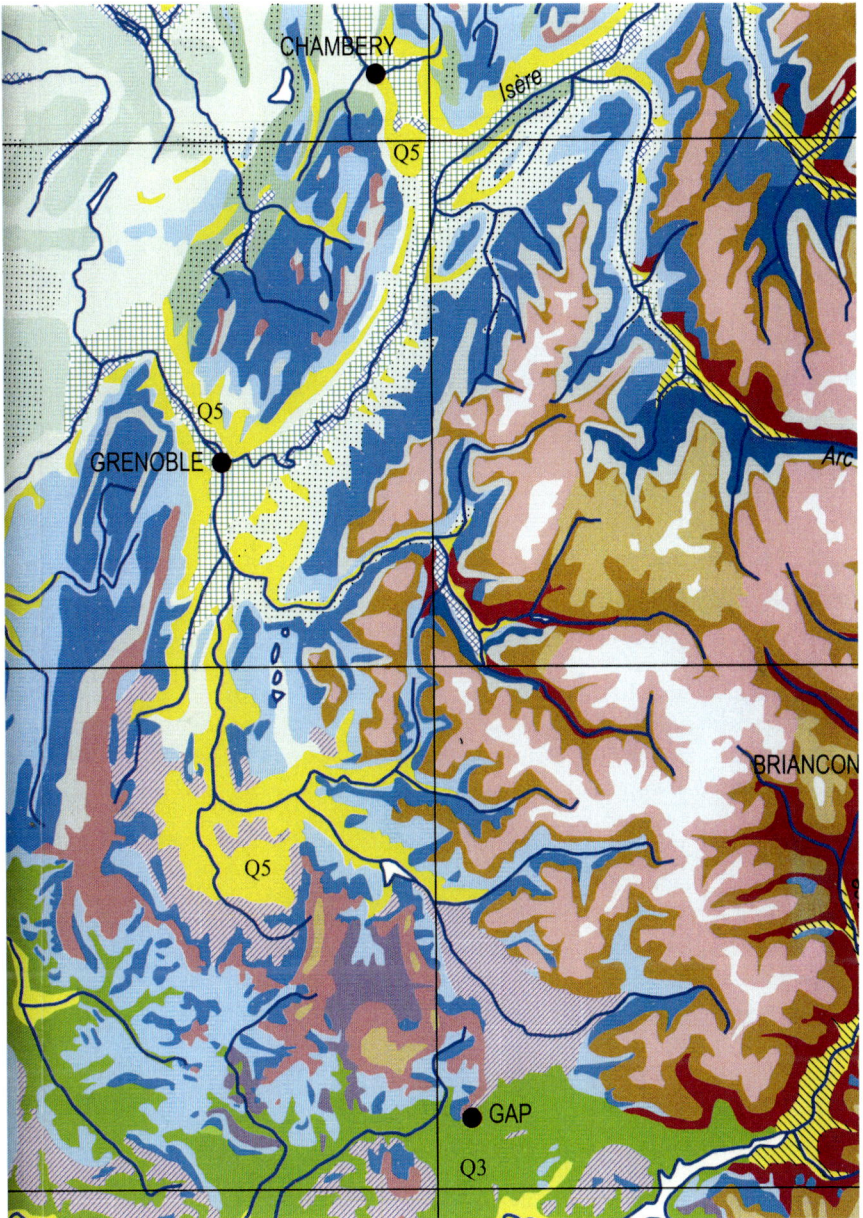

Fig. 6.23 Vegetation map of the western Alps, around Grenoble, France, at scale 1:500,000; each color represents one vegetation series. This territory is in the transition zone between the Mediterranean and middle-European regions. The boundaries between the different vegetation series are not shown on the map (From Ozenda and Borel 2006)

Maps of Vegetation Bands

The word 'zone' is normally used for bioclimatic zones at global scale and the word 'belt' for altitudinal belts in mountains, also bioclimatically conditioned. One also, however, uses the word 'zonation' much more broadly. At the Botanical Congress of Brussels in 1910, the French word *ceinture* was adopted (*Gürtel* in German, *cingolo* in Italian, but with no unequivocal parallel in English) for "bands" of vegetation which may occur in zonation patterns, such as parallel bands crossed by some environmental gradient or in concentric rings around a bog (see Géhu 2006). Here the word "band" is used in this sense of *ceinture*.

Maps of vegetation "bands" are made based on such *ceintures*, as also conceived by Schmid (1961), in the Swiss Alps, for altitudinal bands of vegetation formed by species with similar areas. Outside Switzerland, this type of cartography has been employed in some maps of the vegetation of Piemonte (northwestern Italy) at scale 1:50,000 by Sappa and Charrier (1949) and Sappa (1955), and in Lucania by Famiglietti and Schmid (1969).

Maps of Vegetation Dynamics

Maps of vegetation dynamics aim to show temporal variations in the vegetation, namely its dynamism, and can be made with very diverse criteria according to the different schools.

For the Lauterbrunn Valley in Switzerland, Lüdi (1921) distinguished and mapped initial, transitional and final assemblages. Aichinger (1951, 1967) recognized and described "patterns of development in the vegetation", also mappable, as by Kirchleerau in Switzerland. Gribova and Samarina (1963) mapped autochthonous and secondary plant communities and series resulting from human activity (fire, forest cutting, plowing, etc.); the cartographic representation of these phenomena was obtained by overprinting symbols of various colors on the units that indicate the types of vegetation.

More recently, Faliński (1998a) produced a map of the actual vegetation of a previously cultivated field, parts of which had been abandoned at different times. The author distinguished in the field and on the map eight phases in secondary succession, beginning with the pioneer stage of residual segetal species up to the shrub stage with *Juniperus communis* and *Populus tremula*. As indicated in the legend, one assumes that the final phase (climax) will be represented by pine woods (Peucedano-Pinetum sylvestris) about 140 years after abandonment.

One method for showing the temporal variations of the vegetation is to repeat the cartographic sampling in the same location at successive intervals. One thus obtains a temporal series of maps, from comparison of which it is possible to evaluate the variations in the vegetation occurring through time. A well-known example is that of a brackish pond near Palavas (France), where Braun-Blanquet (1958) made eight successive maps, from 1915 to 1958, that show the changes occurring in the vegetation. Faliński (1986) showed many examples of maps repeated in successive

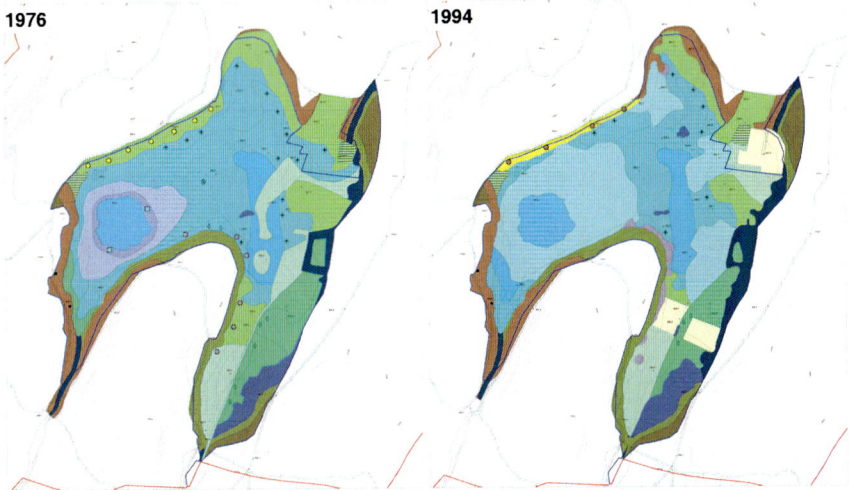

Fig. 6.24 Phytosociological map of the Laghestel di Piné (swamp and lake), Trentino-Alto Adige Region, northern Italy, at scale 1:2,880, made in 1976, 1994 and 2001 (see Fig. 6.45). Comparing the three maps, one can see the following vegetation change: (a) disappearance of the *Caricetum lasiocarpae*, including the variant with *Rhynchospora alba*; (b) strong reduction of the *Caricetum elatae* and *Caricetum rostratae* because of water pollution; (c) development and strong expansion of the *Phragmitetum australis* because of water pollution and cessation of mowing; (d) development and expansion of the *Salicetum cinereae* due to the cessation of mowing in the wet meadows; (e) expansion of the *Corylo-Populetum tremulae* due to cessation of mowing in the mesophylous meadows; and (f) developement of some anthropogenic associations because of plowing (*Galinsogo-Portulacetum*) and dumping rubble (*Tanaceto-Artemisietum vulgaris*) (From Pedrotti 2004a)

years in the Białowieza forest, which were made with very precise methodology and at the same permanent observation areas. At the Reski wetland (15 ha, all within the Białowieza forest), Falińska (2003) monitored vegetation changes over the course of 30 years, from 1972 to 2002, producing four maps of actual vegetation that show the progressive reduction of the wetland associations (Cirsietum rivularis, etc.) and invasion by shrub and tree associations (Fraxino-Alnetum, Salicetum pentandro-cinereae and Ribeso nigri-Alnetum). A final example is for the brackish ponds and dunes of Hon (Netherlands), where monitoring was done with four maps from 1945, 1959, 1969 and 1986; these maps show that the transformation processes have been very fast over the last 50 years, constituting a case of newly forming environments due to the actions of sea and wind (Janssen 2001).

Other maps show man-induced temporal changes in the vegetation, such as those at Laghestel di Pinè in Trentino (northern Italy), where mapping was repeated three times, in 1976, 1994 and 2001 (Fig. 6.24). From this set of maps one can see the progressive disappearance of associations, including the rare Caricetum lasiocarpae, substituted by more commonplace associations such as Phragmitetum australis.

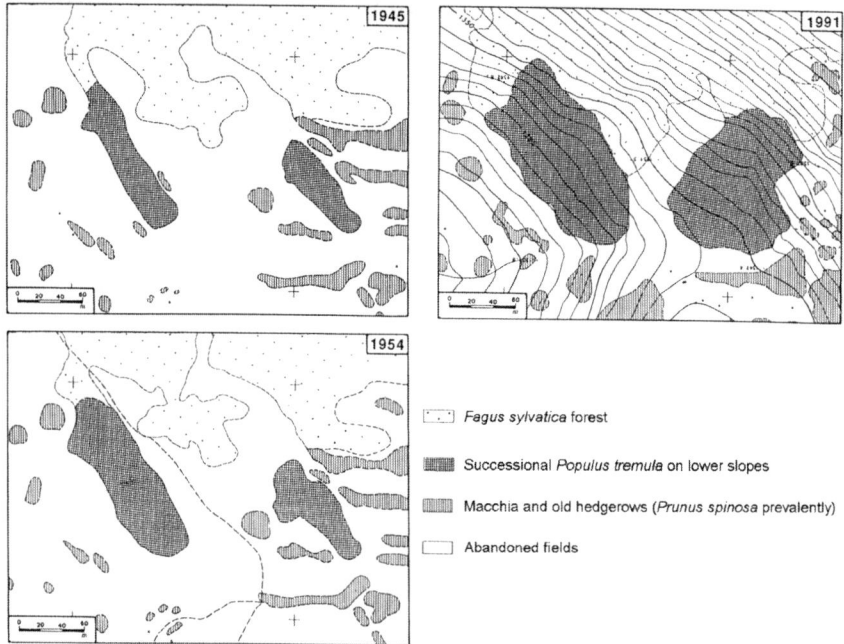

Fig. 6.25 Secondary forest of *Populus tremula* (*Melico uniflorae-Populetum tremulae*) at Gioia Vecchia village, Abruzzo Region, central Italy. This forest was formed from two nuclei that developed after abandonment as cropland and expanded progressively from 1945 to 1991 (From Pedrotti 1996b)

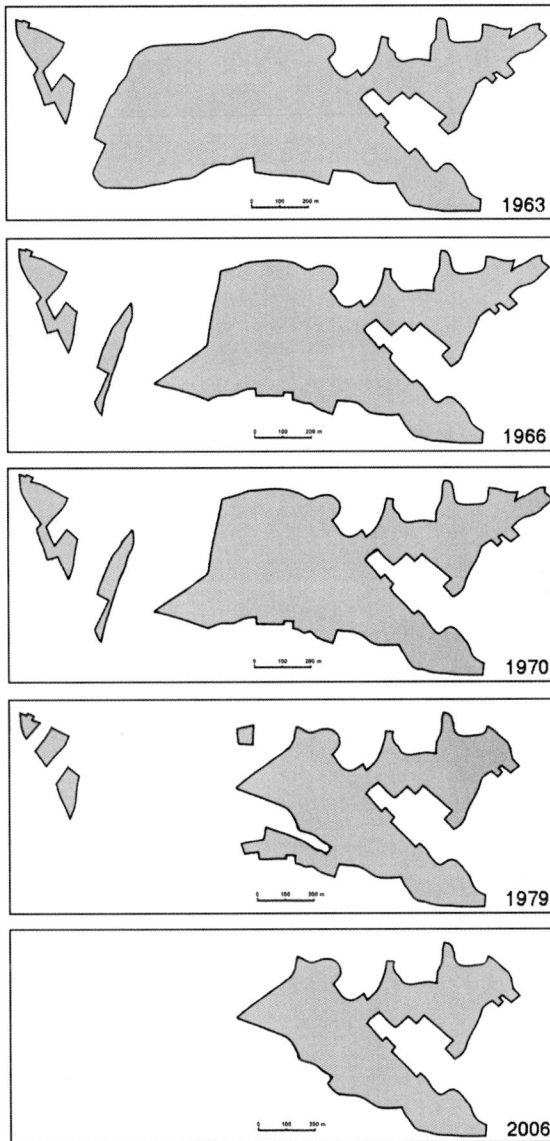

Fig. 6.26 Progressive reduction of wild meadow vegetation (mainly of the order *Trifolio-Hordeetalia*) due to tillage for new crops, in the upper Piano di Montelago, Marche Region, central Italy (From Pedrotti 1979, 2008)

At Gioia Vecchia (Abruzzo), mapping shows the progressive development of a Melico uniflorae-Populetum albae on an abandoned field (Fig. 6.25).

At the Piano di Montelago (near Camerino, central Italy), successive mapping shows, more simply, the progressive reduction of meadow areas as a result of plowing (Fig. 6.26).

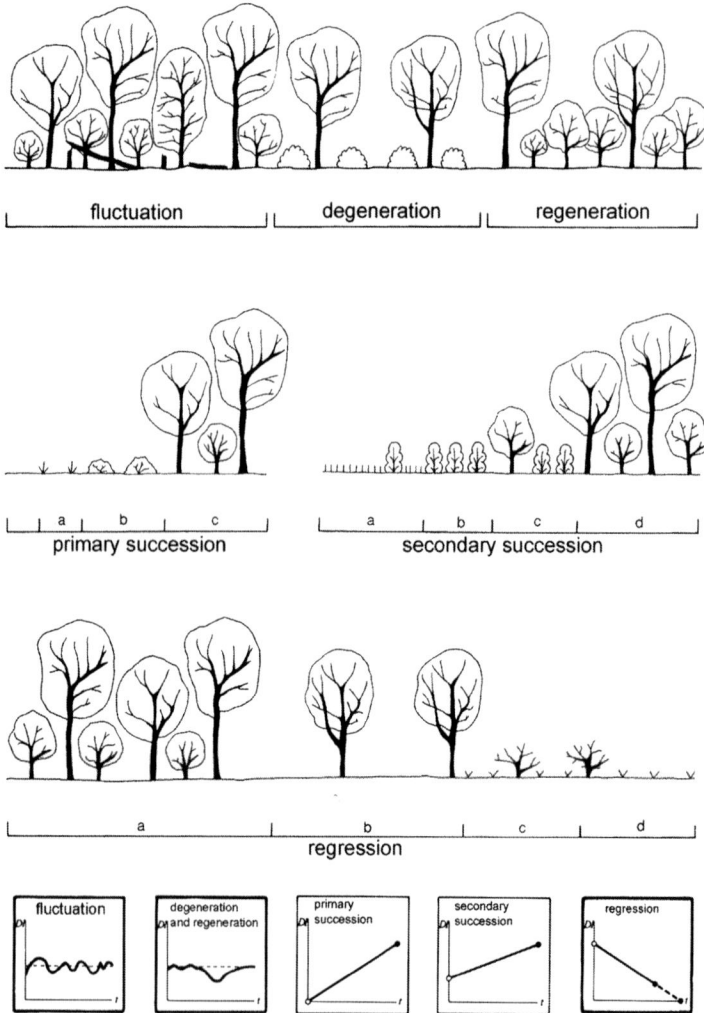

Fig. 6.27 Dynamic tendencies in the vegetation: t = time; DI = dynamisme index; *a, b, c, d* = phases of dynamic processes (From Faliński 1986 and Pedrotti 1995)

Finally, maps of the dynamic tendencies of the vegetation have also been produced in greater detail (Faliński 1986), showing the ecological processes related to the dynamics in the phytocoenoses at the time when they were sampled in the field. One can say then that they represent the dynamic state of the vegetation. These dynamic processes include fluctuation, primary succession, secondary succession, degeneration, regeneration and regression (Fig. 6.27).

Fluctuation consists of all the small, continuous changes that concern the components of a particular phytocoenosis but do not change fundamentally the type of phytocoenosis; these changestake place inside the phytocoenosis and result

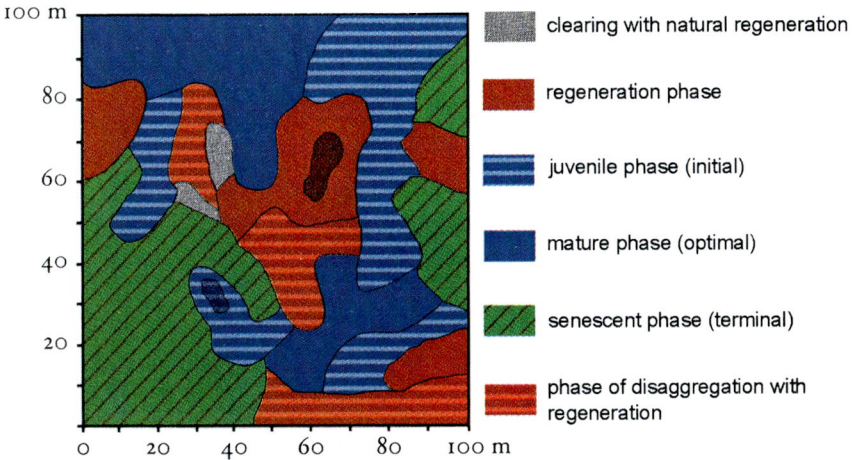

Fig. 6.28 Phases of fluctuation in a subalpine forest of *Pinus cembra* and *Picea abies*, in the Călimani Mountains, Romania, where the different communities are interested by different processes (From Giurgiu et al. 2001)

in a sort of dynamic equilibrium. Such changes often involve gradual exchanges of components, such as the formation of small glades in a forest, with substitution of old individuals by new that arise through natural regeneration, etc. (Fig. 6.28). One can distinguish natural fluctuations that occur in primary associations (forests, high-elevation grasslands, etc.) and anthropogenic fluctuations that appear in secondary associations such as grazed or mowed meadows, etc. Within fluctuation it is possible to identify and map the various phases in natural development that, according to Leibundgut (1959, 1978), are the following: regeneration (see below), juvenile (initial), optimal maturation, senescence and disaggregation. This classification of phases was modified subsequently by Oldeman (1990), who distinguished renovation, aggradation, initial biostasis, biostasis, and decadence; it has also been modified by other authors, but the techniques of cartographic representation remain the same. Within an individual phase it is possible to sample and map, at fine scale, the individual structural components, i.e. trees, shrubs, herbs, standing woody dead, etc. In the Bosco della Fontana, a deciduous forest of the Po Plain (northern Italy) with a prevalence of *Quercus robur*, there was a detailed analysis of the canopy gaps and of the eco-units that fill them (Figs. 6.29 and 6.30).

Degeneration is an ecological process inside phytocoenoses that entails modifications in structure and floristic composition but also without changing the type of phytocoenosis. For structure, degeneration normally involves progressive thinning of the tree stratum and reduction (or excessive development) of the shrub stratum. For floristic composition, degeneration involves penetration and development of foreign species in the phytocoenosis or progressive reduction and eventual disappearance of some or most characteristic species combination that constitute the association. All these phenomena are due to human activities such as tree

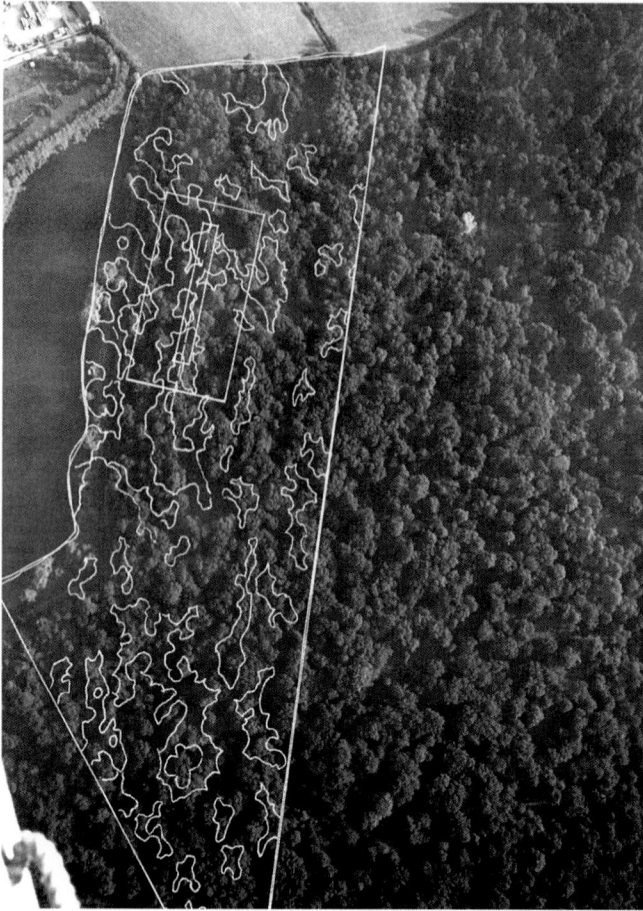

Fig. 6.29 Forest mosaic (mosaico silvatico), in Core Area 1 of the Bosco della Fontana, Mantova, Lombardia Region, northern Italy (From Mason 2002)

cutting in the forest or pasturing of domestic animals in the forest. These activities foster the penetration of nitrophilous and synanthropic species into the forest and change some environmental factors, as by drainage, which leads to the immediate disappearance of hygrophilous species and the establishment of less demanding outside species; this also leads quite quickly to regression phenomena, with the thinning that modifies considerably the forest microclimate. Forms of degeneration were identified by Olaczek (1974) as follow: "monotypization", "fruticization" (expansion of shrubs species), "cespitization" (expansion of caespitose hemicrypto-phytes), "juvenalization" (maintenance of phytocoenoses in initial stages by peri-odic cutting), "neophytization" (invasion by neophytes), and "pinetization" (plantation of conifers, mainly pines) (Fig. 6.31). The substitute phytocoenoses are identified by the way the original phytocoenoses were eliminated and measured

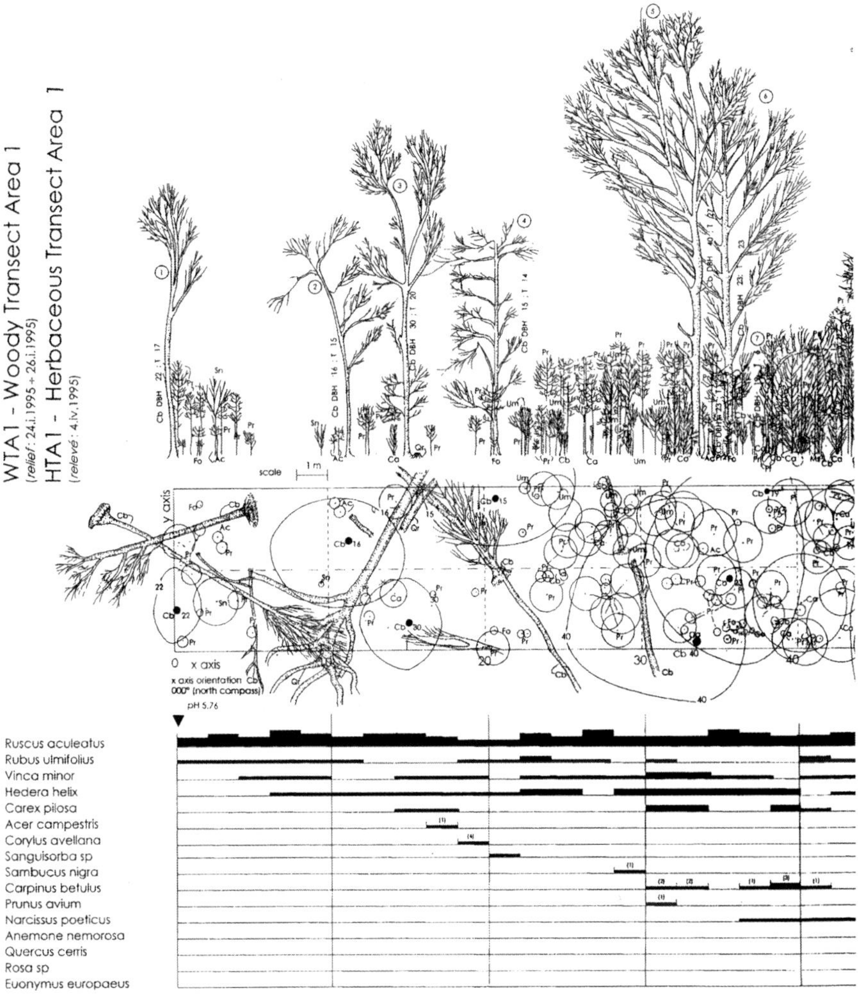

Fig. 6.30 Woody and herbaceous transect areas (Area 1) in the Bosco Fontana, Mantova, northern Italy (From Mason 2002)

by the degree of their destruction (Faliński 1998b). At the Ilawa forest (Poland) degeneration was taking place through reforestation with *Pinus sylvestris* in stands of the forest associations Tilio-Carpinetum and Potentillo albae-Quercetum, resulting in invasion and diffusion of the neophyte *Impatiens parviflora* in various parts of the forest (Fig. 6.32).

Regeneration is the opposite of the previous process, namely the reconstitution of the original situation, of a more closed tree overstorey in openings and in other ways. Regeneration is also a process that takes place inside a phytocoenosis without changing its type. The regeneration process can be examined by mapping the shrub and tree species in test plots, as was done in the beech forest of Gariglione (Sila,

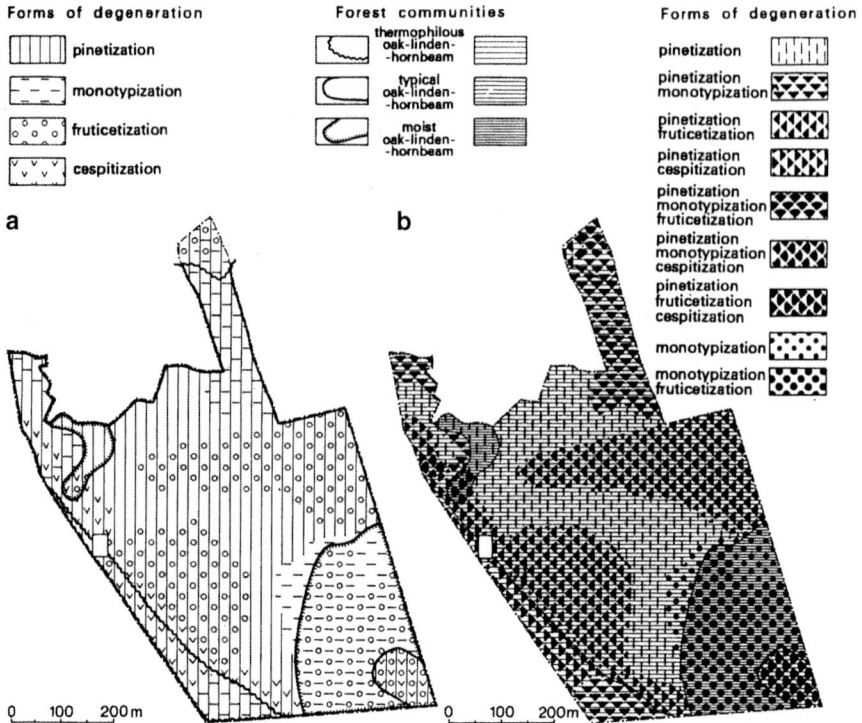

Fig. 6.31 Natural vegetation of the Gaik Nature Reserve near Opoczno, Poland. This represents an attempt to map information on both primary (syntaxonomical) differentiation in oak-hornbeam forest (*Tilio-Carpinetum* and *Potentillo albae-Quercetum petraeae*) and secondary differentiation under the influence of forestry management (forms of degeneration of forests phytocoenoses). The overlap of areal symbols denotes the co-occurrence of degeneration forms (From Faliński 1998a)

southern Italy), where renovation of *Abies alba* was shown by planimetry of its increasing area and along a transect (Canullo et al. 1993). The tree layer was constituted entirely by *Fagus sylvatica*, as a result of selective cutting in the past. Today, though, *Abies alba* seems to be increasing (Fig. 6.33), and when this process is finished, the forest will have regained a more complete structure than it has now, with both *Fagus sylvatica* and *Abies alba* in the tree layer. Since this is a protected area, it will also be possible to show, through observations over many years, eventual dominance cycles involving the two species.

Succession involves the reconstruction *ex-novo* of plant associations over many years, ending with a mature, stable phytocoenosis, as is typical of climax vegetation. In the case of *primary succession* the development of the vegetation starts on a substrate without organic matter, such as alluvial deposits, volcanic lava, or landslide tracks (Fig. 6.34); *secondary succession* starts from situations in which organic matter and living organisms are already present, as on abandoned agricultural fields or meadows no longer grazed or mowed. Secondary succession can then

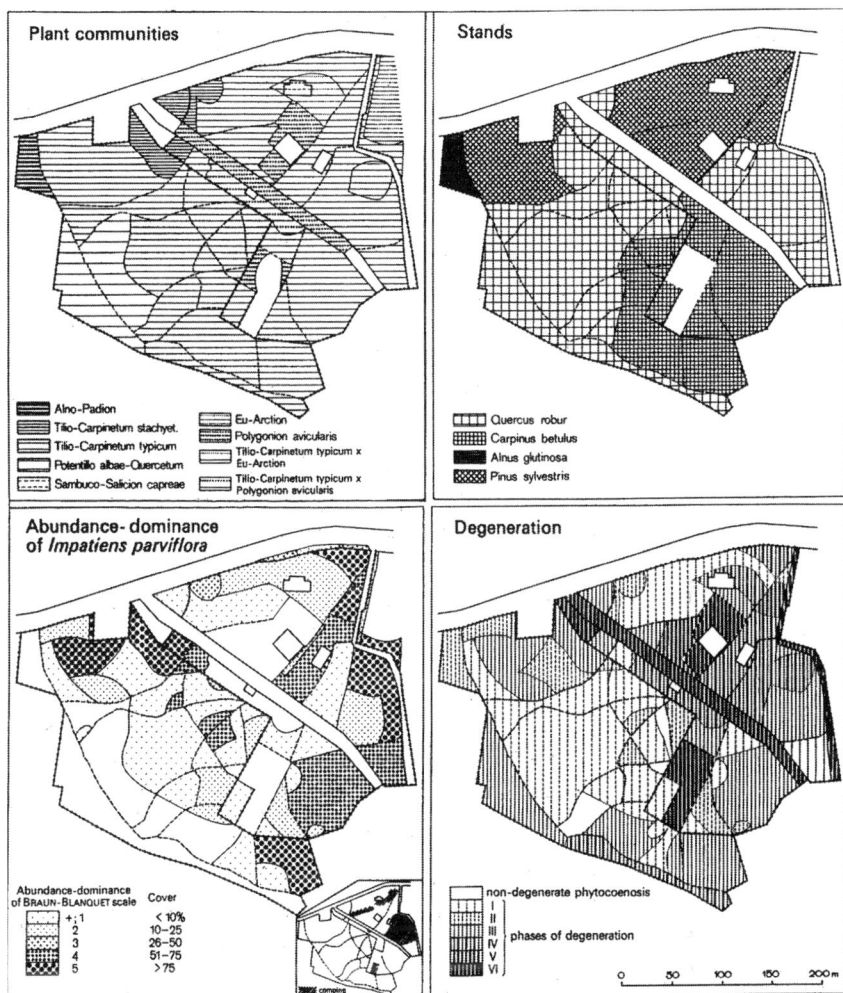

Fig. 6.32 Cartographic presentation and interpretation of phytocoenosis degeneration and neophytism in forest communities of an urban forest in Iława, Poland. The series of four maps allows the distribution and intensity of the community degeneration to be interpreted through the differentiation of vegetation or the causes inducing it (the introduction of a pine stand into an oak-hornbeam forest), as well as in relation to the share taken by the neophyte *Impatiens parviflora* (From Faliński 1998a)

lead to the return of forest in areas where forest vegetation had been eliminated in the past (Fig. 6.35), progressing from one type of phytocoenosis to another, through successional stages (Fig. 6.36).

Regression is a process of gradual simplification of plant associations, under aggressive action by external factors, and may proceed to complete substitution by other plant associations (Fig. 6.37); in extreme cases it may lead to their complete

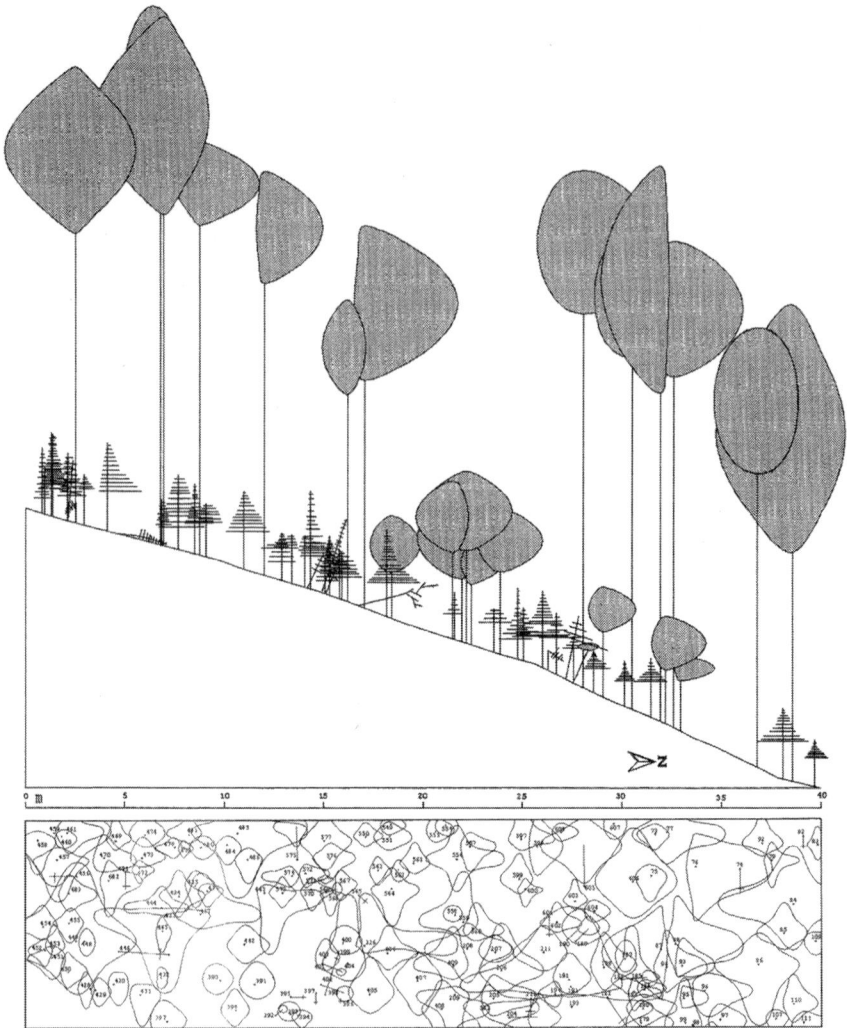

Fig. 6.33 Vertical profile of a 10×40 m transect in the permanent plot area "Gariglione", Calabria Region, southern Italy. The good regeneration of *Abies alba* compared with *Fagus sylvatica* is remarkable; at *bottom*, a map of canopy cover, with *Fagus sylvatica* in *grey* and *Abies alba* in *with symbols* (From Canullo et al. 1993)

disappearance, leaving an environment with only a very sparse plant cover or no plants at all (Fig. 6.38). Regression may start from an initial degeneration phase. Regression can consist of many phases, as is occurring in the Bosco Quarto (Gargano promontory), where 11 phases have been mapped (Fig. 6.39). Maps of dynamic tendency also represent the state of synanthropization (adaptation to human activity) of a phytocoenosis, as in phytocoenoses undergoing secondary succession, degeneration or regeneration; on the other hand, in the case of primary

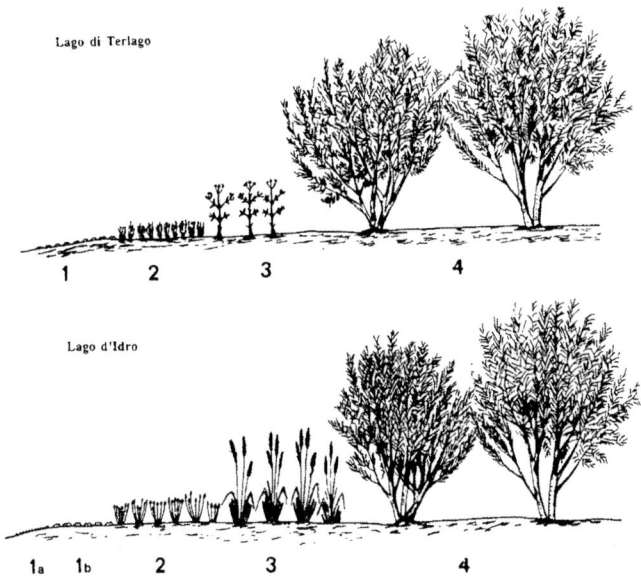

Fig. 6.34 Primary succession on the banks of streams feeding into the Terlago and Idro lakes, Trentino-Alto Adige Region, northern Italy. Terlago Lake: *1 – Riccio cavernosae-Physcomitrelletum*; *2 – Nanocyperion*; *3 – Bidention*; *4 – Salicetum albae*. Idro Lake: *1a – Botrydietum granulati*; *1b – Riccio cavernosae-Physcomitrelletum*; *2 – Eleocharitetum acicularis; 3 – Caricetum gracilis*; *4 – Salicetum albae* (From Cortini Pedrotti 1992)

Fig. 6.35 Val Calamento, Trentino-Alto Adige Region, northern Italy. A big storm has caused a clearing in the *Oxali-Piceetum* forest, which is now being filled in by secondary succession with a *Salicetum capreae* (Photo Franco Pedrotti)

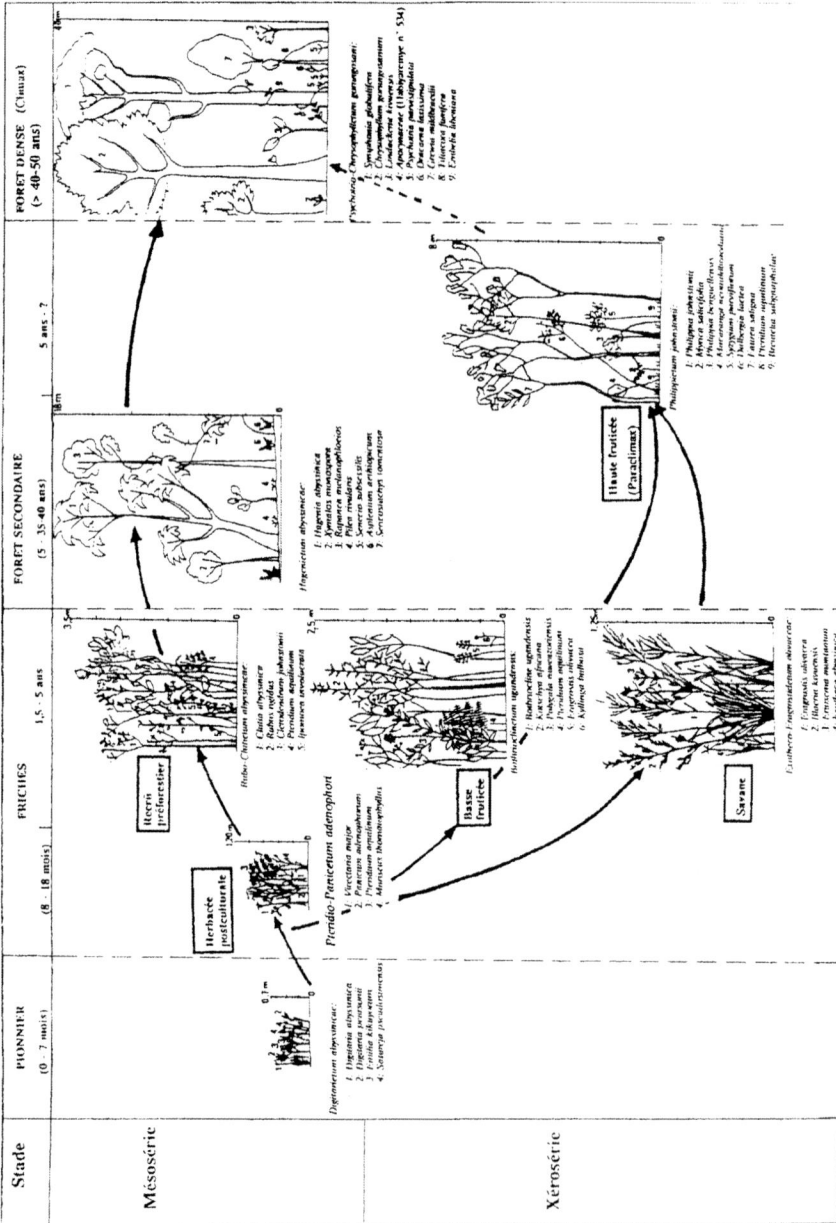

Fig. 6.36 Secondary succession in the forests of Zaire (From Habiyareme 2000)

Fig. 6.37 Regression
process in *Nothofagus
pumilio* forests in Patagonia,
Argentina: *1* forest of tall
trees in fluctuation in the
*Mayteno-Nothofagetum
pumilionis*; *2–4* regression
phases of the forest because
of fire and grazing: *2 Elymo-
Chiliotrichetum* matorral;
3 Stipo-Mulinetum spinosi
pajonal; *4 Triseto-Poёtum
pratensis* pasture meadow
(From Roig et al. 1985; see
also Roig and Faggi 1985)

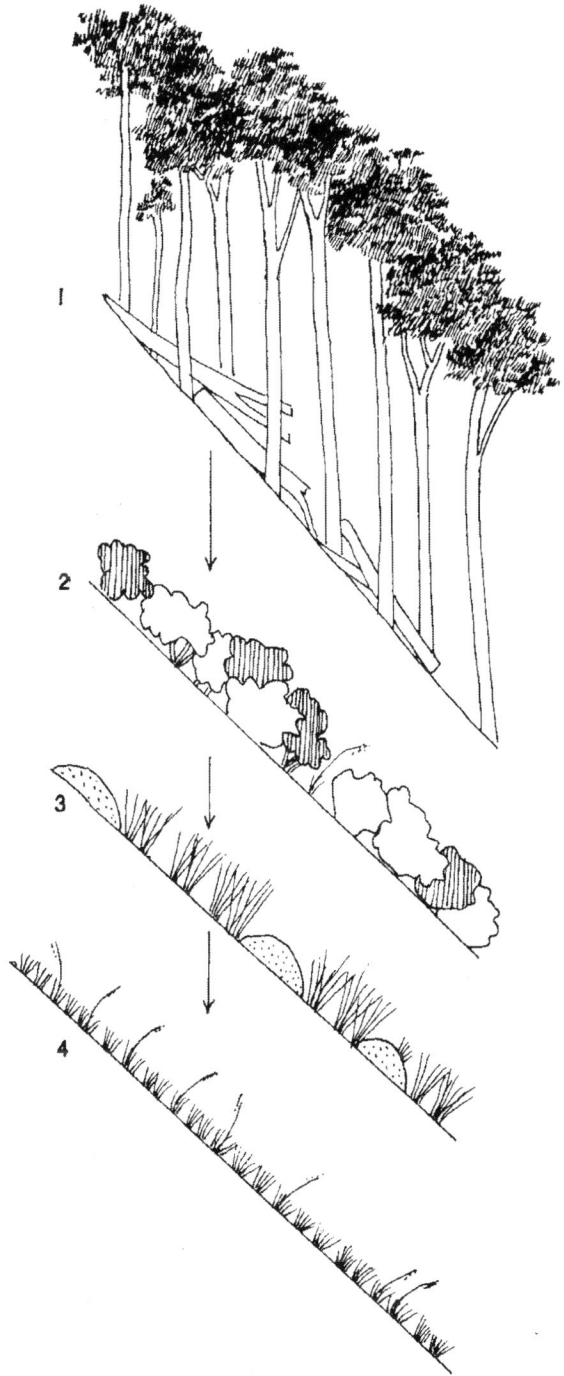

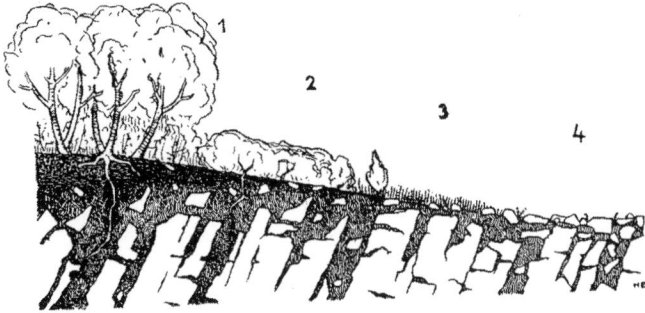

Fig. 6.38 Regression process in forest of *Quercus ilex* developed on carbonate rocks in Languedoc, southern France: *1 – Quercetum ilicis*; *2 – Q. cocciferae*; *3 – Brachypodietum ramosi*; *4* – overgrazing facies with *Euphorbia characias* (From Braun-Blanquet 1936)

succession or natural fluctuation, such maps show the absence of anthropogenic effects.

The map of the dynamic tendency of the Bosco Quarto was made at the scale 1:10,000 (Fig. 6.39); it was obtained from the same mapping units as on the phytosociological map of actual vegetation (Fig. 6.3), for which the dynamic state was also evaluated at the time of field sampling (Figs. 6.40 and 6.41). In practice, during the field sampling, units to be mapped are marked with two symbols, the first representing the vegetation type and the second the dynamic state, as can be noted on the field copy. The mapping units sampled can be subdivided or grouped in order to emphasize the phytocoenoses undergoing the same dynamic processes, each in distinct phases; these units are then shown on the map in various colors according to the dynamic tendency (Faliński and Pedrotti 1990). The Bosco Quarto area (*circa* 5,000 ha) is undergoing processes of regression and degeneration, and only 11.4 % of the forest is subject to natural fluctuations only, which indicates a high degree of synanthropization over a large part of the territory.

A map of vegetation dynamics, with the same criteria, was made for the Val Pagana – Le Forme, in the Abruzzo National Park, at the scale 1:10,000 (Canullo and Pedrotti 1993). Also in this case, the map of dynamic tendencies permits an immediate evaluation of the state of the vegetation: the forests are mostly in a phase of degeneration or regeneration; on north-exposed smaller valleys, because of periodic avalanches in winter and thus permanent pioneering conditions, the vegetation is in a phase that can be defined as recurring primary succession. Primary high-altitude grasslands are characterized by natural fluctuation, secondary grasslands still in use by anthropogenic fluctuation, and those that are abandoned by secondary succession.

The same processes have been surveyed in primary high-altitude and secondary grasslands of Velino Mount, Abruzzo Region (Petriccione and Claroni 1996).

For the map of Oaxaca state in Mexico (Velasquez et al. 2003), both positive processes (regeneration and secondary succession) and negative processes

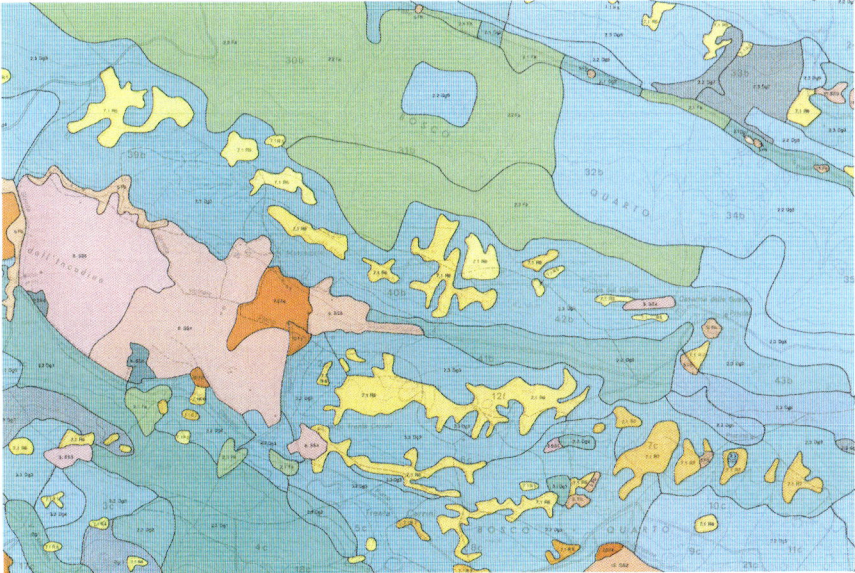

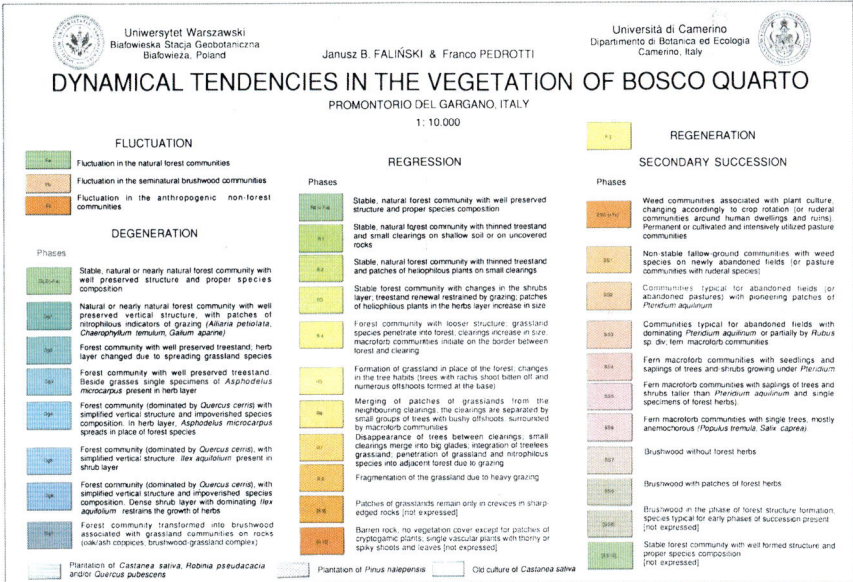

Fig. 6.39 Map of dynamical tendencies in the vegetation of the "Bosco Quarto" forest, Mt. Gargano, Adriatic southern Italy (From Faliński and Pedrotti 1990)

Fig. 6.40 Dynamic tendencies in the vegetation of the "Bosco Quarto": (**a**) natural fluctuation in *Doronico-Carpinetum betuli*; (**b**) natural fluctuation in *Aremonio-Fagetum*; (**c**) anthropogenic fluctuation in segetal vegetation; (**d**) degeneration in *Quercus cerris* forest due to selective cutting and coppicing (From Faliński and Pedrotti 1990)

Fig. 6.41 Dynamic tendencies in the vegetation of the "Bosco Quarto": (**a**) regression process in an overgrazing clearing with the gradual death of trees; (**b**) regression process with the total disappearance of trees vegetation; (**c**) secondary succession with developement of *Pteridium aquilinum* in abandoned croplands on the *bottom* of a big doline; (**d**) secondary succession with developement of new *Quercus ilex* woodland after the *Pteridium aquilinum* phase (From Faliński and Pedrotti 1990)

(deforestation and degradation) have been mapped. On the maps of the Molène Archipelago (France) the initial groupings, processes of superposition and complete substitution were mapped (Bioret et al. 1995). Regardless of the nomenclature employed in these studies, the dynamic processes described are in essence always the same.

Relations Among Different Types of Phytosociological Maps

We can now attempt a comparison of some of the first types of vegetation map presented with the following map types: phytosociological maps of actual vegetation, maps of the dynamic tendency of the vegetation, integrated phytosociological maps and phytosociological maps of potential vegetation.

As a reference area, we take a small protected area of 10 ha on the banks of the Lago di Levico (lake) in Trentino, occupied by moist meadow and swamp forest vegetation (Fig. 6.42). The starting point is a phytosociological map of actual vegetation, with representation of the different plant associations (map A); also identified and mapped are the dynamic tendencies (map B). One proceeds next to identify and delimit the vegetation series, and to represent them on an integrated phytosociological map (map C). With this map it is easy to reconstruct the potential vegetation of the study area (map D) (Fig. 6.43).

An example that treats a larger area is that of Laghestel di Piné (central Italian Alps), a glacial basin excavated in Permian porphyry (Fig. 6.44), where four maps have been made: actual vegetation (phytosociological map), dynamic tendency, vegetation series (integrated phytosociological map), and potential vegetation. The map of actual (natural) vegetation shows 30 plant associations and one subassociation, by cartographic units or by symbols (for very localized units) (Fig. 6.45). The map legend follows phytosociological systematics (Fig. 6.46). The associations are involved in very diverse dynamic processes, as are shown on the map of dynamic tendencies (Fig. 6.47). Understanding these dynamic tendencies permits recognition and understanding of vegetation series; the map of vegetation series (or integrated phytosociological map, see Fig. 6.48) was made in this way; its legend is shown in Figs. 6.49, 6.50, 6.51, 6.52, 6.53, and 6.54. In this legend, the plant associations are listed by vegetation series or sigmeta. Finally, the map of potential vegetation shows only the climax associations that represent the "head associations" of the series on the integrated map (Fig. 6.54).

By examining the four [phytosociological] maps together, one can see easily that the mapping units are always the same on the map of actual vegetation and on the integrated map. On the other hand, the map of dynamic tendencies shows some differences, because an association can be subjected to different development according to the intensity and type of perturbation exerted upon it. On the map of potential vegetation, all the associations that belong to the same series are combined into a single vegetation unit ("head association"), and the map is thus more synthetic than the other three map types.

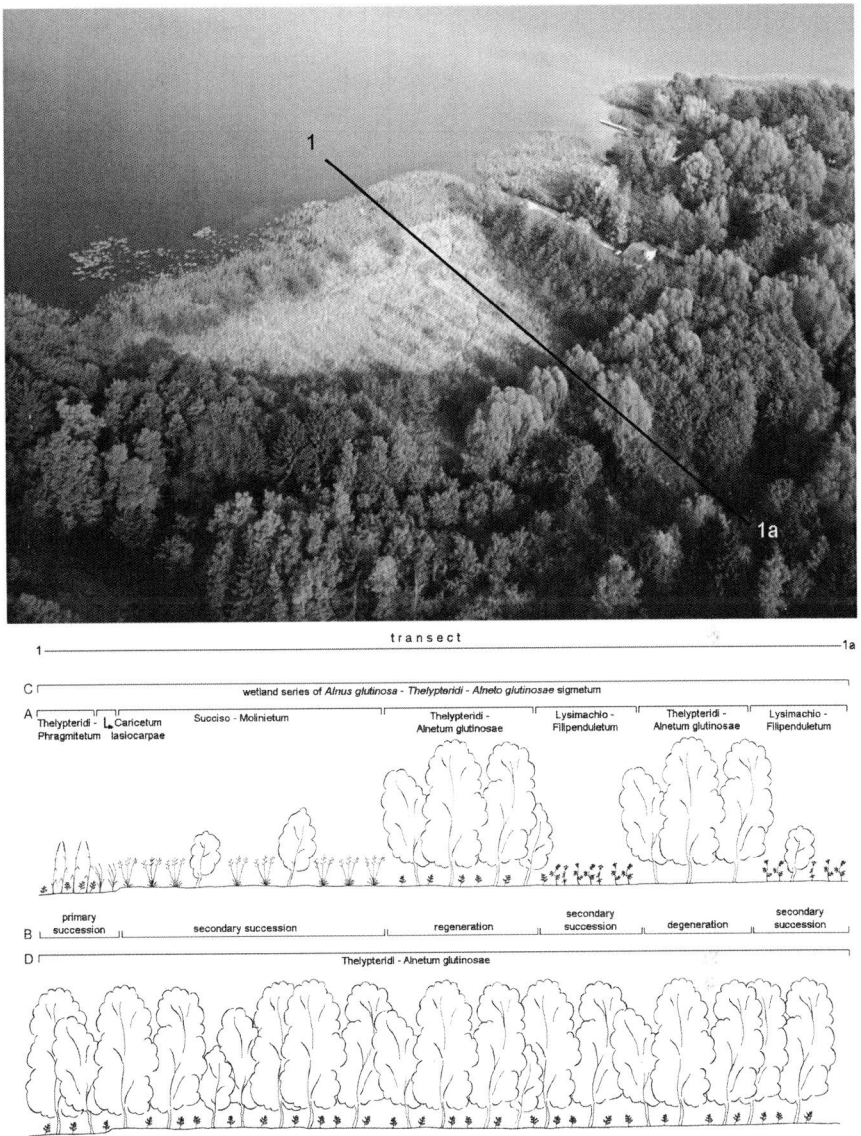

Fig. 6.42 Levico Lake, Trentino-Alto Adige Region, northern Italy, showing bank transects (1-1a) and correlated profiles: *A* – plant associations; *B* – dynamic tendencies in the vegetation; *C* – vegetation series; *D* – potential vegetation (Photo *Servizio Parchi P.A.T., Trento*)

The cartographic units that appear on the four vegetation maps show different proprieties of the vegetation, according to the type of map made; each map type possesses its own particular significance and offers the possibility to know different but complementary characteristics of the vegetation. Even so, the map that contains

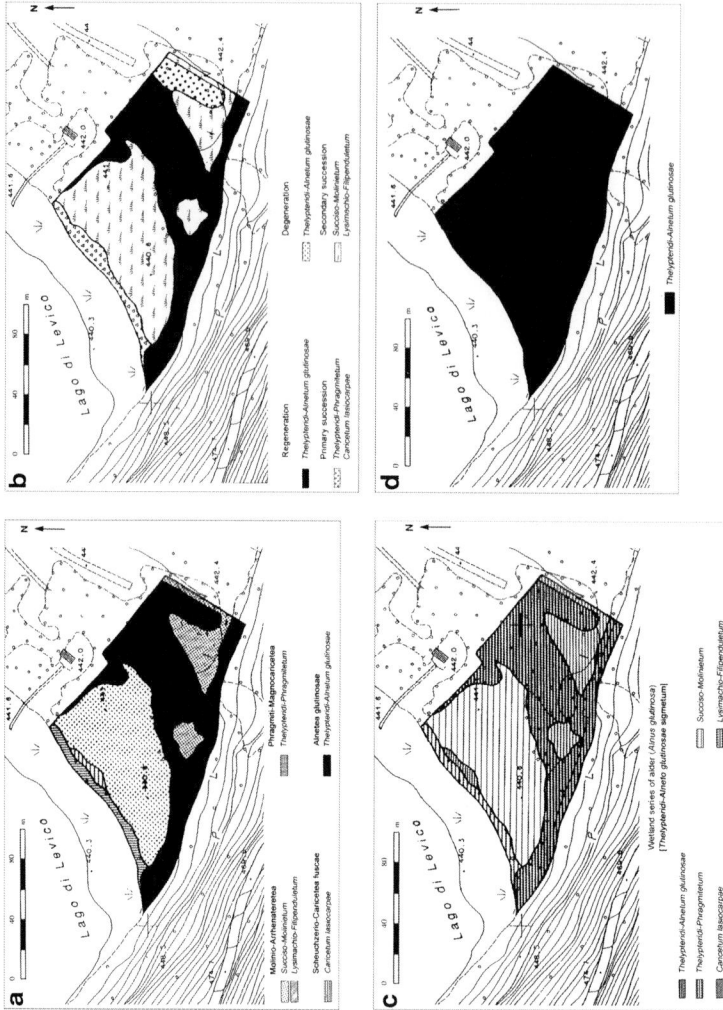

Fig. 6.43 Vegetation maps of the banks littoral of Levico Lake, Trentino-Alto Adige Region, northern Italy: (**a**) map of actual natural vegetation (phytosociological map); (**b**) map of dynamic tendencies in the vegetation; (**c**) integrated phytosociological map; (**d**) map of potential vegetation (From Pedrotti 1998)

Fig. 6.44 The glacial basin of Laghestel di Piné, Trentino-Alto Adige Region, northern Italy. The herbaceous vegetation belongs to the orders *Magnocaricetalia* and *Molinietalia*, the forest vegetation to the order *Vaccinio-Piceetalia* (Photo Franco Pedrotti 2001)

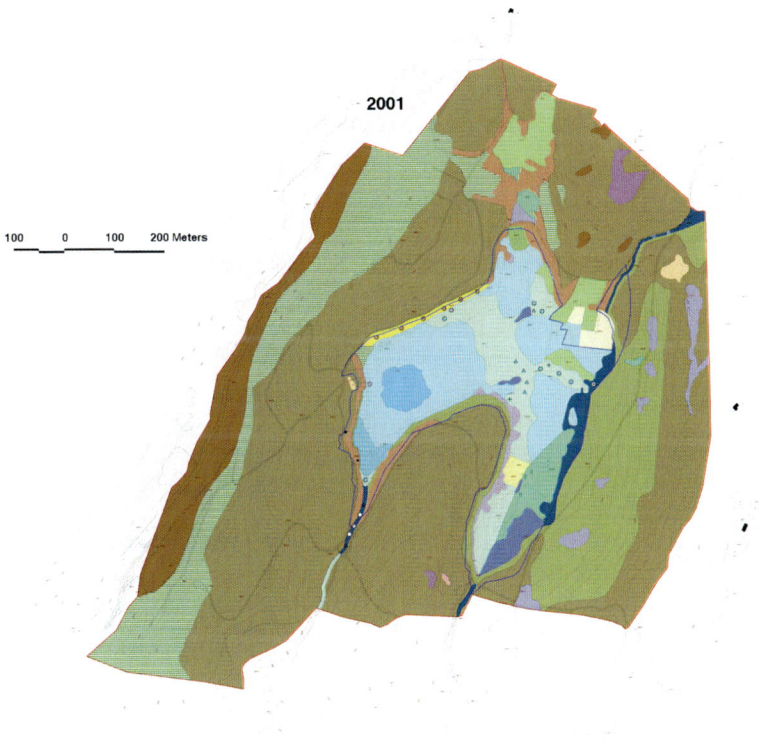

Fig. 6.45 Map of actual natural vegetation of Laghestel di Piné surveyed in 2001 (From Pedrotti 2004a)

Fontinaletea antipyreticae

☆   *Fontinaletum antipyreticae*

Charetea fragilis

         Charetum fragilis

Lemnetea

○   *Lemnetum minoris*

Scheuchzerio-Caricetea fuscae

         Caricetum lasiocarpae

         Caricetum lasiocarpae var. a Rhynchospora alba

◊   *Caricetum fuscae*

Asplenietea trichomanis

         Sileno rupestris - Asplenietum septentrionalis

Calluno-Ulicetea

         Chamaecytiso hirsuti - Callunetum

Rhamno-Prunetea

●   *Corylo - Populetum tremulae*

         Frangulo alni - Viburnetum opuli

Artemisietea vulgaris, Stellarietea mediae

○   *Erigeronetum annui*

         Galinsogo - Portulacetum

         Tanaceto vulgaris - Artemisietum vulgaris

Epilobietea angustifolii

         Rubetum idaei

Phragmiti-Magnocaricetea

○   *Caricetum elatae*

△   *Caricetum rostratae*

□   *Phragmitetum australis*

⊙   *Typhetum latifoliae*

△   *Glycerietum plicatae*

Molinio-Arrhenatheretea

∗   *Scirpetum sylvatici*

         Lysimachio - Filipenduletum

         Junco - Molinietum

         Centaureo - Arrhenatheretum

         Alopecuretum pratensis

⊙   *Juncetum macri*

Alnetea glutinosae

         Carici elatae - Alnetum glutinosae

○   *Salicetum cinereae*

Querco-Fagetea

         Stellario nemorum - Alnetum glutinosae

Vaccinio-Piceetea

         Molinio - Pinetum sylvestris

         Vaccinio - Pinetum sylvestris

Quercetea pubescentis

         Luzulo niveae - Quercetum petraeae

Reforestations

         Picea abies and *Larix tecidua*

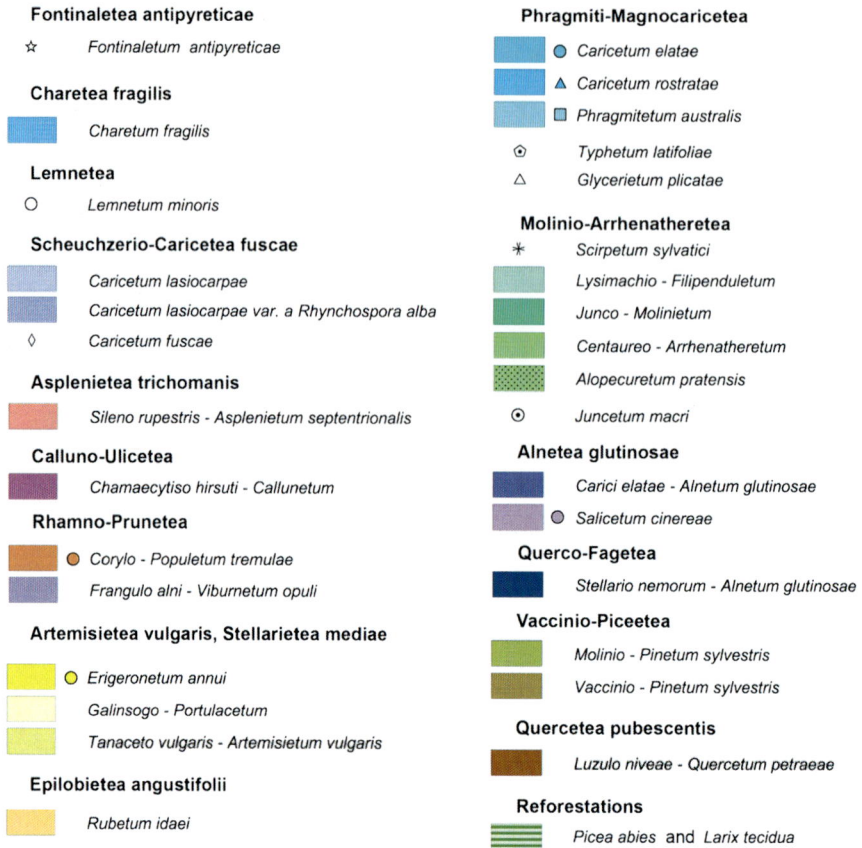

Fig. 6.46 Legend of the previous map (Fig. 6.45); the associations are listed below the vegetation classes (From Pedrotti 2004a)

the most information is the integrated phytosociological map (or synphyto-sociological map).

This methodology has also been applied in other parts of Trentino, as on the biotopes Laghestel di Piné (Pedrotti 2004a), Marocche di Dro and the Fiavé wetland complex (Figs. 7.8, 7.9, 7.10, 7.20, and 7.21) and to all the protected biotopes of Trentino (Pedrotti 2001).

Maps of Plant Ecology

Plant-ecological (phytoecological) maps are maps of ecological aspects rather than vegetation types and generally attempt to show the spatial (or spatio-temporal) variation of:

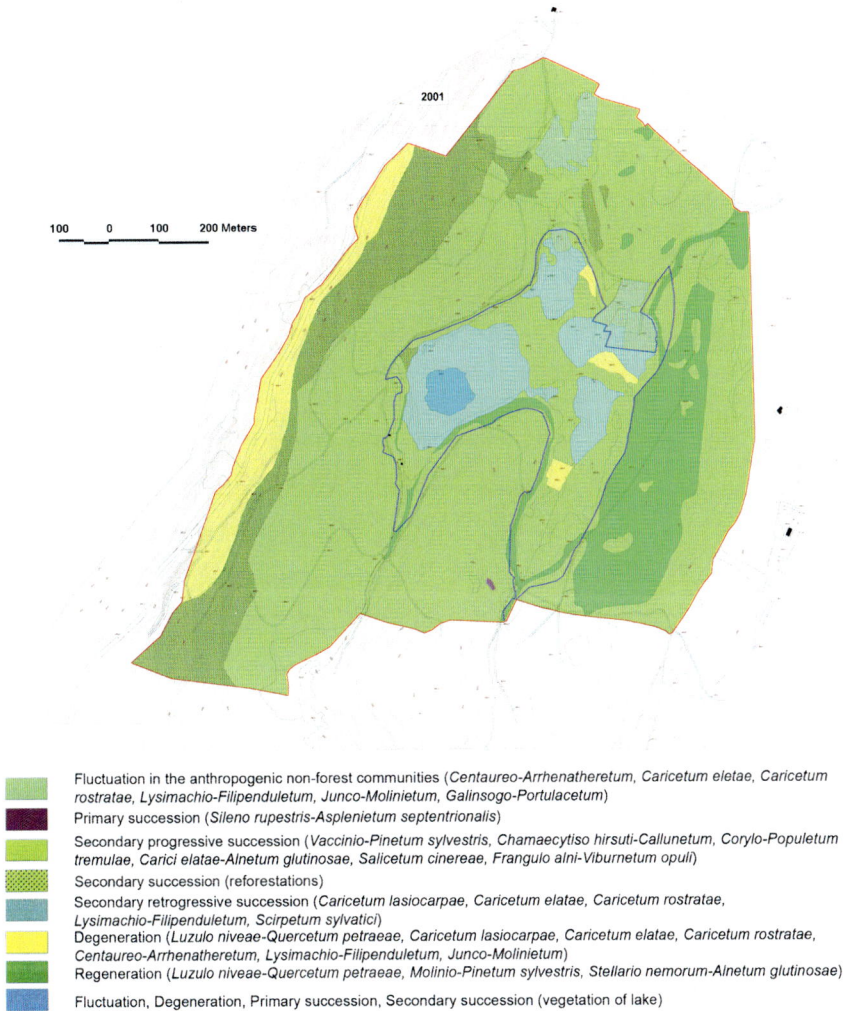

Fluctuation in the anthropogenic non-forest communities (*Centaureo-Arrhenatheretum, Caricetum eletae, Caricetum rostratae, Lysimachio-Filipenduletum, Junco-Molinietum, Galinsogo-Portulacetum*)

Primary succession (*Sileno rupestris-Asplenietum septentrionalis*)

Secondary progressive succession (*Vaccinio-Pinetum sylvestris, Chamaecytiso hirsuti-Callunetum, Corylo-Populetum tremulae, Carici elatae-Alnetum glutinosae, Salicetum cinereae, Frangulo alni-Viburnetum opuli*)

Secondary succession (reforestations)

Secondary retrogressive succession (*Caricetum lasiocarpae, Caricetum elatae, Caricetum rostratae, Lysimachio-Filipenduletum, Scirpetum sylvatici*)

Degeneration (*Luzulo niveae-Quercetum petraeae, Caricetum lasiocarpae, Caricetum elatae, Caricetum rostratae, Centaureo-Arrhenatheretum, Lysimachio-Filipenduletum, Junco-Molinietum*)

Regeneration (*Luzulo niveae-Quercetum petraeae, Molinio-Pinetum sylvestris, Stellario nemorum-Alnetum glutinosae*)

Fluctuation, Degeneration, Primary succession, Secondary succession (vegetation of lake)

Fig. 6.47 Map of dynamic tendencies in the vegetation of Laghestel di Piné (From Pedrotti 2004a)

1. Intrinsic proprieties of phytocoenoses, such as phytomass, structural differences in the vegetation, phenology; or
2. Vegetation differentiation as related to ecological factors, such as climate or soil (usually separately).

Many vegetation maps also have smaller auxiliary maps (called *cartine laterali* in Italian, or "curtains", if shown adjacently) that show hypsometry, geology, pedology, climate, etc. These are maps of ecological factors that influence the

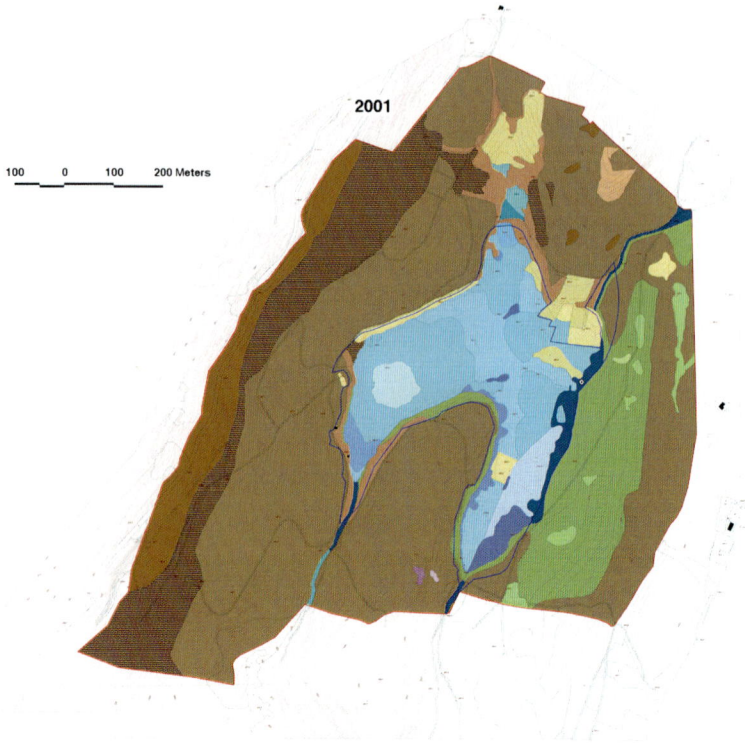

Fig. 6.48 Map of vegetation series of Laghestel di Piné. The legend of this map is given in Figs. 6.49, 6.50, 6.51, 6.52, 6.53, and 6.54 (From Pedrotti 2004a)

vegetation, and the objective is to show the spatial variation of these important factors separately. Climatic maps can in turn show climatic elements such as temperature or precipitation regimes, insolation, duration of snow cover, etc., or can be synthetic climatic (phytoclimate) maps that represent the distribution of climate types, based on bioclimatic indices. An example of this last type is the *Carta fitoclimatica del Trentino-Alto Adige* (Phytoclimatic Map of Trentino-Alto Adige) at scale 1:250,000, on which 26 distinct climate types are shown, based on temperature, precipitation and combined thermal-pluvial continentality (Gafta and Pedrotti 1998).

Plant-ecological maps can differ very much from each other, for example: synphenological maps (Fig. 6.55), maps of the herbaceous understorey layer Fig. 6.56 maps of trees uprooted and fallen to the ground (Figs. 6.57 and 6.58), or maps of structural differentiation in the vegetation based on one or more intrinsic characters or proprieties of phytocoenoses and their populations. Other examples include maps of plant productivity Fig. 6.59 or living biomass (Fig. 6.60). The map

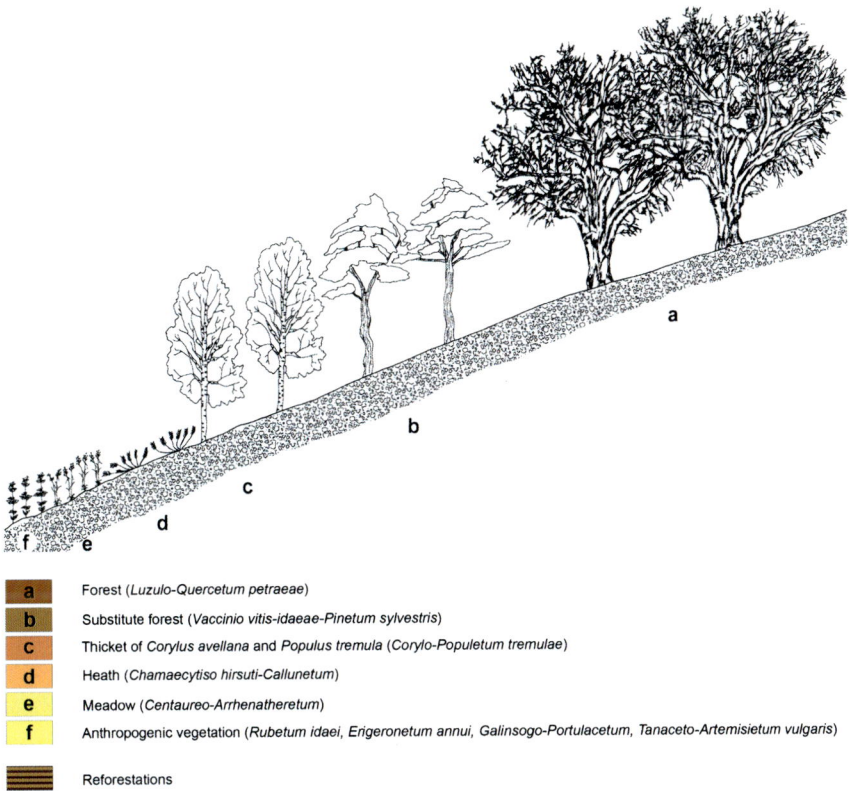

a Forest (*Luzulo-Quercetum petraeae*)

b Substitute forest (*Vaccinio vitis-idaeae-Pinetum sylvestris*)

c Thicket of *Corylus avellana* and *Populus tremula* (*Corylo-Populetum tremulae*)

d Heath (*Chamaecytiso hirsuti-Callunetum*)

e Meadow (*Centaureo-Arrhenatheretum*)

f Anthropogenic vegetation (*Rubetum idaei, Erigeronetum annui, Galinsogo-Portulacetum, Tanaceto-Artemisietum vulgaris*)

Reforestations

Fig. 6.49 Series of *Quercus petraea* [*Luzulo-Querceto petraeae* sigmetum] at Laghestel di Piné, developed on acidic brown soils (Dystric Cambisols) (From Pedrotti 2004a, modified)

of the vegetation of the Monti Sibillini (mountains) obtained from Landsat imagery (Fig. 5.40) also permitted estimation of meadow biomass (Schino et al. 2003).

Ecological maps can also have very practical applications, such as the map of forest damage caused by acid rain shown in Fig. 6.61 and maps of fire risk in Mediterranean France (Boullet and Géhu 1988a, b, c). The latter were constructed with reference to risk factors involving general bioclimate, wind, and the physical structure of the vegetation (e.g. retention of dead lower branches), from which one deduces the overall fire risk. The legend of this map shows seven categories of climatic risk: higher, very high, high, fairly high, fairly low, low, and very low; each in turn is subdivided into four categories of risk due to vegetation structure: very strong, strong, weak and very weak.

An example of an ecological map showing structural vegetation differentiation (cover degree) is that of a sandy area near Jelonka (Poland) being colonized by secondary succession (Fig. 6.62). Cover is shown for the canopy (*Juniperus communis*) and for lichen, bryophyte and grassy communities (Cornniculario-Cladonietum, Spergulo-Corynephoretum, etc.). Particular attention is paid to the ecotones formed around single juniper plants.

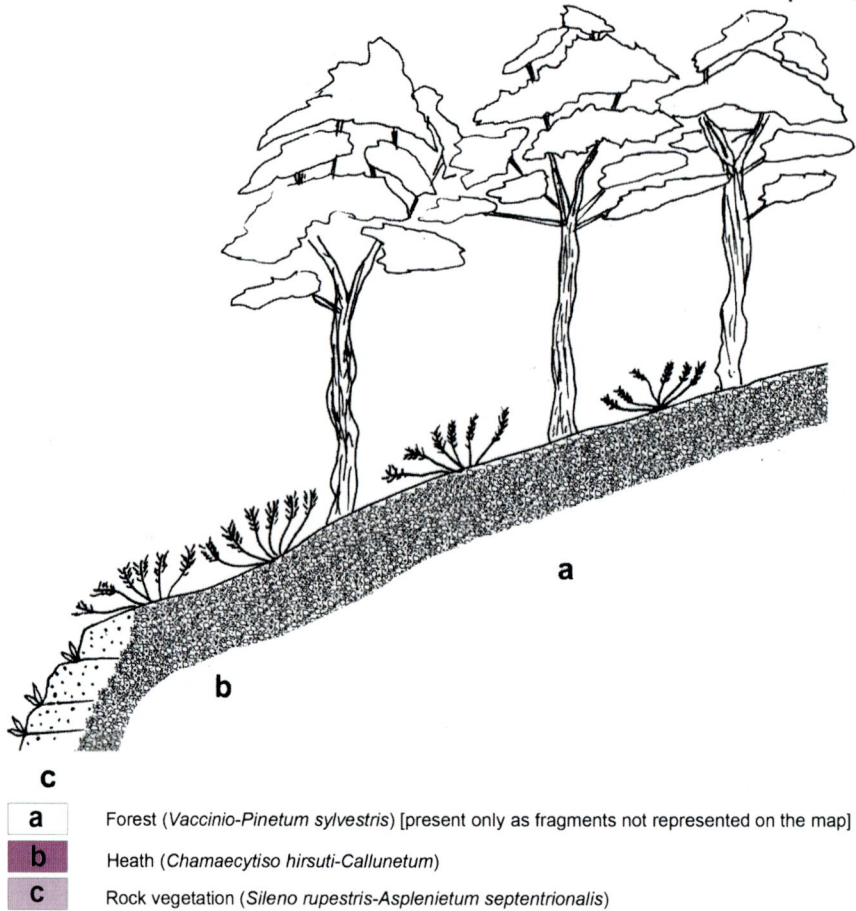

Fig. 6.50 Xerophilous series of *Pinus sylvestris* [*Vaccinio-Pineto sylvestris* sigmetum] at Laghestel di Piné, developed on acidic brown soils (Dystric Cambisols) (From Pedrotti 2004a, modified)

The fundamental maps that show vegetation types also contain ecological information, in that each vegetation type identified and mapped possesses definite ecological characteristics. Representing these types on a map thus gives some indication of their ecology, albeit indirectly. As a result, phytosociological maps also express synecological features.

It must be noted that such vegetation maps are not phytoecological maps in the sense first stated, as can be seen in the vegetation maps of the Gran Sasso d'Italia (mountains) (Biondi 1999) and of the desert around the Caspian Sea (Safronova 2004).

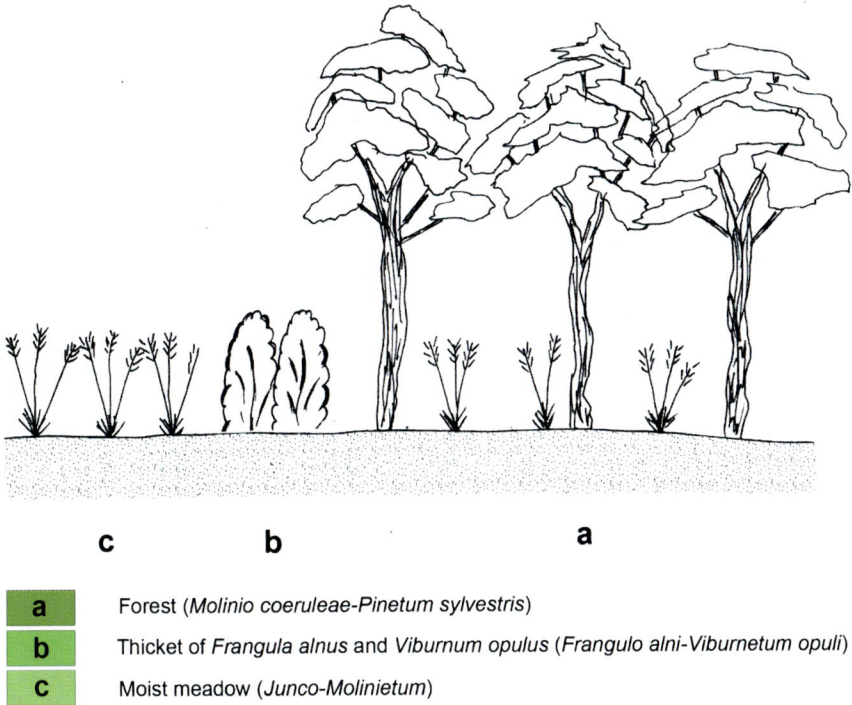

a   Forest (*Molinio coeruleae-Pinetum sylvestris*)

b   Thicket of *Frangula alnus* and *Viburnum opulus* (*Frangulo alni-Viburnetum opuli*)

c   Moist meadow (*Junco-Molinietum*)

Fig. 6.51 Hygrophilous series of *Pinus sylvestris* [*Molinio coeruleae-Pineto sylvestris* sigmetum] at Laghestel di Piné, developed on acidic brown soils (Dystric Cambisols) with groundwater appearing at the surface in some places (From Pedrotti 2004a, modified)

Still, at various times, the need is felt to enrich the maps of vegetation types, as was emphasized by Sochava (1975), stating that maps made by traditional methods "cannot completely satisfy the growing need for phytogeographic information". Analogous observations and proposals have also been made by other authors, such as Ozenda (1975, 1986), who on various occasions has shown the objectives and necessity of an ecological cartography. Finally, Faliński (1993) also spoke about the "ecologization of geobotanical maps", referring to the introduction of ecological data onto maps of vegetation types, with many examples, especially related to the Białowieza forest and other locations in Poland. In a 1-ha sector of the Białowieza forest, for example, occupied by the association Tilio-Carpinetum with three distinct subassociations (calamagrostidetosum, typicum and stachyetosum), one phytosociological sheet of the actual vegetation combines 36 phytoecological sheets that

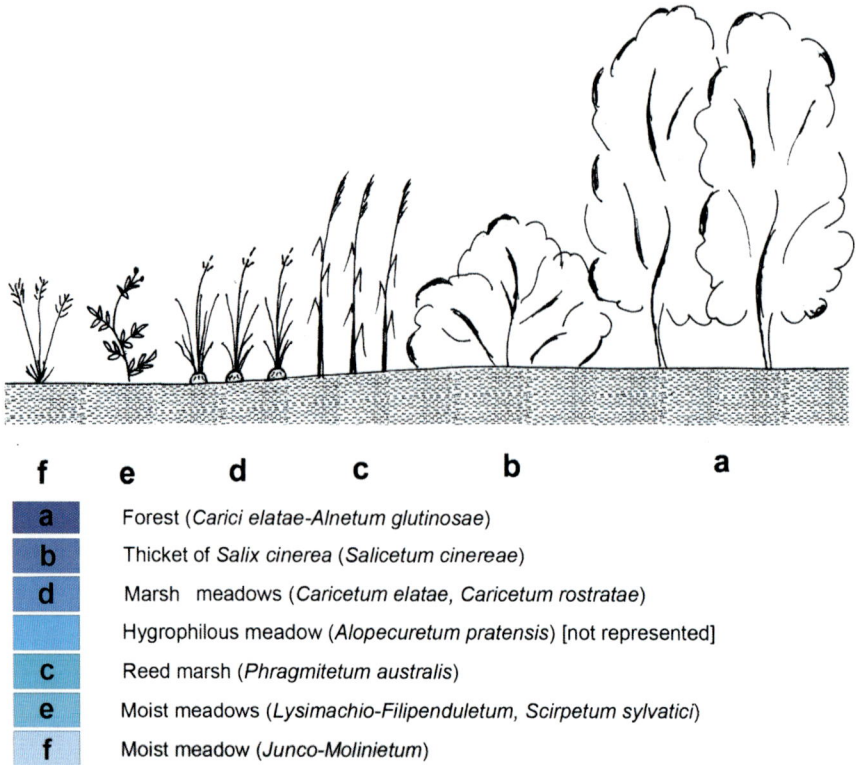

Fig. 6.52 Wetland series of *Alnus glutinosa* [*Carici elatae-Alneto glutinosae* sigmetum] at Laghestel di Piné, developed on hydromorphous and organic soils (Umbric Gleysols and Terric Histosols) (From Pedrotti 2004a, modified)

show ecological aspects such as trees fallen to the ground, decomposition of the uprooted trunks, and forest structure (Faliński 1990–1991).

Synchorological Maps

Synchorological maps show the area of one or more plant associations or other syntaxonomic units in a given geographic area. According to the mode of representation, synchorological maps can be classified as location maps, grid maps, and range maps showing absolute limits (Pop 1977–1979), as already illustrated with chorological maps of species.

Location maps are used usually with associations (or lower units) that combine phytocoenoses which are rare and of quite small areal extent, as may occur on rock

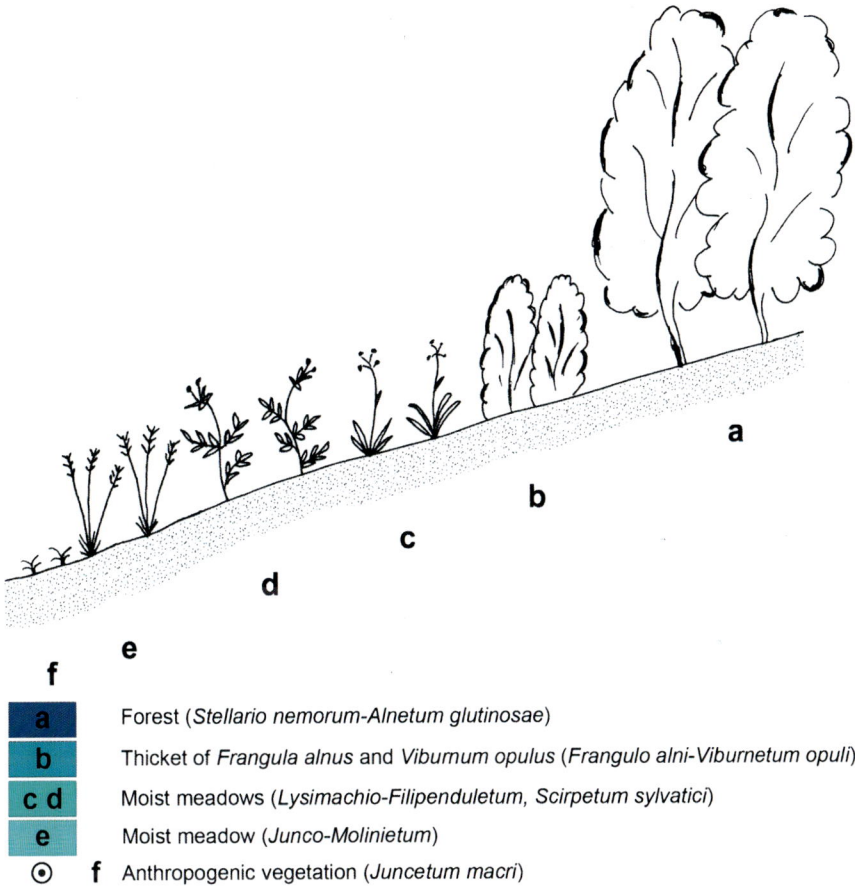

a   Forest (*Stellario nemorum-Alnetum glutinosae*)

b   Thicket of *Frangula alnus* and *Viburnum opulus* (*Frangulo alni-Viburnetum opuli*)

c d   Moist meadows (*Lysimachio-Filipenduletum, Scirpetum sylvatici*)

e   Moist meadow (*Junco-Molinietum*)

⊙   f   Anthropogenic vegetation (*Juncetum macri*)

Fig. 6.53 Riparian series of *Alnus glutinosa* [*Stellario nemorum-Alneto glutinosae* sigmetum] at Laghestel di Piné, developed on alluvial soils (From Pedrotti 2004a, modified)

walls or in wetlands. The maps show, by symbols, only the locations where phytocoenoses belonging to a particular association are developed. A map of this type was used to present the synchorology of three riparian associations in Italy characterized by *Fraxinus oxycarpa* (Fig. 6.63). One variant of such a map shows areas occupied by the associations, as on the synchorological map of *Abies* forests in Trentino (Gafta 1994), which shows the distribution of 9 associations and subassociations. Another map, mixing absolute limits and locations, was used to represent the areas of three associations of alder (*Alnus*) woods in Italy, one quite rare, the Carici brizoidis-Alnetum glutinosae, and two common, the Alnetum incanae and the Aro italici-Alnetum glutinosae (Fig. 6.64).

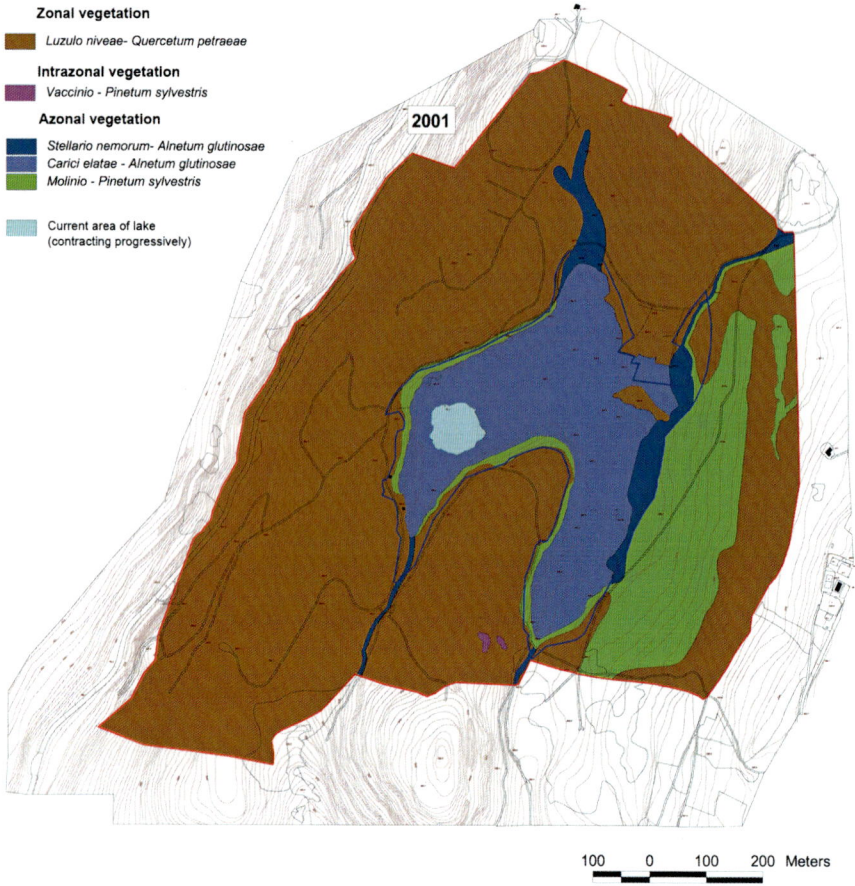

Fig. 6.54 Map of the potential natural vegetation of Laghestel di Piné (From Pedrotti 2004a)

Grid maps use various systems for dividing a territory systematically, such as the UTMG already mentioned in Chap. 4. The presence of an association inside a grid cell is shown by coloration or hatching or by a symbol.

In the atlas of vegetation to be conserved in Switzerland, 118 vegetation types are described and mapped, of which 97 are defined by alliances and the rest by other syntaxonomic units. The mapping was done with great precision, over a network of georeferenced quadrats of 1 km, each of which is referenced to the national map of Switzerland (Hegg et al. 1993); Fig. 6.65 shows the distribution of the Carpinion alliance, as an example.

In some cases, above all when the vegetation is quite rare and its distribution very localized, it is possible to make a small synchorological point map to indicate the locations where an ecotope (patch) with all the stages comprising the series is

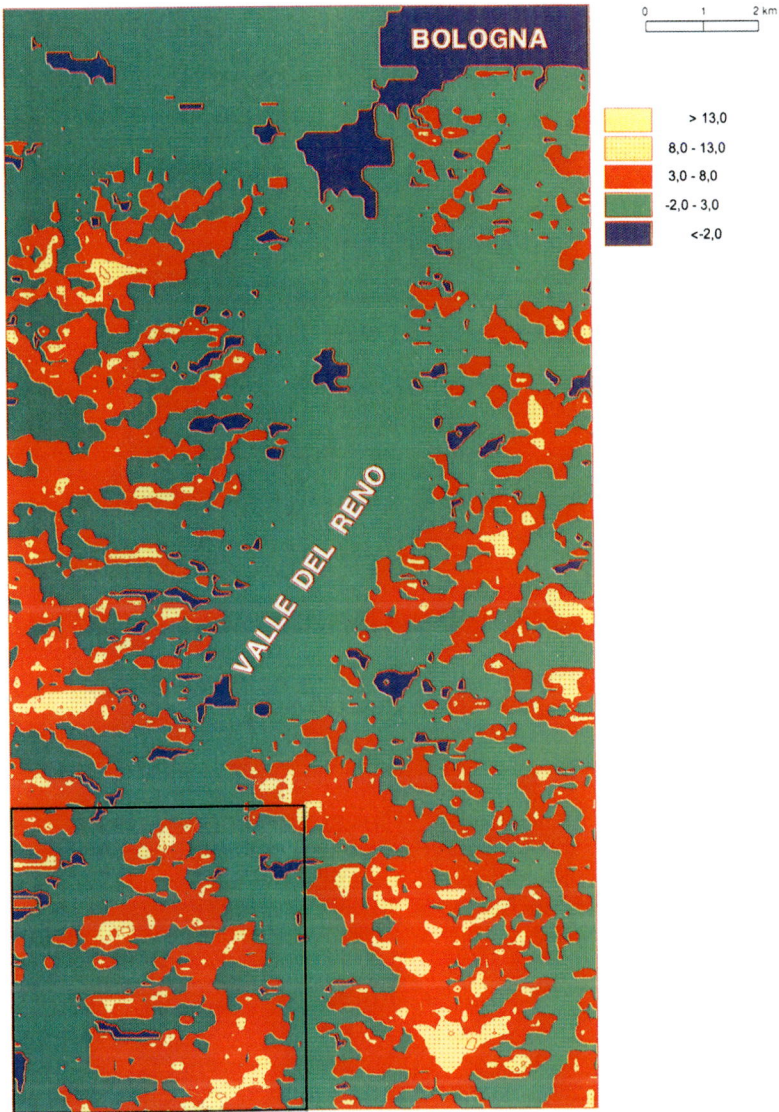

Fig. 6.55 Synthetical phenological map of springtime flowering in the lower Reno river valley, Bologna, Emilia-Romagna Region, north-central Italy. The *numbers* and *colors* show delays in flowering (in days) relative to the reference site (From Puppi Branzi and Zanotti 1989)

present, as in the case of the *Quercus petraea* series (Hieracio racemosi-Querceto petraeae sigmetum) in Umbria, present only in two locations (Fig. 6.66).

Vegetation series can also be mapped in this way, with a gridded reference framework, as was done for all the vegetation series in Trentino-Alto Adige (data unpublished). An example for five series with *Pinus sylvestris* is shown here (Fig. 6.67).

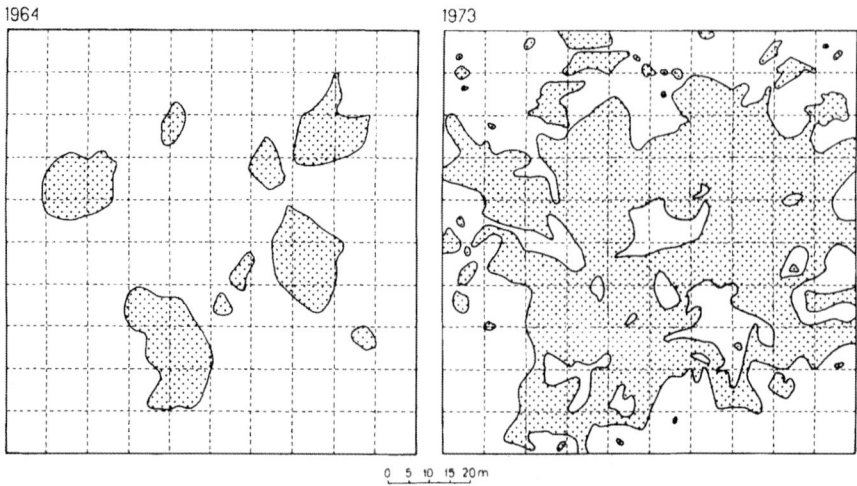

Fig. 6.56 Changes in the abundance of heliophilous species (especially *Urtica dioica*) in the *Tilio-Carpinetum* of Białowieza primeval forest (Poland) as a consequence of natural thinning during 1964–1974 (From Faliński 1978)

Fig. 6.57 Trees downed by strong wind, in the *Tilio-Carpinetum* of Białowieza primeval forest, Poland (From Faliński 1978)

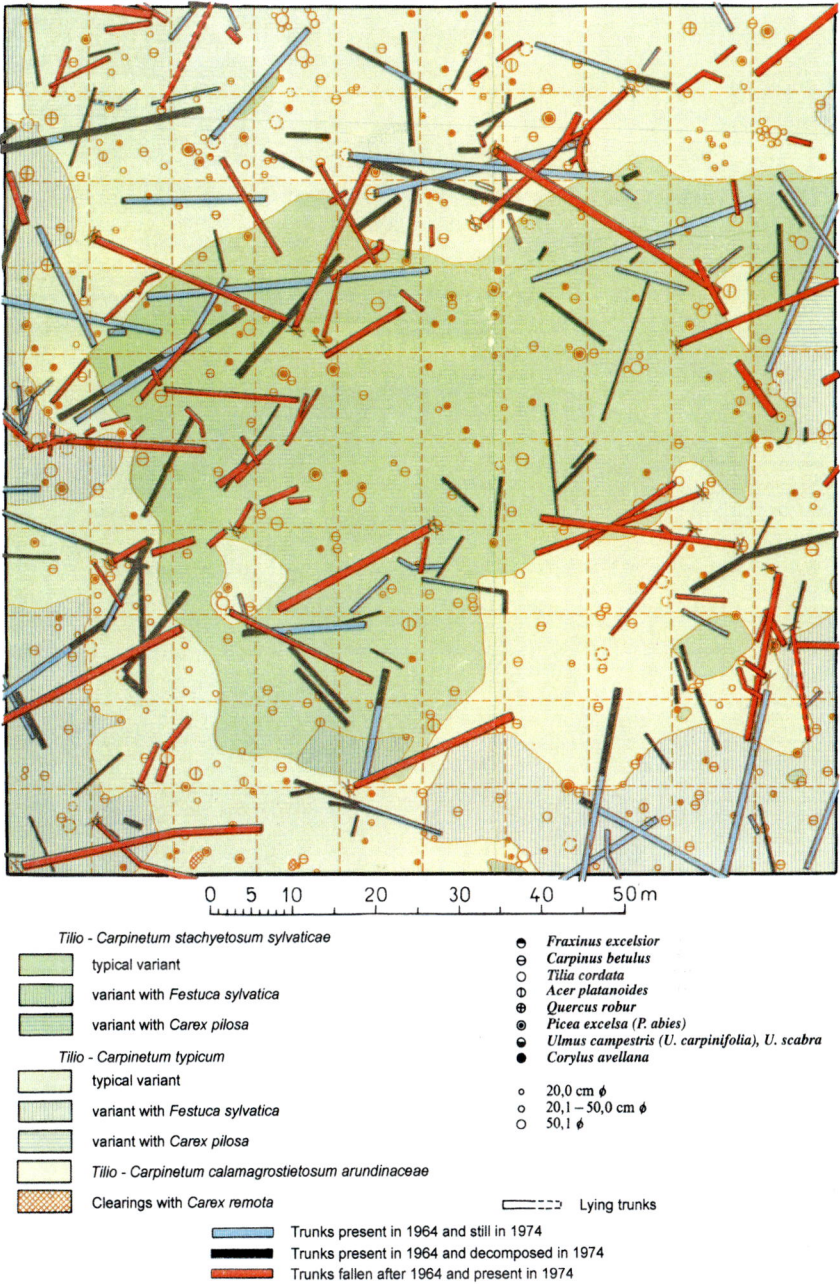

Fig. 6.58 Map of fallen tree trunks in a permanent survey area in the Białowieza primeval forest, Poland (From Faliński 1976)

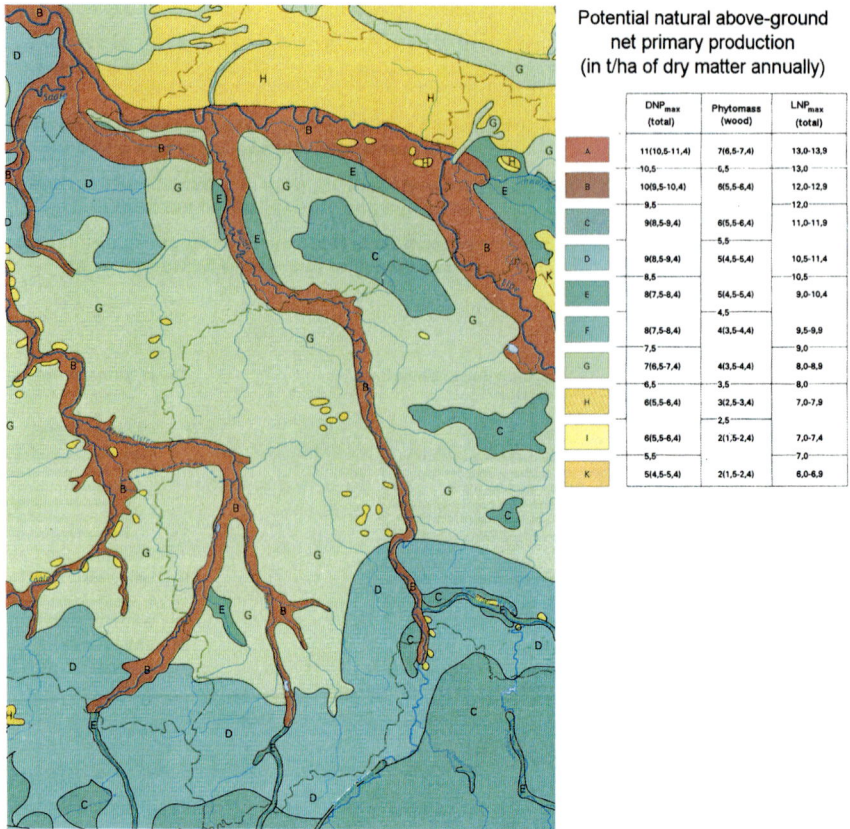

Potential natural above-ground
net primary production
(in t/ha of dry matter annually)

	DNP_{max} (total)	Phytomass (wood)	LNP_{max} (total)
A	11(10,5-11,4)	7(6,5-7,4)	13,0-13,9
	-10,5-	-6,5-	-13,0
B	10(9,5-10,4)	6(5,5-6,4)	12,0-12,9
	-9,5		-12,0
C	9(8,5-9,4)	6(5,5-6,4)	11,0-11,9
		-5,5-	
D	9(8,5-9,4)	5(4,5-5,4)	10,5-11,4
	-8,5		-10,5-
E	8(7,5-8,4)	5(4,5-5,4)	9,0-10,4
		-4,5-	
F	8(7,5-8,4)	4(3,5-4,4)	9,5-9,9
	-7,5		-9,0
G	7(6,5-7,4)	4(3,5-4,4)	8,0-8,9
	-6,5	-3,5-	-8,0
H	6(5,5-6,4)	3(2,5-3,4)	7,0-7,9
		-2,5-	
I	6(5,5-6,4)	2(1,5-2,4)	7,0-7,4
	-5,5		-7,0
K	5(4,5-5,4)	2(1,5-2,4)	6,0-6,9

Fig. 6.59 Distribution of potential net primary production classes of natural ecosystems from eastern Germany, at scale 1:750,000: DNP = net primary production as related to median age; LNP = current annual net primary production (From Hoffmann 1985)

Range maps of absolute distributional boundaries are generally made at broad scale to represent wider areas, by means of one or more closed curves, as on the maps of some vegetation orders in North America (Fig. 6.68). Synchorological maps can also be made for urban environments; in metropolitan Rome, for example, Fanelli (2002) used points to map 104 associations and other assemblages, largely synanthropic but some natural (remnants of natural vegetation), such as oak-ash forests (Fraxino orni-Quercetum ilicis, Fig. 6.69).

Synchorological maps are very useful for showing the distribution of associations that are rare or threatened due to human influences, such as the Mentho-Caricetum pseudocyperi floating mats in the lakes of central Italy (Fig. 6.70). The red book of phytocoenoses of the French coastline by Géhu (1991b) shows a distribution map of all the associations in danger (Fig. 6.71).

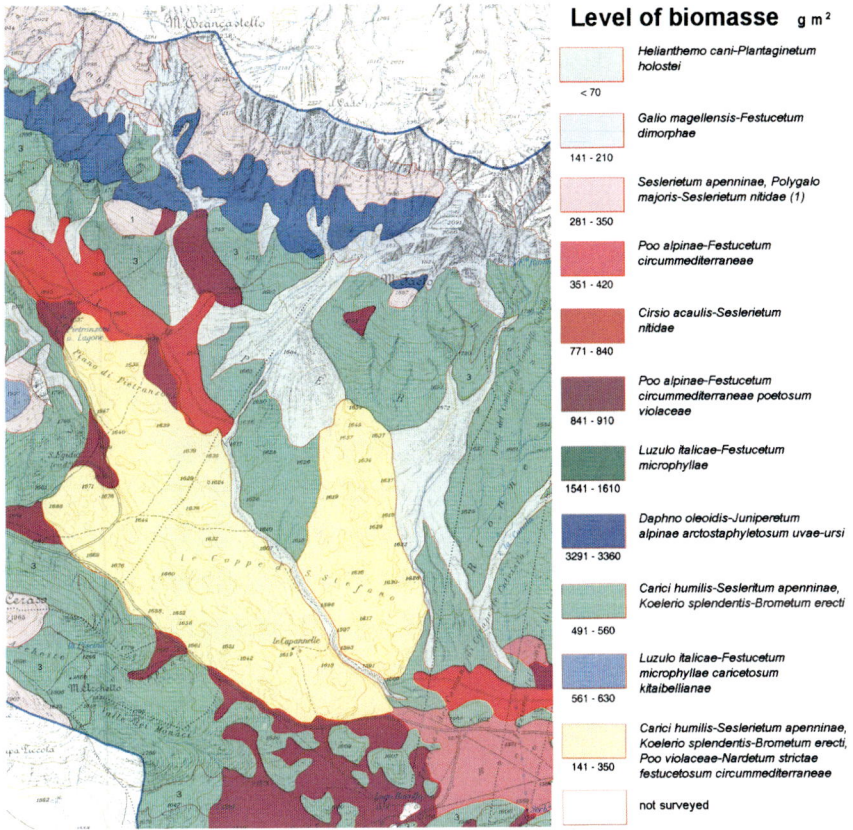

Level of biomasse g m²

< 70	Helianthemo cani-Plantaginetum holostei
141 - 210	Galio magellensis-Festucetum dimorphae
281 - 350	Seslerietum apenninae, Polygalo majoris-Seslerietum nitidae (1)
351 - 420	Poo alpinae-Festucetum circummediterraneae
771 - 840	Cirsio acaulis-Seslerietum nitidae
841 - 910	Poo alpinae-Festucetum circummediterraneae poetosum violaceae
1541 - 1610	Luzulo italicae-Festucetum microphyllae
3291 - 3360	Daphno oleoidis-Juniperetum alpinae arctostaphyletosum uvae-ursi
491 - 560	Carici humilis-Seslerietum apenninae, Koelerio splendentis-Brometum erecti
561 - 630	Luzulo italicae-Festucetum microphyllae caricetosum kitaibellianae
141 - 350	Carici humilis-Seslerietum apenninae, Koelerio splendentis-Brometum erecti, Poo violaceae-Nardetum strictae festucetosum circummediterraneae
	not surveyed

Fig. 6.60 Plant biomass of the Campo Imperatore highland pastures of the Gran Sasso d'Italia, Abruzzo Region, central Italy, at scale 1:25,000 (From Gratani et al. 1999)

Maps of the Conservation Status of Vegetation

Vegetation is the most obvious expression of differing environmental conditions, and each type of environment houses different plant associations. Through its various characteristics, vegetation reveals at each moment its state of conservation or alteration and the levels of influence to which it has been subjected by human activities. Vegetation thus constitutes a biological indicator of the general state of the environment, and it follows that the "vegetation approach" is very useful in the general evaluation of the environment.

Two types of problems are then posed:

(a) That of evaluating the state of the vegetation as such; and
(b) That of estimating the state of environmental conservation in this complex, making use also of vegetation parameters.

DAMAGE TO FORESTS IN GERMANY IN 1984
1: 200,000

Damaged Forest Area

up to 30% damaged
30 - 40%
40 - 50%
50 - 60%
more than 60%

Fig. 6.61 Map of damages due to acid rain in Germany in 1984 (*Bundesamt f. Naturschutz, Bonn, Germany*)

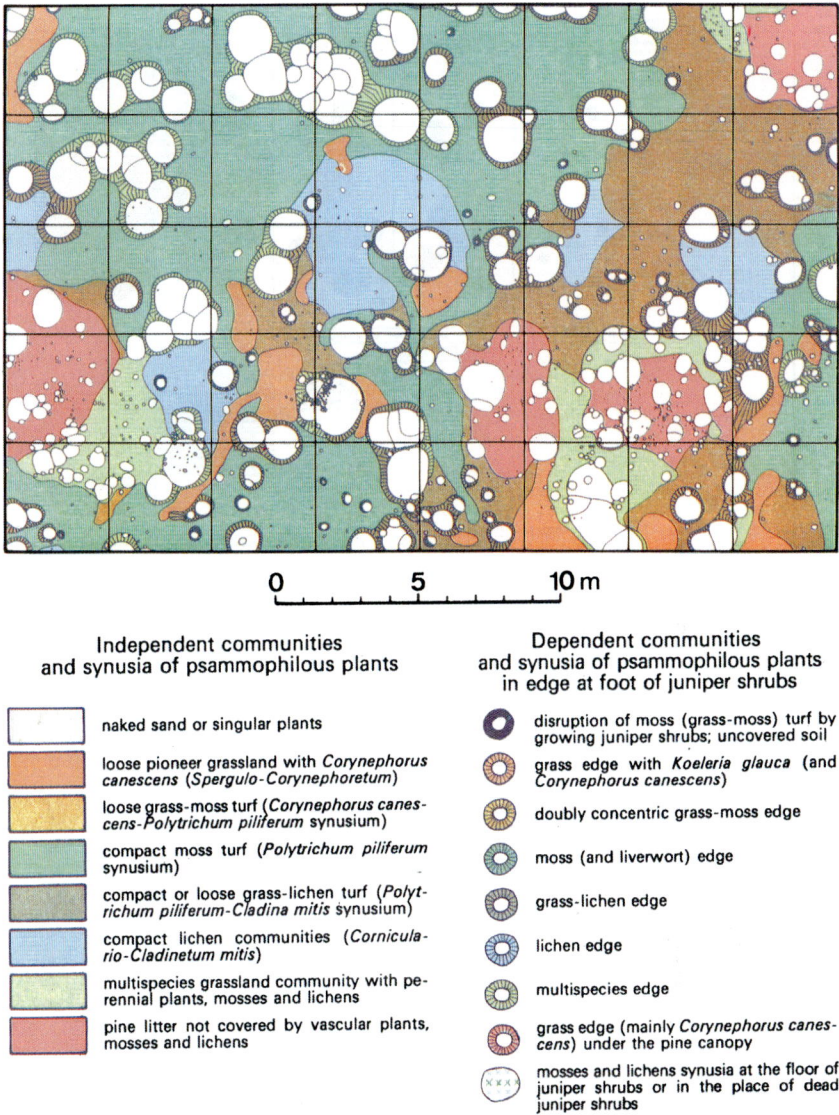

0   5   10 m

Independent communities and synusia of psammophilous plants

☐ naked sand or singular plants

☐ loose pioneer grassland with *Corynephorus canescens* (*Spergulo-Corynephoretum*)

☐ loose grass-moss turf (*Corynephorus canescens-Polytrichum piliferum* synusium)

☐ compact moss turf (*Polytrichum piliferum* synusium)

☐ compact or loose grass-lichen turf (*Polytrichum piliferum-Cladina mitis* synusium)

☐ compact lichen communities (*Corniculario-Cladinetum mitis*)

☐ multispecies grassland community with perennial plants, mosses and lichens

☐ pine litter not covered by vascular plants, mosses and lichens

Dependent communities and synusia of psammophilous plants in edge at foot of juniper shrubs

⊙ disruption of moss (grass-moss) turf by growing juniper shrubs; uncovered soil

⊙ grass edge with *Koeleria glauca* (and *Corynephorus canescens*)

⊙ doubly concentric grass-moss edge

⊙ moss (and liverwort) edge

⊙ grass-lichen edge

⊙ lichen edge

⊙ multispecies edge

⊙ grass edge (mainly *Corynephorus canescens*) under the pine canopy

⊙ mosses and lichens synusia at the floor of juniper shrubs or in the place of dead juniper shrubs

Fig. 6.62 Map of communities and synusiae of psammophilous plants developed around the trunks and between shrubs of *Juniperus communis*, at Jelonka, Poland, at scale 1:200 (From Faliński 1998b)

According to the various conceptions one may speak of naturalness and synanthropization of the vegetation, with many nuances. Among the numerous diagnostic possibilities offered by geobotanical cartography, two are illustrated here:

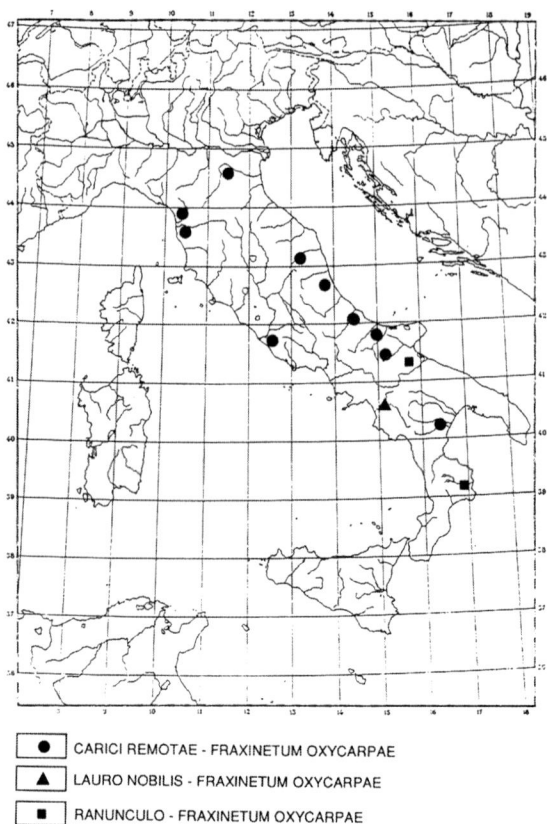

Fig. 6.63 Synchorological map of *Fraxinus oxycarpa* associations in Italy (From Pedrotti and Gafta 1996)

(a) The first considers areas that are relatively homogeneous geomorphologically, climatically and in terms of vegetation and attempts to evaluate the vegetation comprehensively;

(b) The second considers the plant stands that compose the vegetation and attempts to determine where the phytocoenoses of a single cartographic unit occur as concrete units.

A map of vegetation naturalness illustrates the first case. Naturalness is defined beginning with the process of "synanthropization", considered to be a process that alters the primary phytocoenoses; it is manifested in two main ways, by modification of the original floristic composition and by alteration of the vegetation structure.

Modification of the species combination characteristic of a given association occurs, as Faliński (1998b) stated, when autochthonous (native, or "natural")

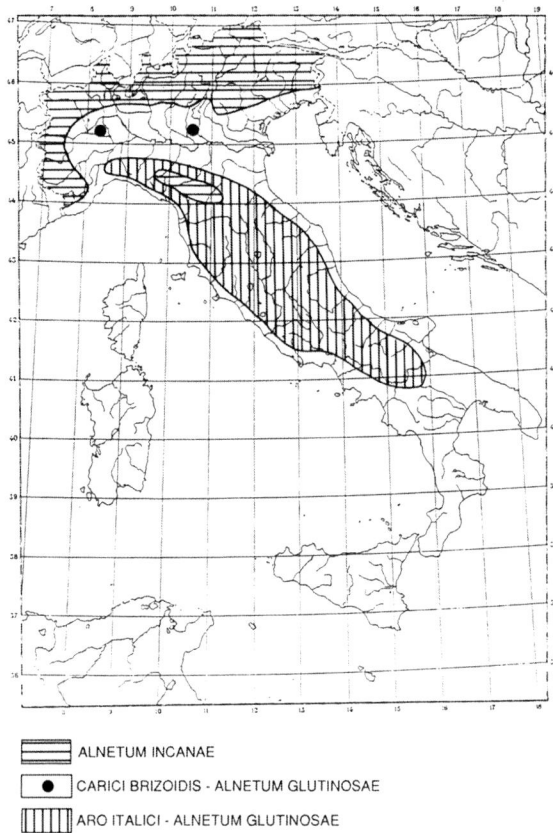

ALNETUM INCANAE

CARICI BRIZOIDIS - ALNETUM GLUTINOSAE

ARO ITALICI - ALNETUM GLUTINOSAE

Fig. 6.64 Synchorological map of some associations of *Alnus glutinosa* and *A. incana* in Italy (From Pedrotti and Gafta 1996)

species in a phytocoenosis are replaced by allochthonous (foreign, or "exotic") species; or when species of restricted ecological amplitude are replaced by species of wide ecological amplitude; or when species with small natural ranges (steno-topic) are replaced by cosmopolitan species. Structural alteration involves changes, usually simplification, in the general stratification of a phytocoenosis. The process of synanthropization (species substitution and change of structure) occurs in differ-ent intensities. This may not compromise the identity of the primary phytocoenosis, or may lead, on the other hand, to a complete substitution of the primary phytocoenosis by phytocoenoses less complex, often with less phytomass, through the process of regression.

Producing a map of naturalness is based on a preliminary subdivision of the territory into areas relatively homogeneous in terms of geomorphology and climate, inside which the overall naturalness is estimated in various degrees. On the map of

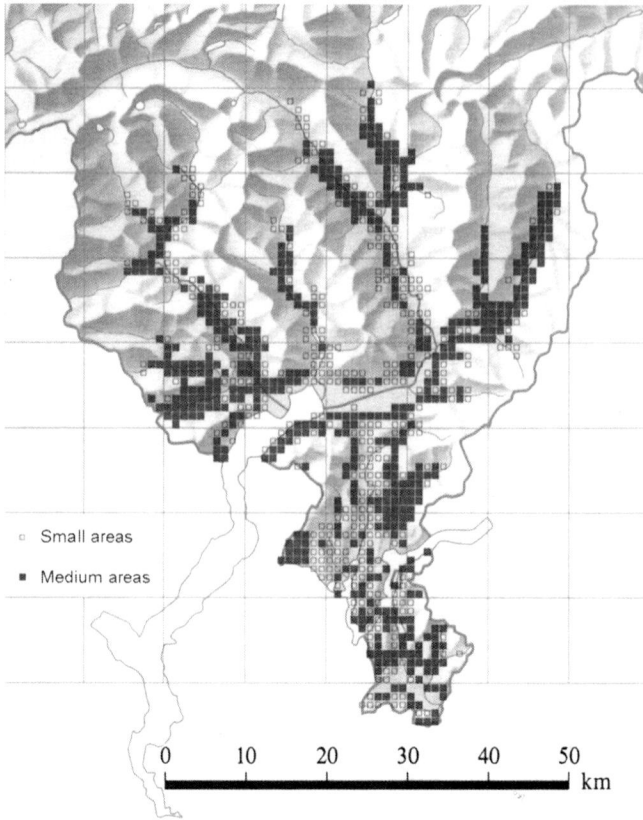

Fig. 6.65 Map of the distribution of *Carpinion* forests in Canton Ticino, Switzerland, based on a network of 1 km² squares (From Hegg et al. 1993)

the vegetation naturalness in Trentino-Alto Adige (Figs. 6.72 and 6.73) there were six distinct levels, with naturalness decreasing from level I to level VI (Pedrotti and Minghetti 1997):

I.   Areas with primary natural vegetation constituted by phytocoenoses undergoing processes of natural fluctuation and primary succession (in high mountains);

II. Areas with natural vegetation represented by forest phytocoenoses with their structure modified by man and undergoing processes of degeneration and regeneration (conifer forests);

III. Areas with vast complexes of natural vegetation, formed by forest phytocoenoses with their structure modified by man and in phases of degeneration and regeneration, plus remnant hedgerows and other units of semi-natural and synanthropic vegetation undergoing processes of secondary succession and anthropogenic fluctuation (coppice woods of broad-leaved trees alternating with grassy areas);

Acidophilous series of *Quercus petraea*
Hieracio racemosi-Querceto petraeae sigmetum

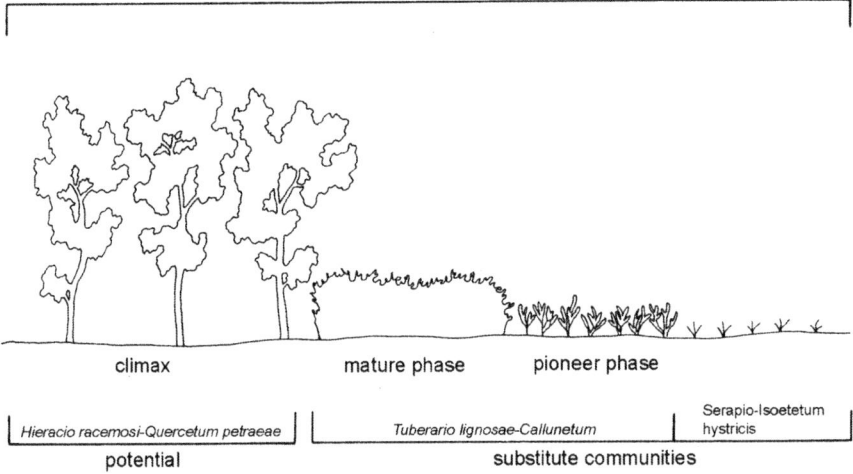

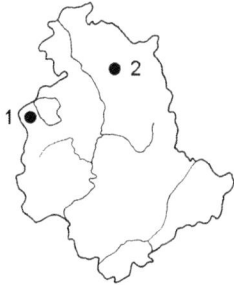

climax   mature phase   pioneer phase

Hieracio racemosi-Quercetum petraeae	*Tuberario lignosae-Callunetum*	Serapio-Isoetetum hystricis
potential	substitute communities	

1 - Ferretto (Trasimeno Lake)

2 - Gubbio

Fig. 6.66 Synchorological map of acidophilic series of *Quercus petraea*, Ferretto, Umbria Region, central Italy; also of heaths (*Calluna vulgaris*) of the West Trasimeno hills at Ferretto (From Pedrotti 1982, 1995)

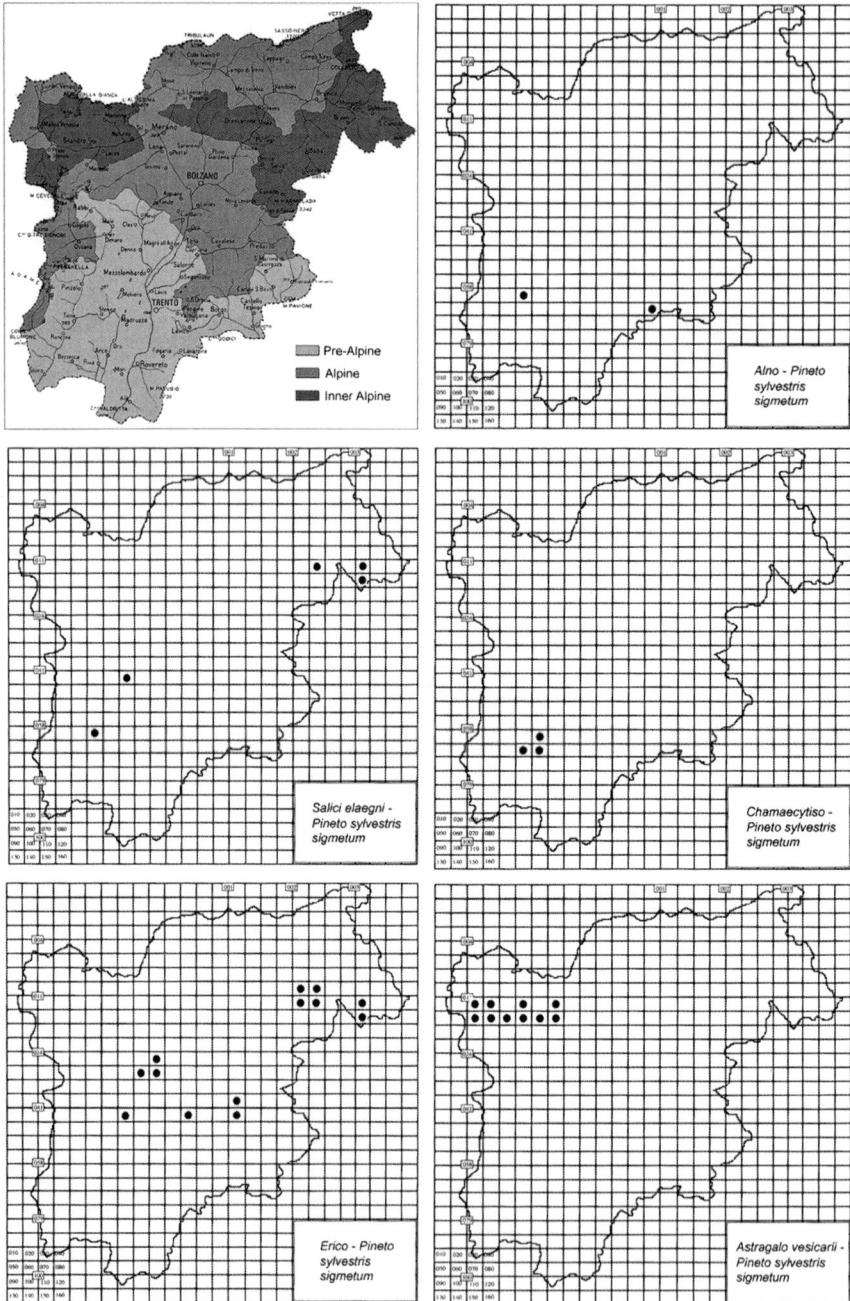

Fig. 6.67 Distribution of five *Pinus sylvestris* – sigmeta in the Trentino-Alto Adige Region, referred to the UTM network. The bioclimatic map shows the subdivision into pre-Alpine, Alpine and endo-Alpine sectors. The *Pinus sylvestris* sigmeta are formed from the following associations: the azonal associations *Alno-Pinetum sylvestris* and *Salici elaeagni-Pinetum sylvestris*; intrazonal associations: *Chamecytiso-Pinetum sylvestris* in the pre-Alpine sector and *Erico-Pinetum sylvestris* in the Alpine and endo-Alpine sectors; and the zonal association *Astragalo vesicarii-Pinetum sylvestris* in the endo-Alpine sector

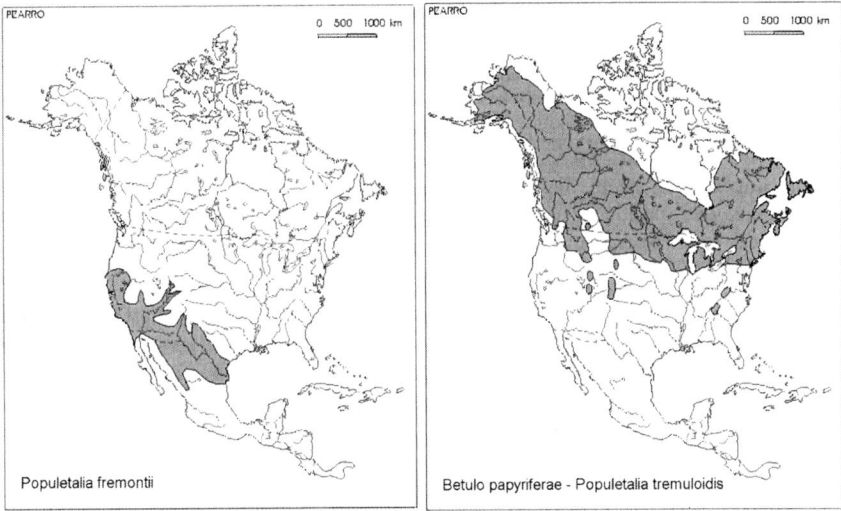

Fig. 6.68 Distribution of the vegetation orders *Populetalia fremontii* and *Betulo papyriferae-Populetalia tremuloidis* in North America (From Rivas-Martínez et al. 1999)

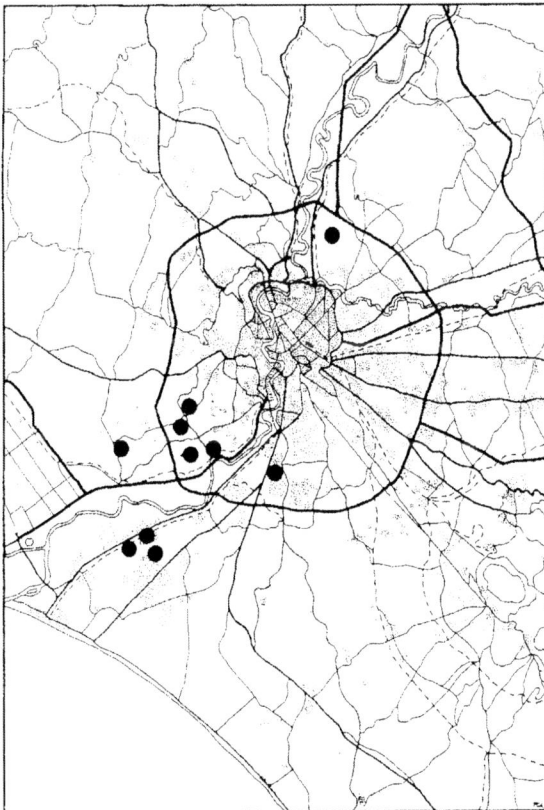

Fig. 6.69 Synchorological map of *Fraxino orni-Quercetum ilicis* in Rome, Latium Region, central Italy (From Fanelli 2002)

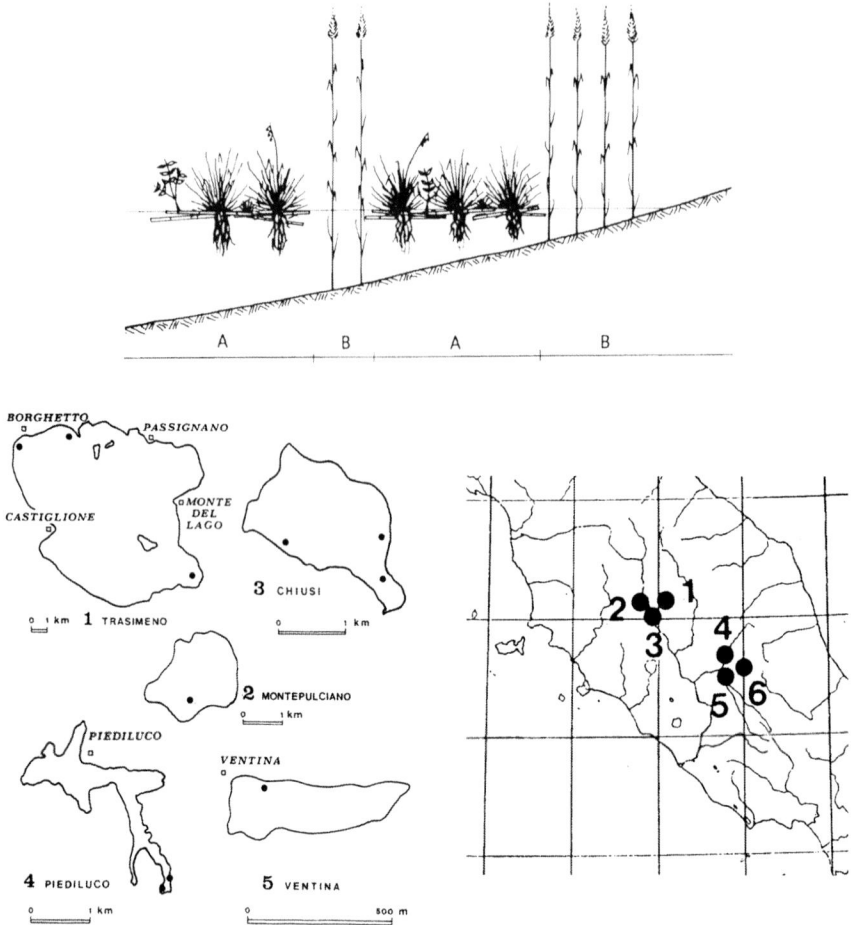

Fig. 6.70 Synchorological map of the association *Mentho aquaticae-Caricetum pseudocyperi*, at different scales for the banks of the Trasimeno lake and for lakes of central Italy (*numbers* identify the different lakes). This map is also an example of scaling in cartographical representation (From Orsomando and Pedrotti 1986)

IV. Areas with semi-natural vegetation represented mainly by grassy phytocoenoses undergoing processes of anthropogenic fluctuation and secondary succession, plus units of synanthropic vegetation (grassy areas) around human settlements;

V. Areas with cultivated synanthropic vegetation (segetal vegetation) and nuclei of synanthropic vegetation (cultivated areas) around human settlements; and

VI. Areas with human-introduced synanthropic vegetation only (areas in urban centers).

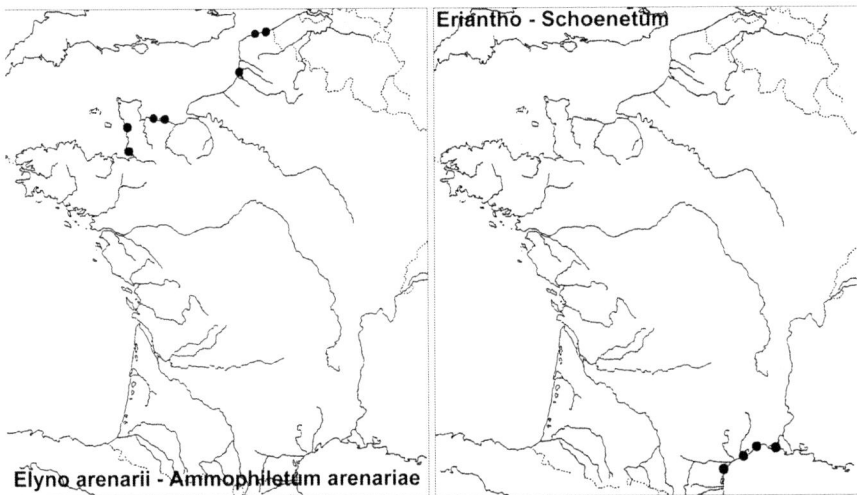

Fig. 6.71 Distribution in France of two dune associations (From Géhu 1991b)

In the second diagnostic approach, phytocoenoses may be evaluated by their degree of degeneration, distance from the climax, or degree of naturalness. *Hemeroby* is the sum of the effects on ecosystems that result from invasive human actions, intentional or not, and is expressed through the following levels (Jalas 1955; Sukopp 1972): meta-hemeroby (limited impact, as concerns destruction of life), meso-hemeroby (moderate impact), oligo-hemerobia (weak impact), and ahemeroby (without impact).

The distance from the climax involves evaluating the position of each phytocoenosis in the dynamic series and thus its distance from (or proximity to) the climax (Pirola 1982).

Two examples are given in Figs. 6.74 and 6.75, the first showing a map made by traditional methods and the second a map made by more complex methods. The map of the degree of anthropogenic modification of Polish vegetation was constructed by Faliński (1975, 1998b) with reference to geographical meso-regions, to each of which the author attributed a value for the degree of "synanthropization" from I to VI. The introduction of neophytes into new areas affects both the flora and the autochthonous vegetation in far-reaching ways. The ratio of alien to autochthonous species was called the "level of invasion" by Chytrý et al. (2009), who produced a map of the level of invasion of Europe (Fig. 6.75). On the plains and

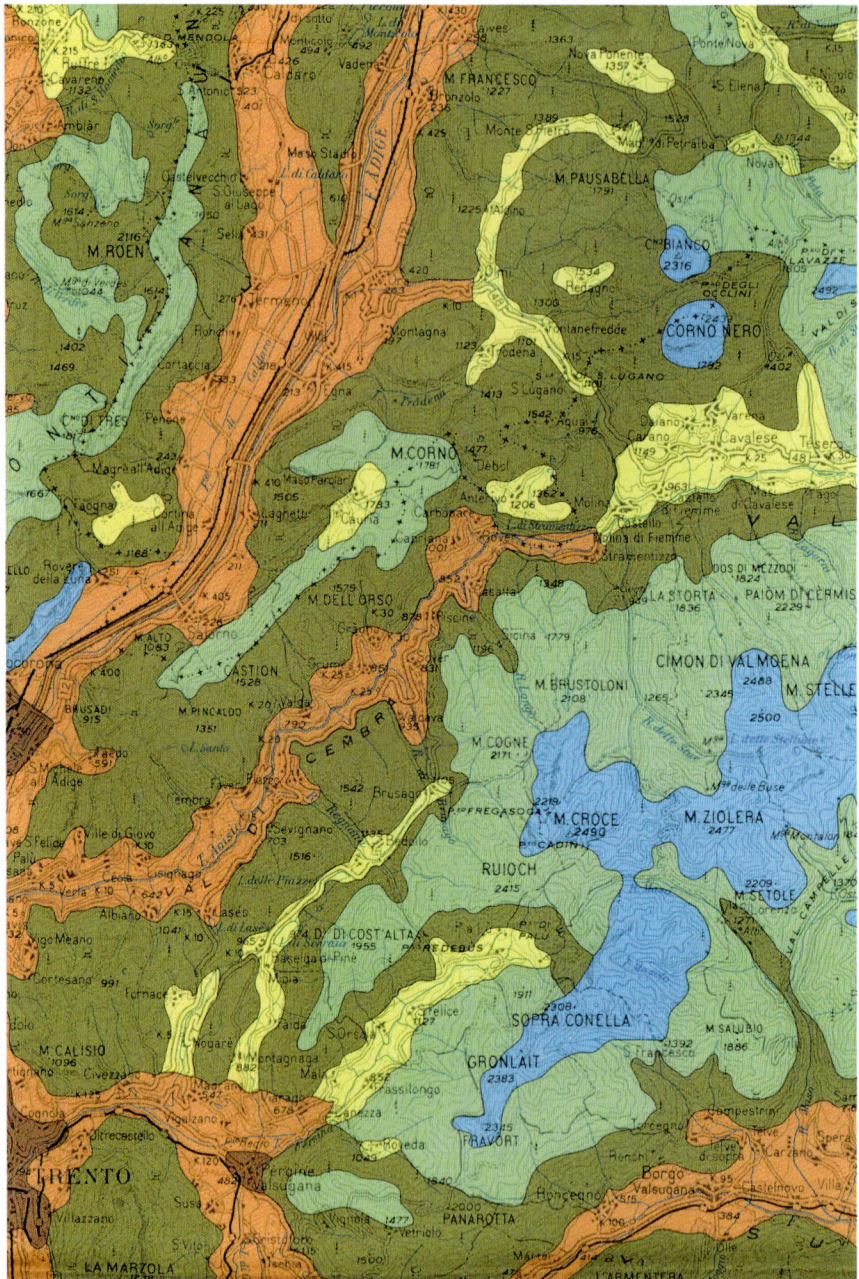

Fig. 6.72 Map of the naturalness of the vegetation of the Trentino-Alto Adige Region (From Pedrotti and Minghetti 1997)

I - *Areas with primary natural vegetation consisting of plant communities affected by processes of fluctuation and primary succession.*
Nival, alpine and subalpine belt areas characterized by pioneer rock and debris vegetation, late snow-lies , lakes and peat-bogs vegetation, primary grasslands, dwarf shrub vegetation and patches of coniferous forest occasionally with small secondary meadows and synanthropic vegetation near summer cowsheds and mountain huts; also areas in the mountain and hill belts with pioneer vegetation on rocky cliffs, coarse debris and very steep slopes.

II - *Areas with natural vegetation consisting of forest communities structurally altered by man and affected by processes of degeneration and regeneration.*
Subalpine and mountain belt areas with coniferous forests and patches of old-growth deciduous forests, including occasional scattered small clearings occupied by secondary meadows; possibly including small areas of natural vegetation of peat-bogs and lakes and synanthropic vegetation near summer cowsheds and mountain huts.

III - *Areas with large complexes of natural vegetation consisting of forest communities structurally simplified by man and affected by processes of degeneration and regeneration and with patches, sometimes extensive, of semi-natural and synanthropic vegegation affected by processeses of secondary succession and man-caused fluctuation.*
Subalpine, mountain and hill belt areas with deciduous coppices and coniferous forests interspersed with large secondary semi-natural grasslands; small patches of natural vegetation of peat-bogs and lakes are also present, together with patches of synanthropic vegetation near summer cowsheds and mountain huts.

IV - *Areas with semi-natural vegetation mainly consisting of grassland communities affected by processes of man-caused fluctuation and secondary succession, including patches of synanthropic vegetation of human settlements.*
Valley floor and middle-slope areas, generally in the mountain belt, characterized by secondary semi-natural grasslands; there are also patches of synanthropic vegetation in villages and of natural vegetation in fens, lakes, floodplains and other stands unsuited to crops.

V - *Areas with synanthropic crop vegetation and patches of synanthropic vegegation of human settlements.*
Valley floor and middle-slope areas, generally in the hill belt, dominated by crop vegetation (in orchads, sowing crops, forest plantations) together with ruderal or nitrophilous vegetation in towns; there are also small areas of natural vegetation in fens, lakes, floodplains and other stands unsuited to crops.

VI - *Areas with synanthropic vegetation of human settlements.*
Urban areas with synanthropic vegetation on roads, walls, gardens, parks.

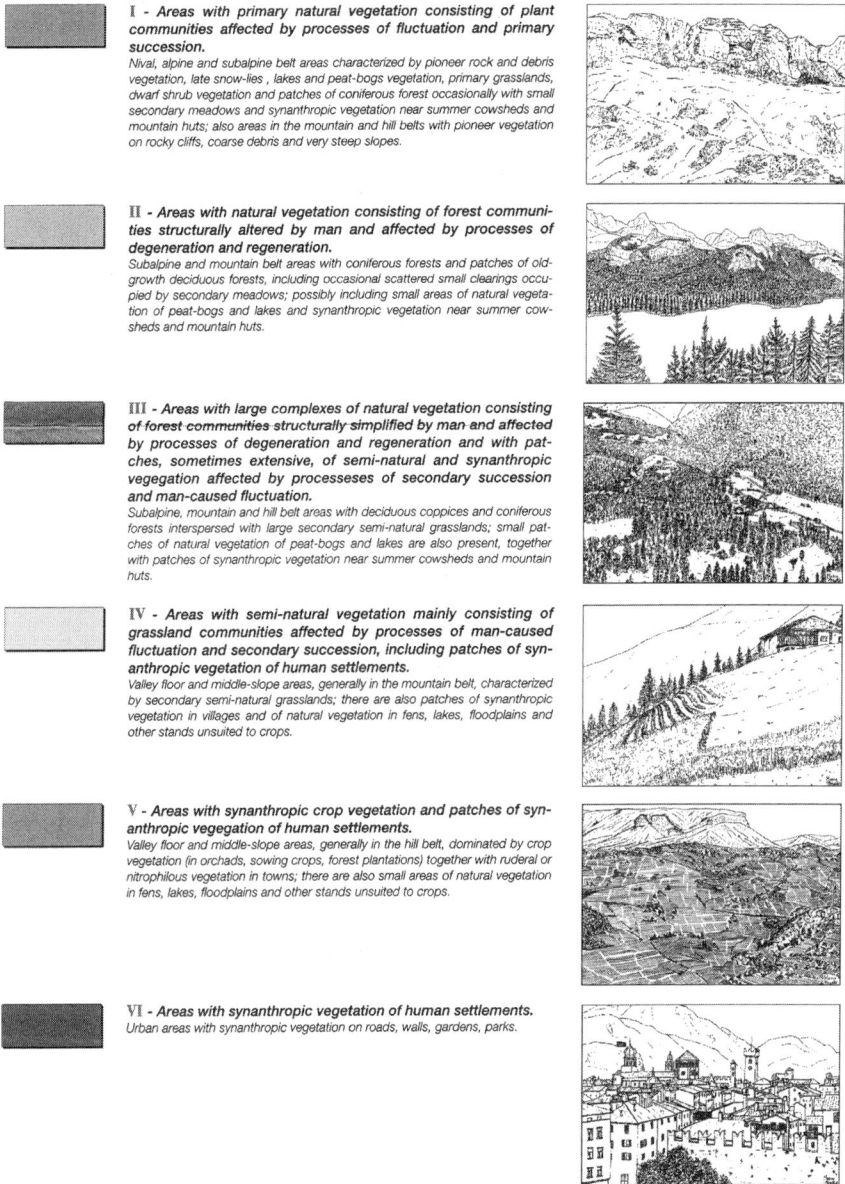

Fig. 6.73 Legend of the previous map (Fig. 6.72) (From Pedrotti and Minghetti 1997 and Pedrotti 1999b)

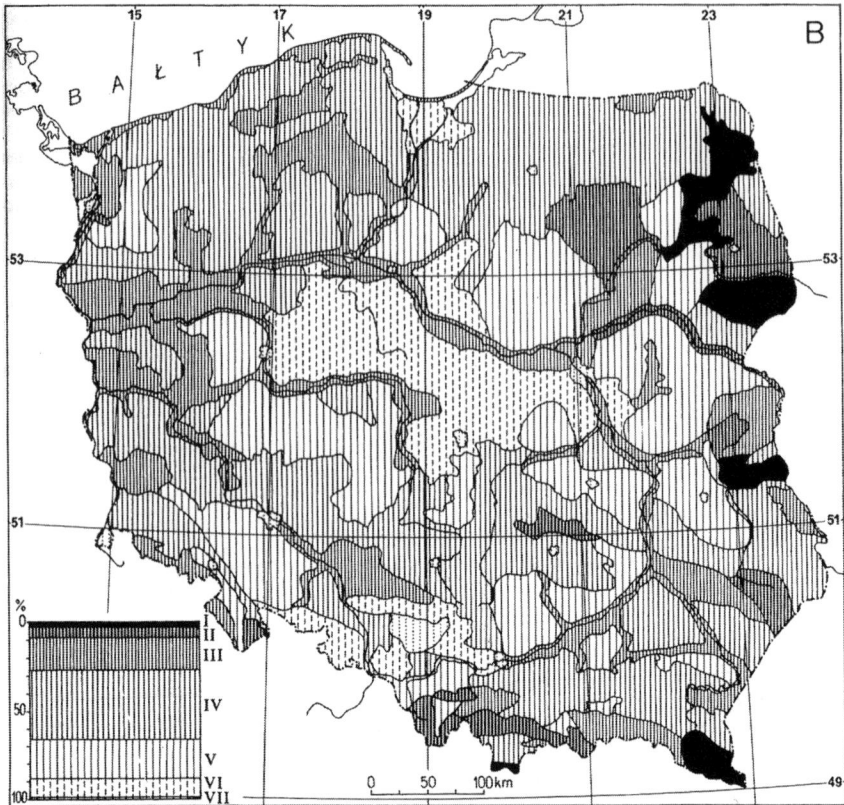

Fig. 6.74 Synthetic diagnosis of the degree of anthropogenic transformation of the vegetation in the geographical meso-regions in Poland: *I–VII* represent stages of synanthropisation (From Faliński 1998a)

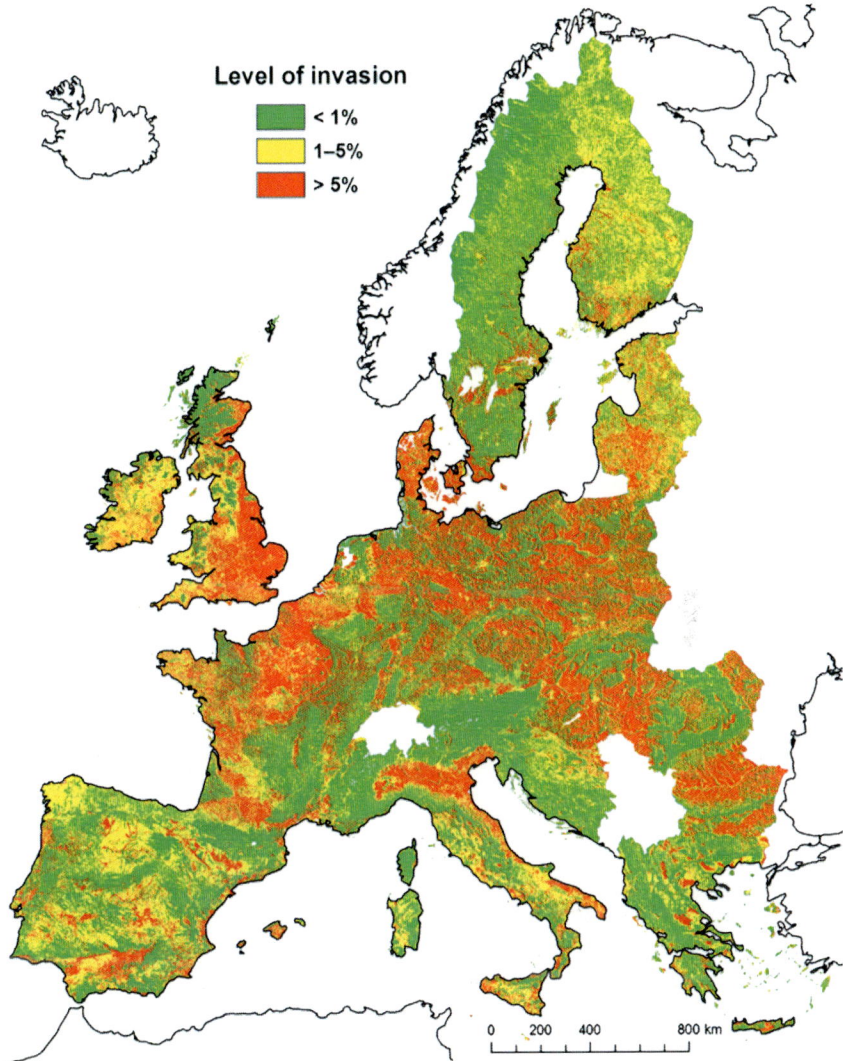

Fig. 6.75 Estimated level of invasion by alien plants, caused by human influence, in Europe. This map is based on the percentage of neophythes in vegetation plots corresponding to individual CORINE land-cover classes (From Chrytý et al. 2009)

foothills, the level of invasion is more than 5 %, while on mountain ranges it falls to less than 1 %.

Such mapping can be very diversified. For the region of Castiglione dei Pepoli (northern Appennino), Puppi et al. (1980) produced a phytosociological map of the

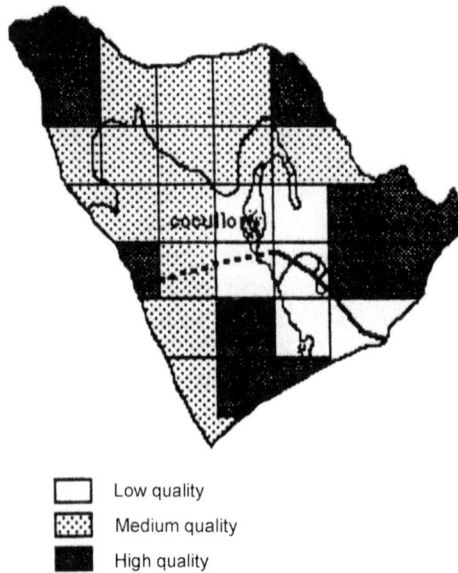

Low quality
Medium quality
High quality

Fig. 6.76 Synthetic map of environmental quality: the squares are grouped by similarity in values of an environmental quality index (From Greco and Petriccione 1988–1989)

actual vegetation (scale 1:25,000), from which they derived a map of the degree of "artificialization" of the vegetation, with distinct levels on a 1–5 scale: no or almost no artificialization, weak, medium, strong or fairly strong, and very strong. For the Lago di Vico (lake), Blasi et al. (1989) made a phytosociological map of actual vegetation, at scale 1:12,500, which served as the basis for a map of naturalness, at the same scale; the naturalness of the vegetation units was distinguished as very high, high, medium, low and very low.

A synthetic map of "environmental quality" with three classes (low, medium and high quality) was made for the region of Cocullo in Abruzzo (central Apennines) by Greco and Petriccione (1988–1989; cf Greco et al. 1991). This map was derived from a phytosociological map of actual vegetation and employed a quantification based on 11 indices, among them indices of floristic richness, vegetation diversity, phytocoenotic naturalness, environmental stress, vegetation naturalness, anthropogenic flora and environmental tampering (Fig. 6.76).

Other aspects relative to the state of vegetation conservation that can be mapped include reduction, fragmentation and rarity of the vegetation for anthropogenic reasons (Fig. 6.77).

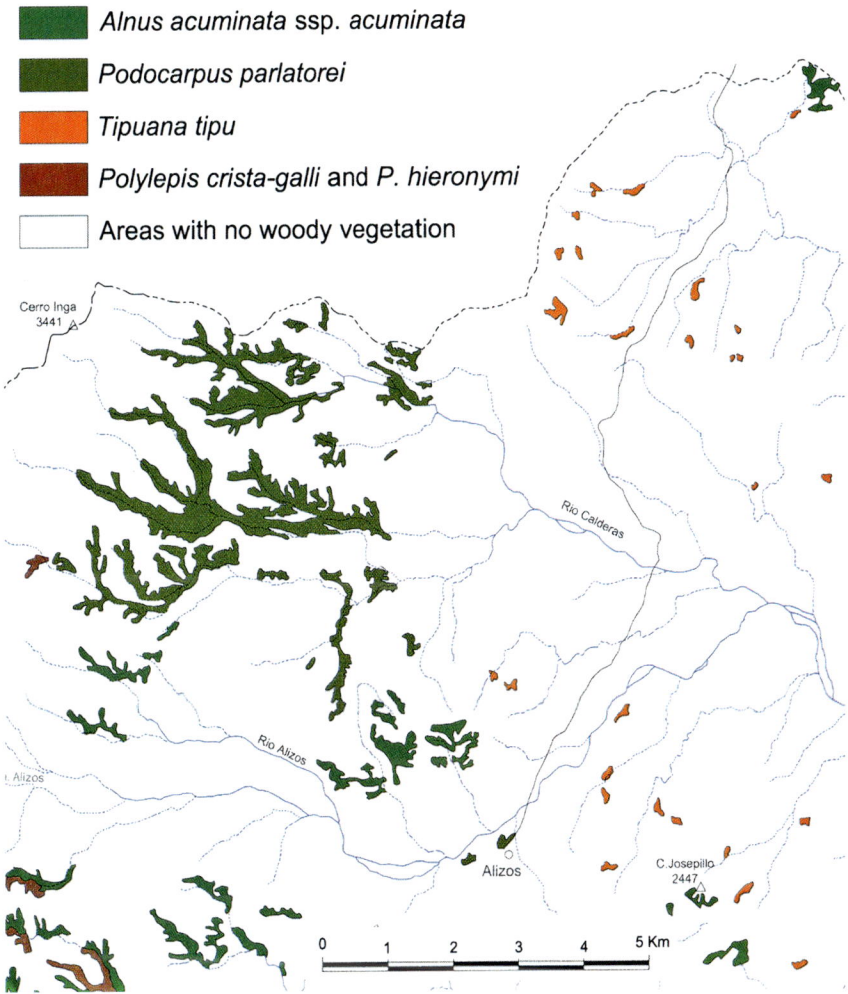

Fig. 6.77 Forest fragmentation in the Rio Camacho basin, Tarija, Bolivia: after deforestation, forests now cover only the 4.3 % of the total surface (From Liberman Cruz and Pedrotti 2006)

Examples of Vegetation Maps

7

This chapter presents examples for a wide range of environment types, including mountain ranges, valleys, volcanos, islands, dunes, lakes, bogs, rivers, and cities. Each map is identified by title, scale, number of vegetation units shown, and bibliographic reference, followed by a brief description.

Mountain Ranges and Valleys

Problems involved in mapping vegetation on mountain ranges and in valleys are examined in Chap. 10. Mapping vegetation in altitudinal belts is illustrated at broad scale (Fig. 7.1), medium scale (Figs. 7.2 and 7.3) and fine scale (Fig. 5.8). At fine scale, the vegetation belts are resolved into the associations housed in the belt, as far as the altitudinal bands are still recognizable. A fair and balanced representation in the combined mapping of plant associations, vegetation series and altitudinal belts is obtained with integrated phytosociological maps (see Figs. 6.7 and 6.8).

Etages de végétation de la Châine Alpine (Vegetation Belts in the Alps), 1:2,250,000, 16 units (vegetation belts), Ozenda (1984, 2002) (Fig. 7.1).

This map shows the vegetation of the Alps by means of distinct colline (foothill), montane, subalpine, alpine and nival vegetation belts; also shown is the piedmont vegetation. The mountain massifs that form the Alpine range influence the distribution of vegetation in different altitudinal bands. In the large valleys that separate the main individual Alpine chains, the vegetation of the foothill (colline) belt can extend further into the range, as seen for example in the vegetation with *Ostrya carpinifolia* in the upper Adige valley (northeastern Italy). Also, the gradient from the plains outside the true Alps to the dividing ridgelines with glaciers (within the Alps) can be seen well on the map, in spite of the broad scale.

Borgo Valsugana sheet (*Carta del Foglio Borgo Valsugana*) of a vegetation map at 1:50,000, 50 units (46 plant associations and 4 physiognomic formations, with

F. Pedrotti, *Plant and Vegetation Mapping*, Geobotany Studies,
DOI 10.1007/978-3-642-30235-0_7, © Springer-Verlag Berlin Heidelberg 2013

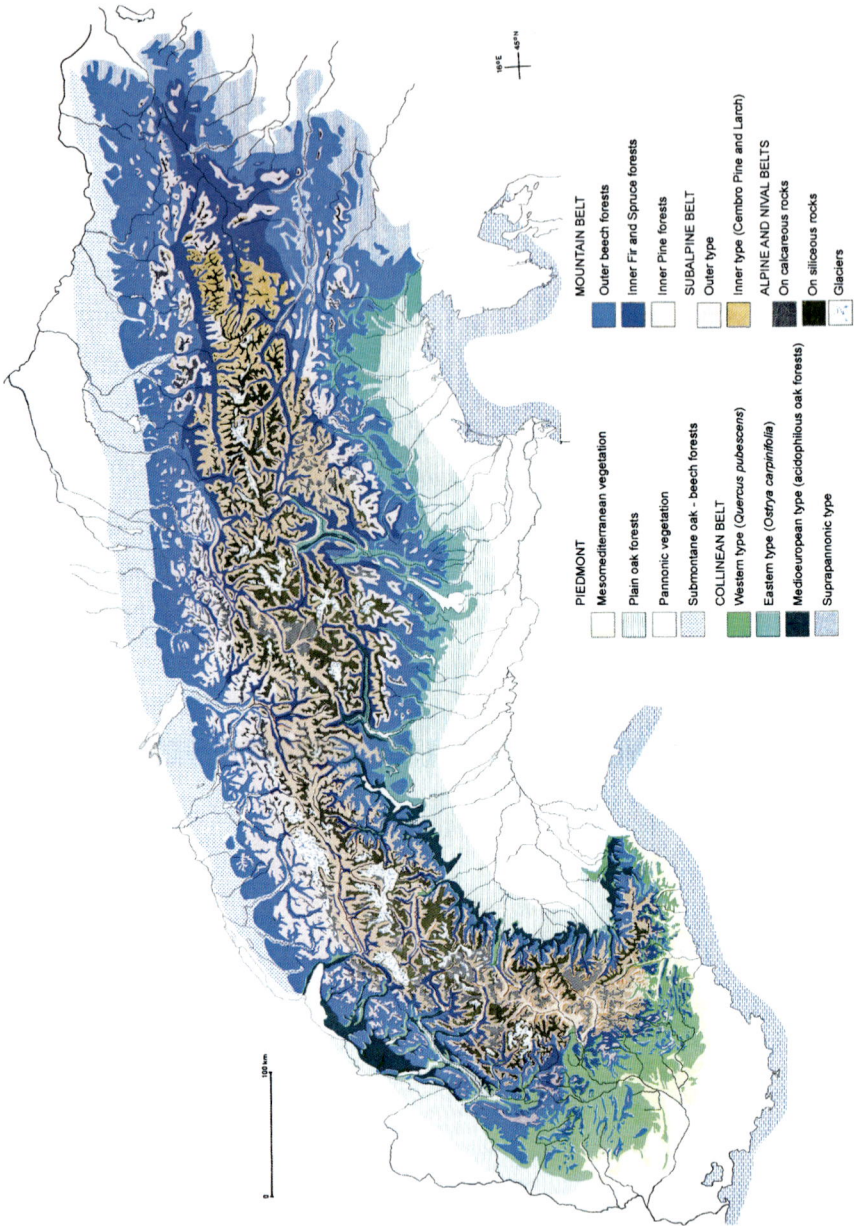

Fig. 7.1 Vegetation belts in the Alps, at scale 1:2,250,000 (From Ozenda 1984)

presence of tree species indicated by symbols); survey by F. Pedrotti, M. Liberman Cruz and C. Movia; Pedrotti (1988a) (Fig. 7.2).

This is a phytosociological map of the actual vegetation of the mountain groups Lagorai and Valsugana (Trentino, northeastern Italy), at the limit between the

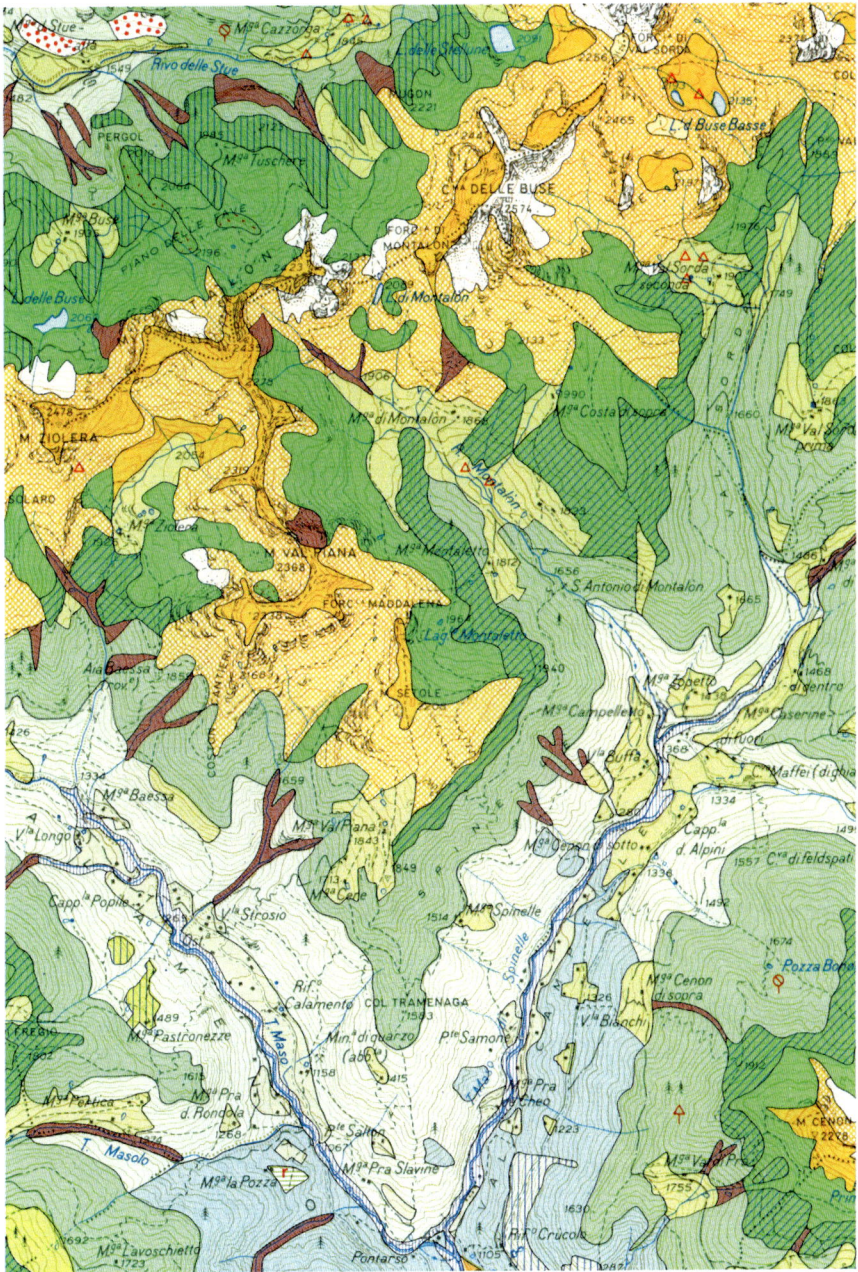

Fig. 7.2 The Borgo Valsugana sheet of the vegetation map of Trentino-Alto Adige Region, northern Italy, at scale 1:50,000. The plant associations occur in altitudinal belts as follow: alpine belt with primary meadows of *Caricetum curvulae*, *Festucetum halleri* and *F. variae*; subalpine belt with *Rhododendro-Vaccinietum* and *Homogyno-Piceetum*; and montane belt with *Oxali-Piceetum* and Abietetum albae s.l. Clearing in the montane belt are occupied by *Sieversio montanae-Nardetum* and *Arrhenatheretum elatioris* s.l.; in ravines there is an *Alnetum viridis*, and along the rivers the *Alnetum incanae* (From Pedrotti 1988a)

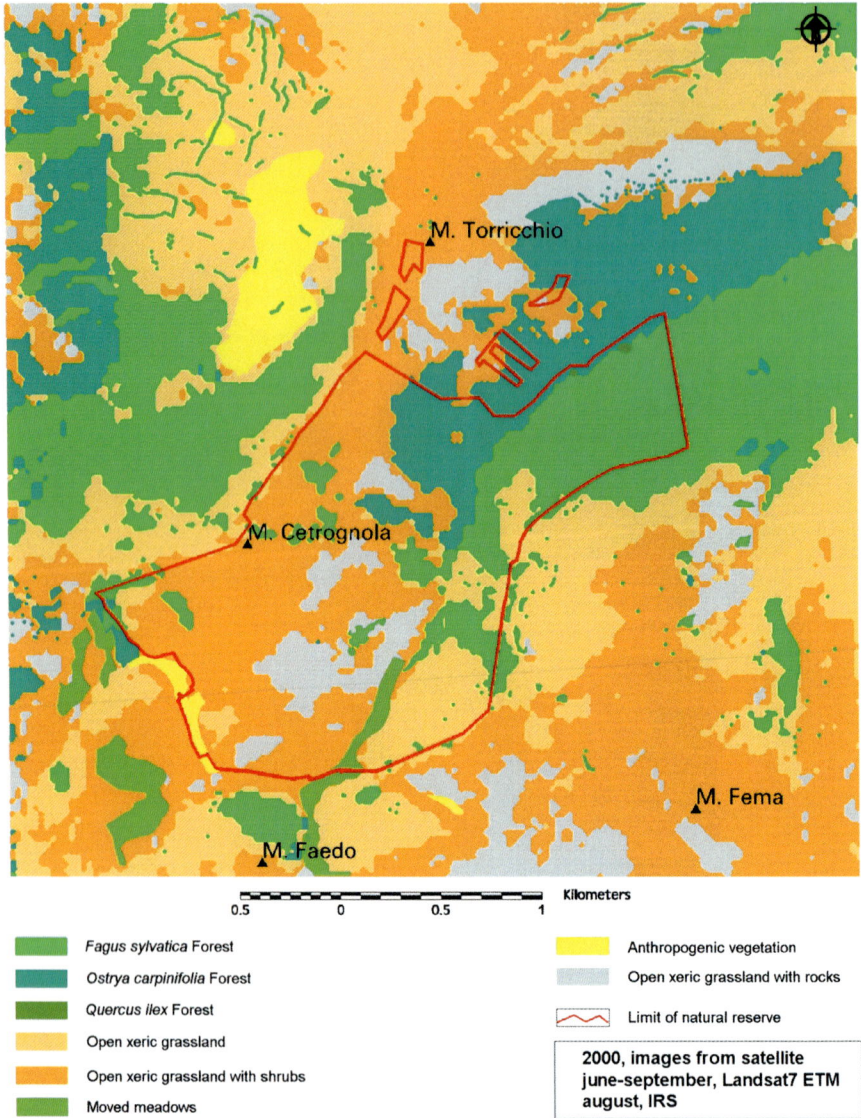

Fagus sylvatica Forest		Anthropogenic vegetation
Ostrya carpinifolia Forest		Open xeric grassland with rocks
Quercus ilex Forest		Limit of natural reserve
Open xeric grassland		
Open xeric grassland with shrubs		**2000, images from satellite**
Moved meadows		**june-september, Landsat7 ETM**
		august, IRS

Fig. 7.3 Vegetation map of the Torricchio Nature Reserve, Marche Region, central Italy, obtained from Landsat satellite imagery (From Borfecchia et al. 2003)

pre-Alpine band and the inner, continental zone. Shown on the map are numerous associations peculiar to the two different phytoclimatic zones. With the map of actual vegetation there is also a map of the potential vegetation that it represents: in the alpine belt Caricetum curvulae, Festucetum halleri and Festucetum variae; in the subalpine belt Rhododendro-Vaccinietum, Rhododendro-Vaccinietum cembretosum, Rhododendro-Vaccinietum laricetosum, and Piceetum subalpinum; in the montane belt Piceetum montanum, Abietetum albae, Abieti-Fagetum, Aceri-Tilietum, Luzulo-Fagetum, Carici-Fagetum, and Erico-Pinetum sylvestris; and in the colline belt Salvio-Fraxinetum and Fraxino orni-Ostryetum carpinifoliae.

Vegetation Map of the Nature Reserve "Montagna di Torricchio" (Italy), 1:25,000, 8 units (each representing a single association or group of related associations), Borfecchia et al. (2003) (Fig. 7.3).

This is a map of the actual vegetation of the Torricchio Reserve (Marche, Adriatic central Italy) and surrounding territory, obtained from satellite imagery (Landsat 7 ETM). The map obtained by this methodology corresponds to one made by traditional methods (cf. Venanzoni et al. 1999) for the large units such as beech forest, *Fraxinus-Ostrya* forest, dry grasslands and dry steppe.

Vegetation Map of the Tovel Valley, in the Adamello-Brenta National Park (Trentino) 1:10,000, 28 units (plant associations, assemblages and physiognomic units) (Pedrotti 1997c, 2004b) (Figs.7.4 and 7.5).

This is a phytosociological map of actual vegetation. From the valley bottom to the upper forest limit, the zonal forest vegetation is formed by *Fraxinus ornus-Ostrya* forest (Fraxino orni-Ostryetum), beech forest (Carici albae-Fagetum and Cardamino pentaphylli-Fagetum), fir forest (Adenostylo glabrae-Abietetum) and spruce forest (Adenostylo glabrae-Piceetum). Also present, on calcareous debris, are the associations Listero cordatae-Piceetum and Erico-Pinetum sylvestris (both considered intrazonal). The subalpine belt is characterized by various shrub associations and the alpine belt by primary meadows (Caricetum firmae and Seslerio-Caricetum sempervirentis). The lacustrine vegetation around the Lago di Tovel (lake) belongs to various assemblages and associations (Charetum asperae, an assemblage with *Ranunculus trichophyllus* ssp. *trichophyllus,* an assemblage with *Ranunculus trichophyllus* ssp. *eradicatus,* etc.), mapped as a single unit because of the scale employed.

Volcanos

Mapping vegetation of volcanos presents problems similar to those for mountain chains, but because of their morphology, the vegetation on volcanos tends to occur in concentric bands, as can be seen on the examples shown here (Nevado Sajama and Mt. Etna).

Vegetation Map of the Nevado Sajama (Bolivia), 1:130,000, 7 units (plant formations), Liberman Cruz (1986) (Fig. 7.6).

This is a physiognomic map that shows plant formations such as matorral and steppe. A forest and matorral (scrub) of *Polylepis tarapacana* have developed on the slopes of the volcanic cone, up to 4,400 m, followed upward by herbaceous

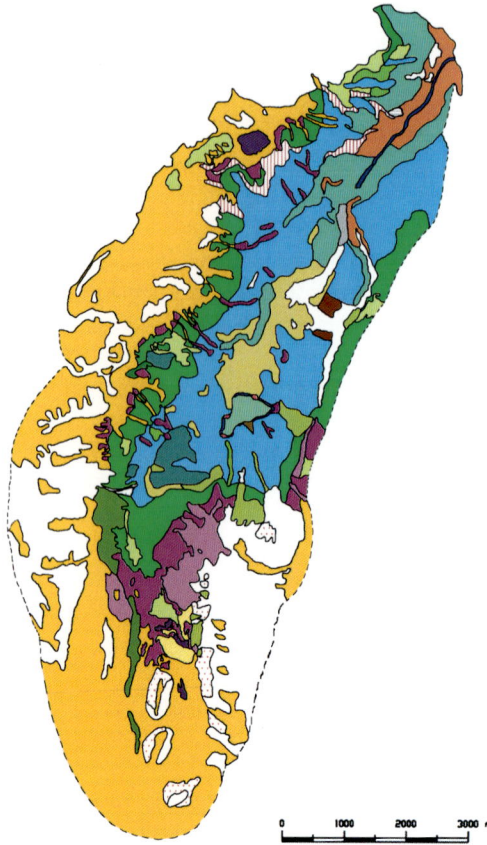

Fig. 7.4 Vegetation map of the Tovel Valley, Trentino-Alto Adige Region, northern Italy (From Pedrotti 2004b)

alto-Andine vegetation (Fig. 7.7). Other formations present are a matorral of *Parastrephia lepidophylla* and, on the level area at the base of the volcano, the "bofedales" (moist meadows) and a steppe of *Festuca orthophylla*, *Stipa icchu* and other grasses. The vegetation develops on the volcanic cone in concentric rings, interrupted by lava flows and deposits of volcanic scoria, analogous to what can be seen on other volcanos, such as Mt. Etna (Poli 1965; Poli et al. 1981). The last edition of the map of Etna is particularly significant in its representation of the vegetation of volcanos (Poli Marchese and Patti 2000).

Disturbed Environments

For debris environments, the initial development phases of the vegetation are particularly important for the dynamics and phases in the primary succession.

Pioneer, sparse or no vegetation

Thlaspietum rotundifolii, Androsacetum helveticae, Arabidetum caeruleae (alpine belt), *Festucetum spectabilis* (colline and montane belt)

Lacustrine vegetation

Charetum asperae, Grouping of *Ranunculus trichophyllus* ssp. *trichophyllus* and ssp. *eradicatus*, Algal bentonic vegetation

Primary meadows

Dryadetum octopetalae

Seslerio - Caricetum sempervirentis, Caricetum firmae

Secondary meadows

Laserpitio - Festucetum alpestris

Poion alpinae

Sieversio montanae - Nardetum strictae

Montane shrub stands

Amelanchiero - Pinetum mugo

Salicetum capreae

Subalpine Krummholz

Alnetum viridis

Rhododendretum hirsuti

Rhododendro ferruginei - Laricetum

Erico carneae - Pinetum prostratae

Rhododendro hirsuti - Pinetum prostratae

Sorbo chamaemespili - Pinetum mugo

Conifer Forests

Adenostylo glabrae - Abietetum albae

Adenostylo glabrae - Piceetum

Erico - Pinetum sylvestris

Salici elaeagni - Pinetum sylvestris

Listero cordatae - Piceetum

Deciduous broad-leaved Forests

Calamintho grandiflorae - Aceretum pseudoplatani

Cardamino pentaphylli - Fagetum

Carici albae - Fagetum

Ostryo carpinifoliae - Fraxinetum orni

Riparian Vegetation

Salicetum incano-purpureae

Anthropogenic Vegetation

Lolietum perennis

Echio - Melilotetum, Poo compressae - Tussilaginetum and fragments of other ruderal and nitrophilous associations

Reforestations

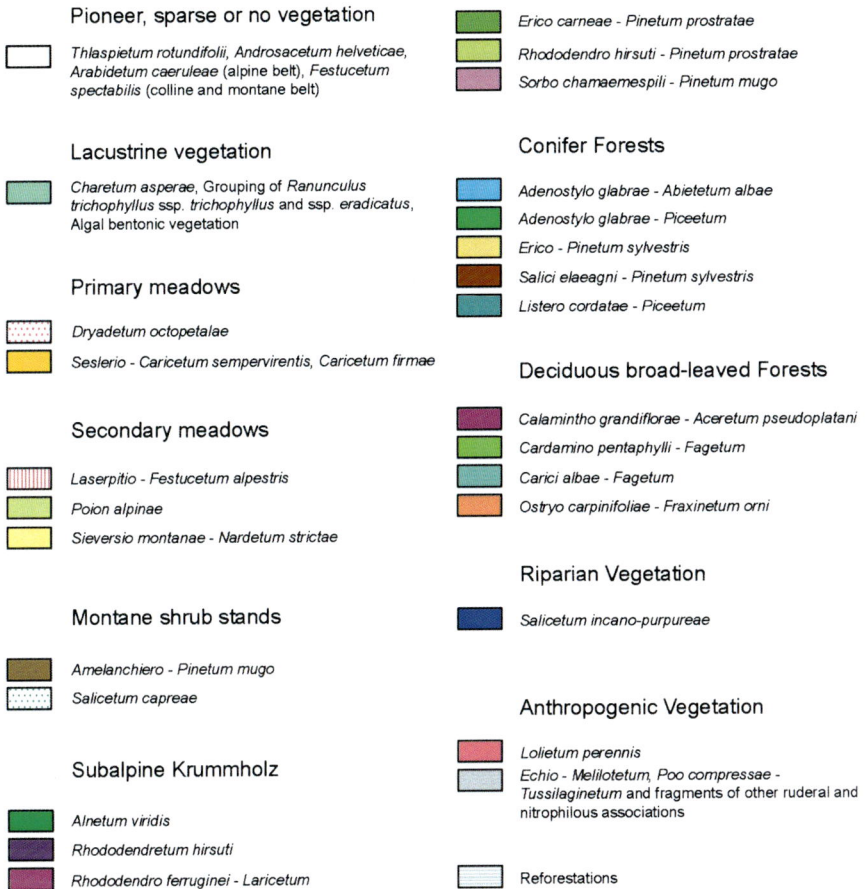

Fig. 7.5 Legend of the previous map (Fig. 7.4)

Marocche di Dro, in the Sarca Valley (Trentino, northern Italy; Fig. 7.8), 1:25,000, 11 units (plant associations) and Map of the Dynamic Tendencies of the Vegetation, 1:25,000, (ecological processes), Pedrotti and Minghetti (1994); Pedrotti et al. (1996) (Fig. 7.9).

The first of these is a phytosociological map of actual vegetation that belongs to three vegetation series, based respectively on *Ostrya carpinifolia*, *Fagus sylvatica* and *Quercus ilex*. The map of dynamic tendencies involves a relatively recent debris environment, in which the main process is that of primary succession (Fig. 7.10). This begins with pioneer stages (such as a Stipetum calamagrostidis) and then proceeds with meadows (Euphrasio tricuspidatae-Seslerietum) and scrub (Cotino-Amelanchieretum), ending with an *Ostrya carpinifolia-Fraxinus ornus* forest (Fraxino orni-Ostryetum carpinifoliae). There is also secondary succession, in the openings created by grazing or in some areas once cultivated (both now abandoned). The relations among the actual vegetation, dynamic tendencies, vegetation series and potential vegetation are evident in Fig. 7.10.

Vegetaciòn herbacea altoandina (High-Andean herbaceous vegetation)		Estepa de Gramìneas, producto de incendios (grass steppe, after fire)	
Matorral y bosque de *Polylepis tarapacana* (scrub and forest)		Vegetaciòn de areas salinas (saline vegetation)	
Vegetaciòn herbacea de "bofedal" (groundwater marsh vegetation)		Areas sin vegetaciòn, campos de nieve (snow cover)	
Matorral de *Parastrephia lepidophylla* (scrub)		Areas sin vegetaciòn, rocas (rocky areas without vegetation)	
Estepa de Gramìneas y Arbustos (steppe with grass and shrubs)		Areas sin vegetaciòn, acumulaciòn de desechios orgànico (hyper-nitrophic areas without vegetation)	

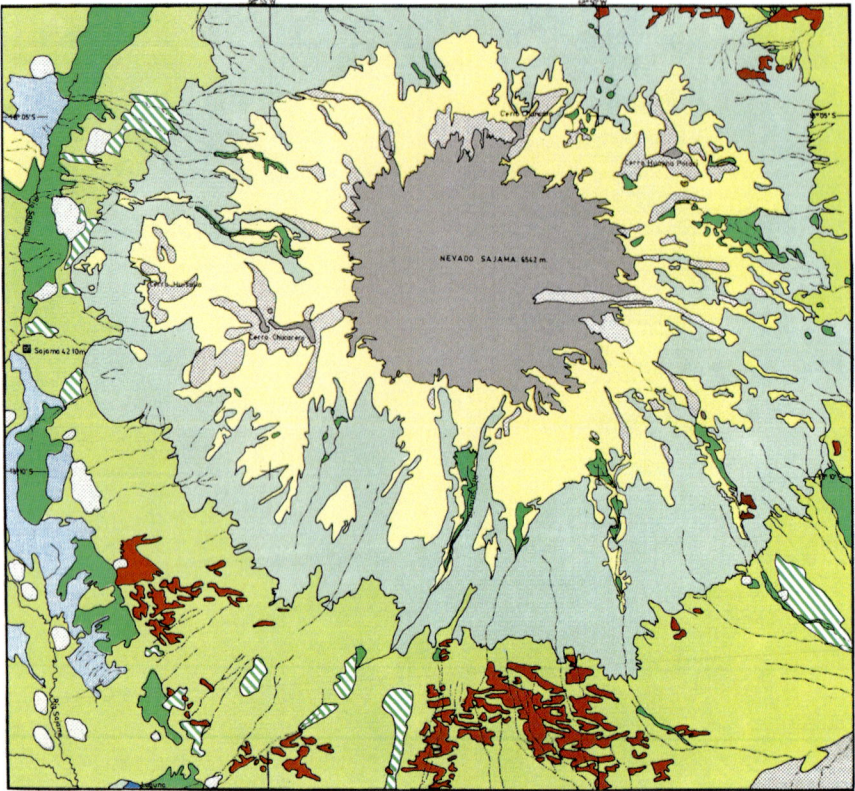

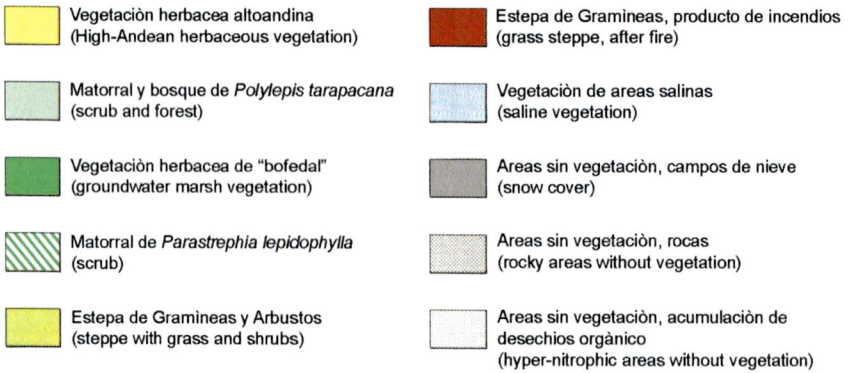

Fig. 7.6 Vegetation map of the Nevado Sajama volcano in Bolivia, at scale 1:130,000 but re-sized to 1:150,000 for printing (From Liberman Cruz 1986)

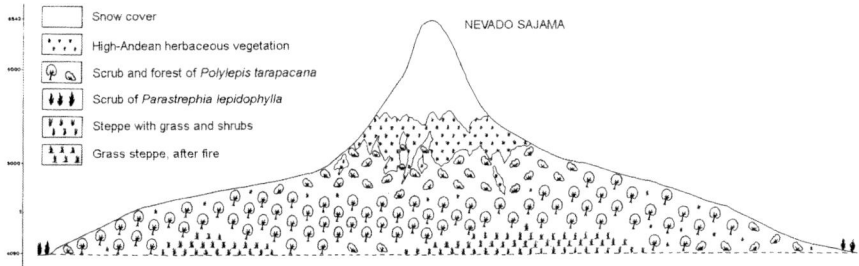

Fig. 7.7 Nevado Sajama volcano (6,542 m) with altitudinal distribution of vegetation

Fig. 7.8 The "marocche" (typical debris landslide) at Dro village in the Sarca Valley, Trentino-Alto Adige Region, northern Italy (Photo Franco Pedrotti)

Plains

While vegetation boundaries in mountains are due mainly to climatic factors varying with altitude, in flat terrain the limits are due above all to various types of alluvial deposits and to the underlying rock types.

Map of Actual Vegetation in a sector of the Białowieza Forest (eastern Poland), 1:10,000, 30 units (plant associations and groupings), Pedrotti and Venanzoni (1994a) (Figs. 7.11 and 7.12).

This is a map of the actual vegetation of a 3 × 2 km area of the Białowieza Forest (Poland), on an alluvial plain drained by the river Narewka and two of its tributaries. The forest is formed by hornbeam (Tilio-Carpinetum) and to a lesser extent by pinewoods (Peucedano-Pinetum) and *Quercus robur* (Pino-Quercetum

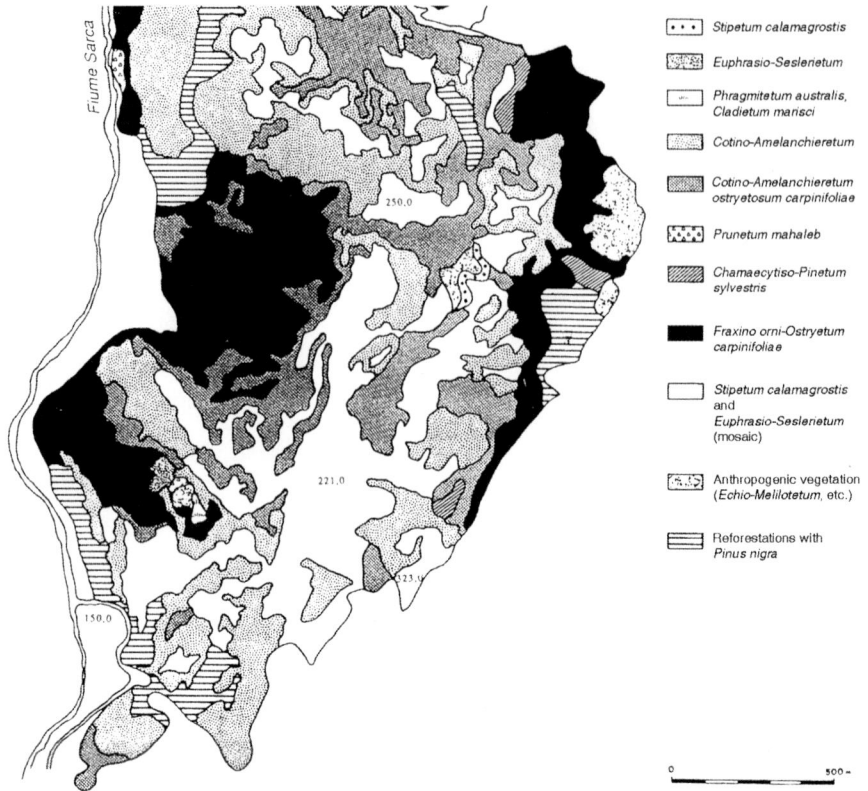

Fig. 7.9 Vegetation map of southern "marocche" from Dro (From Pedrotti et al. 1996)

and Querco-Piceetum); in some depressions there are also patches of swamp forest of Carici elongatae-Alnetum glutinosae and Carici elongatae-Quercetum. Riparian vegetation is formed by a Circaeo-Alnetum glutinosae. The part of the map on the left side of the river Narewka is not included in the national park, for which reason the forest has been cut periodically, creating artificial clearings with secondary vegetation of the Epilobio-Salicetum capreae. The wetland vegetation, developed on level areas along water courses, is constituted by various associations of tall forbs (Filipendulo-Geranietum, Cirsietum rivularis, etc.) and *Carex* sedges (Caricetum elatae, Caricetum appropinquatae, Caricetum gracilis, etc.). In the waterways, finally, there is a Sagittario-Sparganietum emersi and, in some abandoned meanders, a Nupharo-Nymphaeetum and a Hydrocharidi-Stratiotetum.

Islands

Mapping vegetation on islands does not present particular problems; they are the same as those seen with the maps of mountain relief, plains and volcanos. Many times, for example, an island is in fact a volcanic cone partly submerged in the sea.

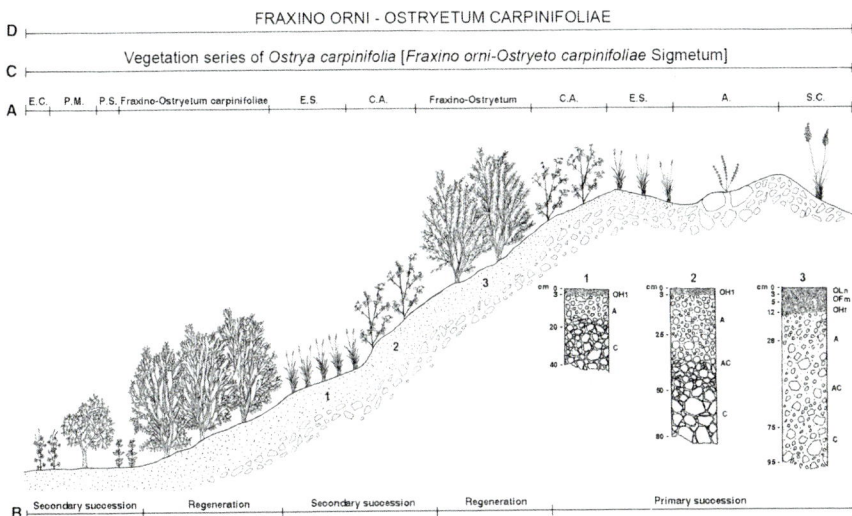

Fig. 7.10 Vegetation transect of "marocche" from Dro with four profiles: A plant associations; B dynamic tendencies in the associations; C – vegetation series; D – potential vegetation. Particular communities: E.C. – *Echio-Melilotetum*; P.M. – *Prunetum mahaleb*; P.S. – *Panico-Setarion*; E.S. – *Euphrasio tricuspidatae-Seslerietum albicantis*; C.A. – *Cotino-Amelanchieretum*; A – *Asplenietum trichomano-rutae-murariae*; S.C. – *Stipetum calamagrostidis*. Soil profiles: *1* – thin "pierrique"and "vide" peyrosol, on stabilized debris; *2* – somewhat deeper "pierrique" and "vide" peyrosol; *3* – deeper "rendosol" (amphimull) (From Pedrotti et al. 1996 and Pedrotti 1999b)

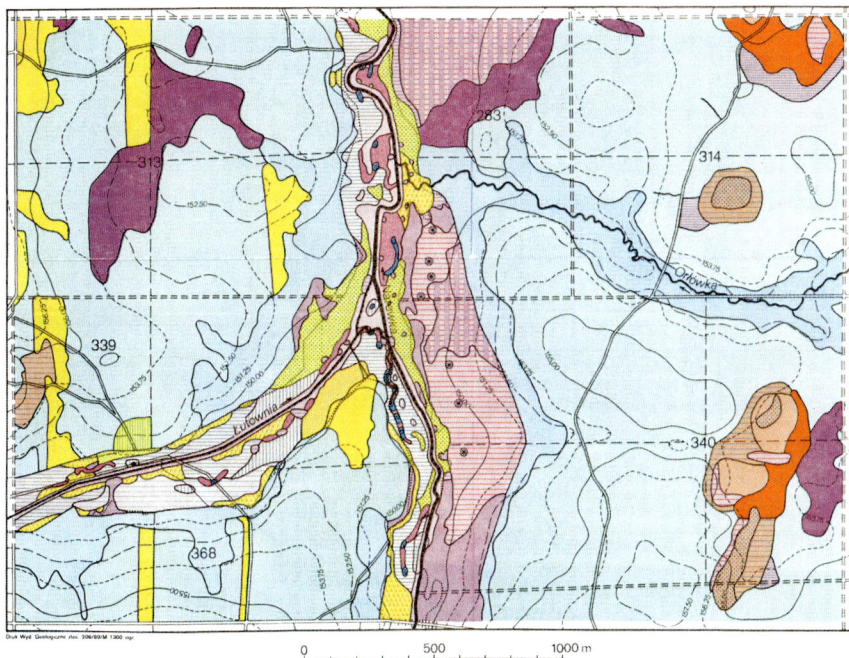

Fig. 7.11 Vegetation map of Białowieza primeval forest, Poland, showing a rectangular plot area of 2 × 3 km^2 (From Pedrotti and Venanzoni 1994a)

A. STROMSIDE FLOOD-PLAIN FOREST COMMUNITIES

Circaeo-Alnetum

B. DECIDUOUS FOREST COMMUNITIES

Tilio-Carpinetum

C. CONIFEROUS FOREST COMMUNITIES

Pino-Quercetum s.s. (= Serratulo-Pinetum)

Peucedano-Pinetum

Querco-Piceetum

Vaccinio uliginosi-Pinetum

D. WET AND BOG ALDER-BIRCH FOREST AND BRUSHWOOD COMMUNITIES

Thelypteridi-Betuletum

Community with Alnus glutinosa

Carici elongatae-Alnetum s.s. (= Ribo-Alnetum)

Carici elongatae-Quercetum

Salicetum pentandro-cinereae

Salici rosmarinifoliae-Betuletum humilis

E. SUBSTITUTE CLEARING COMMUNITIES

Epilobio-Salicetum capreae

F. MACROFORBS, GRASSLANDS AND SIMILAR COMMUNITIES

Filipendulo-Geranietum

Lysimachio-Filipenduletum

Cirsietum rivularis

Community with Carex cespitosa

Epilobio-Juncetum effusi

Community from alliance Arrhenatherion

Lolio-Plantaginetum

G. SEDGE AND REED COMMUNITIES

Caricetum elatae

Caricetum appropinquatae

Caricetum gracilis

Community with Calamagrostis neglecta

Phalaridetum arundinaceae

Phragmitetum communis

Sagittario-Sparganietum emersi

H. WATER PLANT COMMUNITIES

Nupharo-Nymphaeetum

Hydrocharidi-Stratiotetum

Lemno-Spirodeletum polyrrhizae

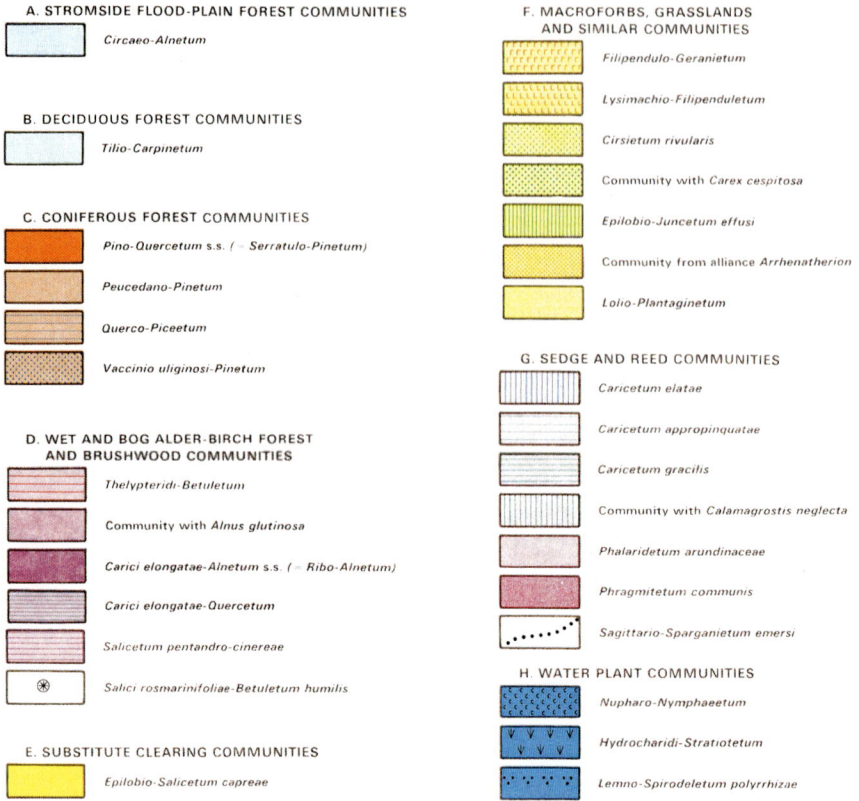

Fig. 7.12 Legend of the previous map (Fig. 7.11) (From Pedrotti and Venanzoni 1994a)

The nice aspect of vegetation mapping on islands is that the territory is complete and clearly defined geographically.

Vegetation Map of Pantelleria (Sicilian Channel), 1:120,000, 9 units (plant associations), Gianguzzi (1999) (Fig. 7.13).

This is a map of potential vegetation, distinct in infra-mediterranean, thermo-mediterranean and meso-mediterranean belts; azonal vegetation is also shown. The map was made from a map of the actual vegetation of Pantelleria, at scale 1:20,000.

Vegetation Map of the Isla del Sol (an island in Lake Titicaca, Bolivia), 1:10,000, 23 units (plant communities), Liberman Cruz et al. (1995) (Fig. 7.14).

This is a map of actual vegetation showing how the vegetation of the Isla del Sol, seat of ancient Incan culture, has been influenced strongly by human activity (Fig. 7.15). In fact, the entire island is today covered by secondary vegetation, including matorral of *Baccharis incarum* and *Satureja boliviana*, and matorral of *Puya mollis*. Nuclei of residual *Polylepis incana* forests, which in the past must

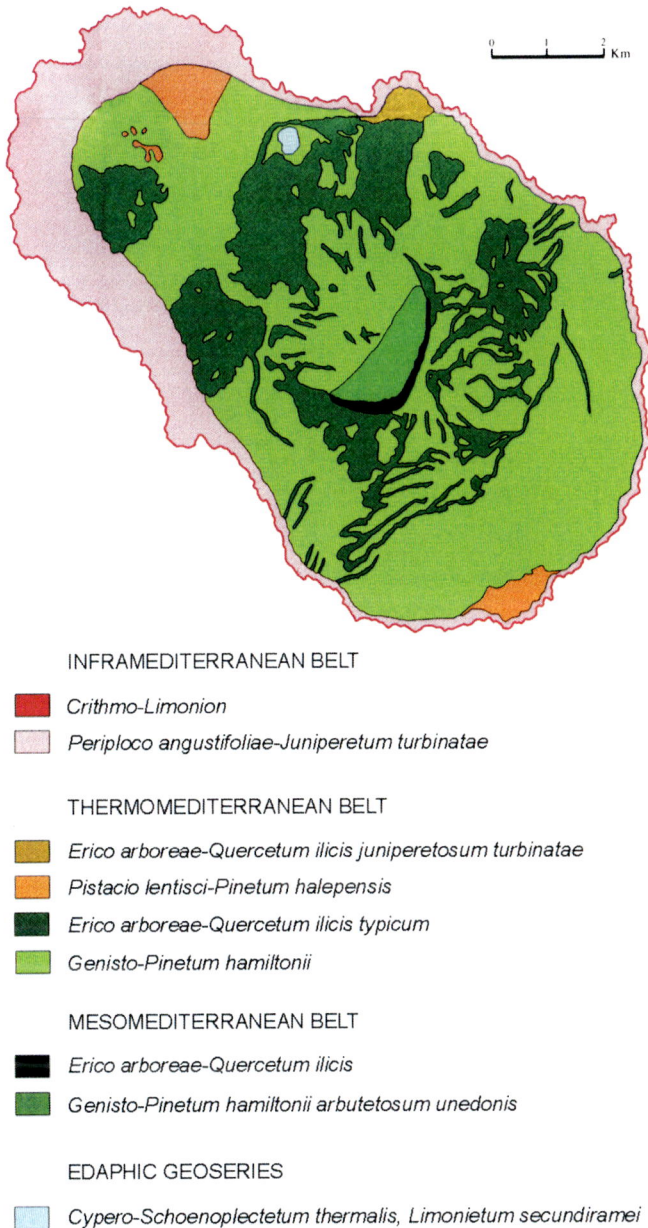

INFRAMEDITERRANEAN BELT

■ *Crithmo-Limonion*

□ *Periploco angustifoliae-Juniperetum turbinatae*

THERMOMEDITERRANEAN BELT

■ *Erico arboreae-Quercetum ilicis juniperetosum turbinatae*

■ *Pistacio lentisci-Pinetum halepensis*

■ *Erico arboreae-Quercetum ilicis typicum*

□ *Genisto-Pinetum hamiltonii*

MESOMEDITERRANEAN BELT

■ *Erico arboreae-Quercetum ilicis*

■ *Genisto-Pinetum hamiltonii arbutetosum unedonis*

EDAPHIC GEOSERIES

□ *Cypero-Schoenoplectetum thermalis, Limonietum secundiramei*

Fig. 7.13 Map of the potential vegetation of Pantelleria Island, off Sicily in southern Italy (From Gianguzzi 1999)

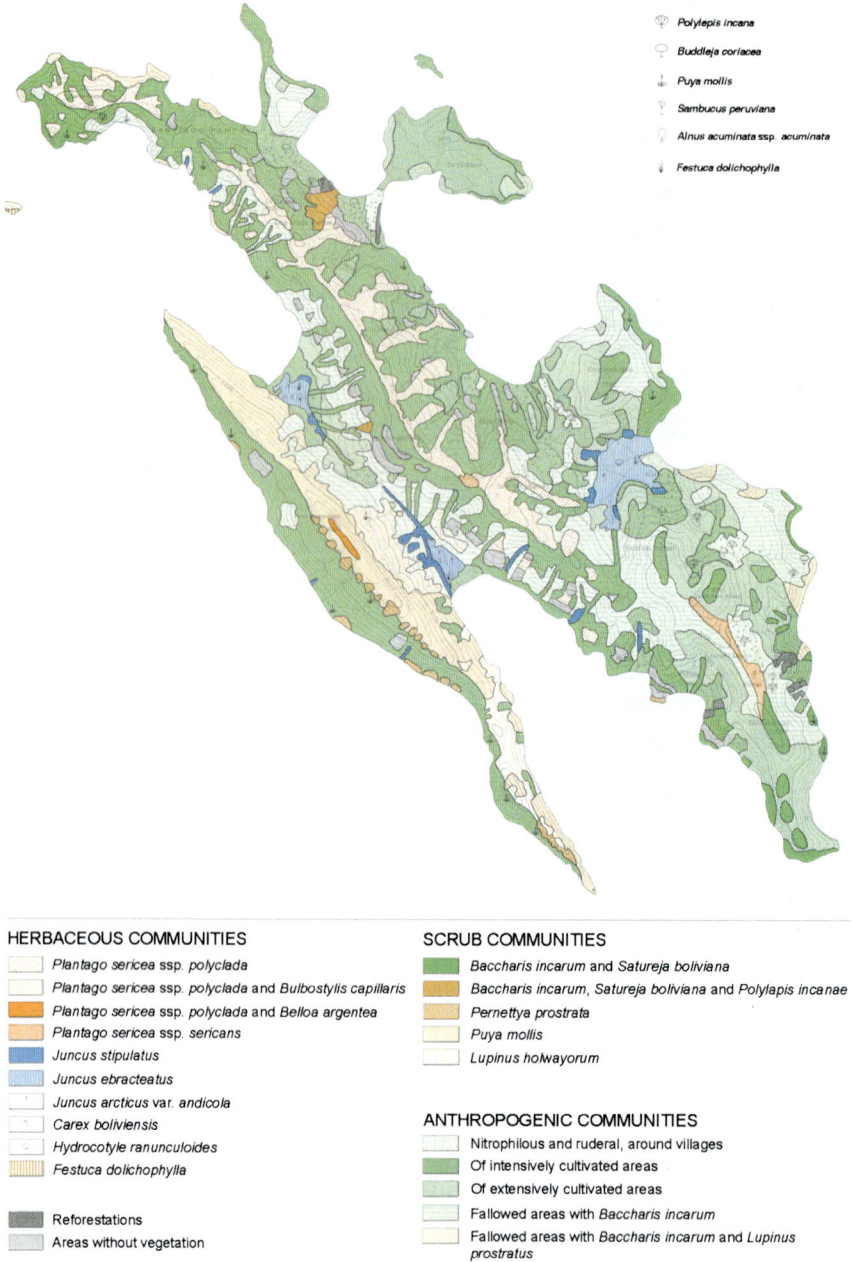

Polylepis incana

Buddleja coriacea

Puya mollis

Sambucus peruviana

Alnus acuminata ssp. acuminata

Festuca dolichophylla

HERBACEOUS COMMUNITIES

Plantago sericea ssp. polyclada
Plantago sericea ssp. polyclada and *Bulbostylis capillaris*
Plantago sericea ssp. polyclada and *Belloa argentea*
Plantago sericea ssp. sericans
Juncus stipulatus
Juncus ebracteatus
Juncus arcticus var. *andicola*
Carex boliviensis
Hydrocotyle ranunculoides
Festuca dolichophylla

Reforestations
Areas without vegetation

SCRUB COMMUNITIES

Baccharis incarum and *Satureja boliviana*
Baccharis incarum, *Satureja boliviana* and *Polylapis incanae*
Pernettya prostrata
Puya mollis
Lupinus holwayorum

ANTHROPOGENIC COMMUNITIES

Nitrophilous and ruderal, around villages
Of intensively cultivated areas
Of extensively cultivated areas
Fallowed areas with *Baccharis incarum*
Fallowed areas with *Baccharis incarum* and *Lupinus prostratus*

Fig. 7.14 Vegetation map of the "Isla del Sol", in Lake Titicaca in the Bolivian altiplano, at scale 1:16,000 (reduced for printing to 1:60,000) (From Liberman Cruz et al. 1995)

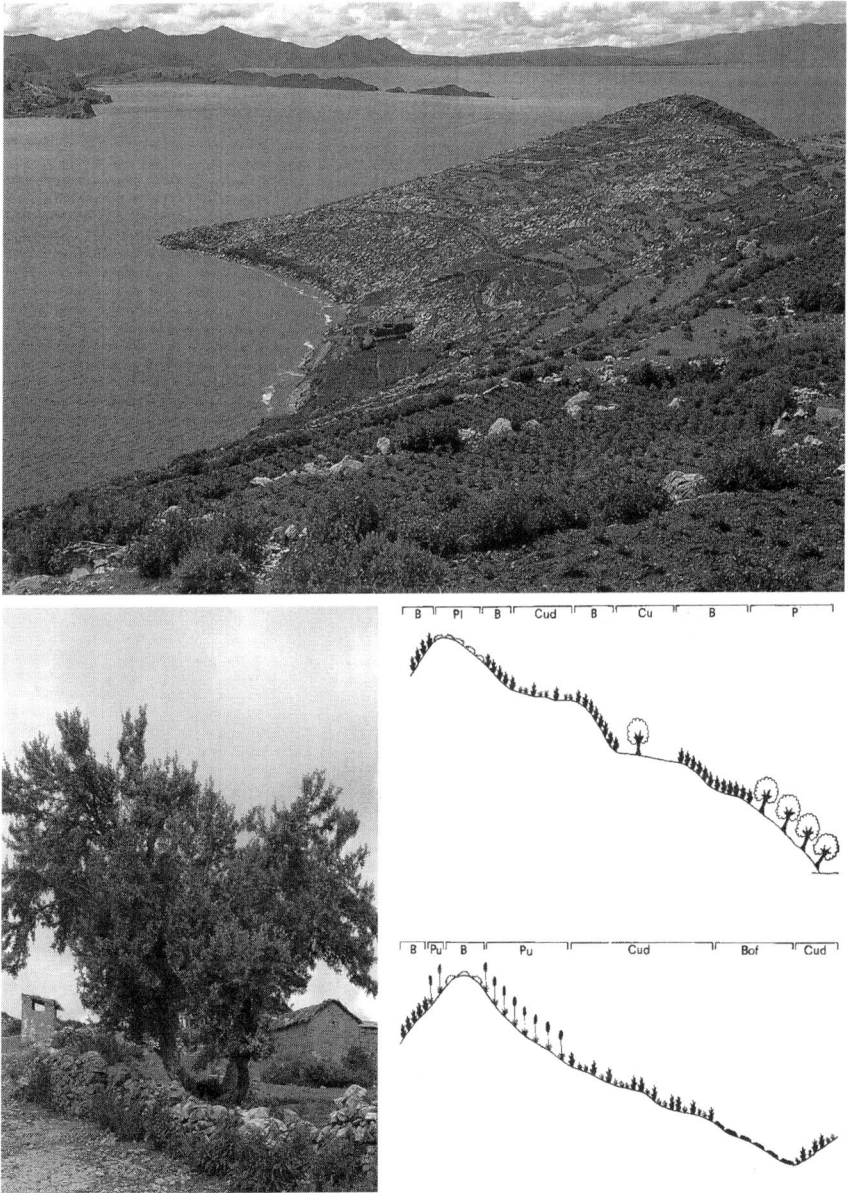

Fig. 7.15 Vegetation of the "Isla del Sol": the slopes are mostly cultivated. At *lower left* is a tree of *Polylepis incana*; at *lower right* are vegetation profiles: B – matorral of *Baccharis incarum* and *Satureja boliviana*, Pl – *Plantago sericea*, Cud – abandoned or fallow agricultural field, Cu crop areas with isolated trees of *Polylepis incana*, P – residual forest of *Polylepis incana*, Pu – *Puya mollis*, and Bof – bofefedales (humid meadows) (Photo Franco Pedrotti)

have covered most of the island, are today present in only two places. The presence of six species is also mapped, by symbols, including *Polylepis incana* and *Buddleja coriacea*, a shrub that has disappeared from nature and is present only in human-modified areas, planted near houses and along streets.

Dunes

Vegetation Map of the Lago di Burano (lake) **and the Capalbio dunes** (Grosseto, Italy), 1:5,000 and Map of one dune sector, at 1:2,000; the latter (Fig. 5.32) shows eight mapped vegetation units (plant associations), Pedrotti et al. (1975, 1979); Pedrotti and Cortini Pedrotti (1976).

This is a phytosociological map of actual vegetation; on dune areas near the shoreline there is a macchia (dense shrub vegetation, cf maquis, matorral) of Juniperetum macrocarpae-phoniceae, interrupted by blowouts which have been colonized by a Crucianelletum. On the seaward side of the dunes is the association Agropyretum mediterraneum; the interior part of the dunes, on the other hand, is occupied by macchia of Oleo-Lentiscetum, interrupted by secondary clearings with vegetation of the order Helianthemetalia guttati.

Karst Basins

Karst basins in the central Apennines are more or less large depressions in which water stagnates during the autumn and winter months; in spring the water is absorbed by sinkholes. For this reason, the vegetation tends to occur in concentric rings following the general geomorphology, the slope of the area immediately surrounding the karst basin, and the sinkholes. Every karst basin has its particular characteristic physics, reflected in the distribution of vegetation (Pedrotti 1985).

Phytosociological Vegetation Map of the Piani di Montelago (plains, near Camerino, central Italy), 1:3,000, 30 vegetation units (plant associations and groupings) (Pedrotti 1967) (Fig. 7.16).

This is a phytosociological map of actual vegetation of an upper and a lower karst plain (Fig. 7.17). The lower plain has: meso-xerophilic meadows (Mesobrometum) around the outside border of the basin, where inundation by water occurs only exceptionally; moist meadows inside the area of the preceding association, where water remains shallow and stays for only a short period (Hordeo-Ranunculetum velutini and Deschampsio-Caricetum distantis); transitional meadows related to the sinkholes (assemblage with *Ranunculus sardous*); wet meadows (*Caricetum gracilis*) in the central part of the basin; and a Caricetum gracilis being invaded progressively by *Phragmites australis* in some central areas mowed infrequently.

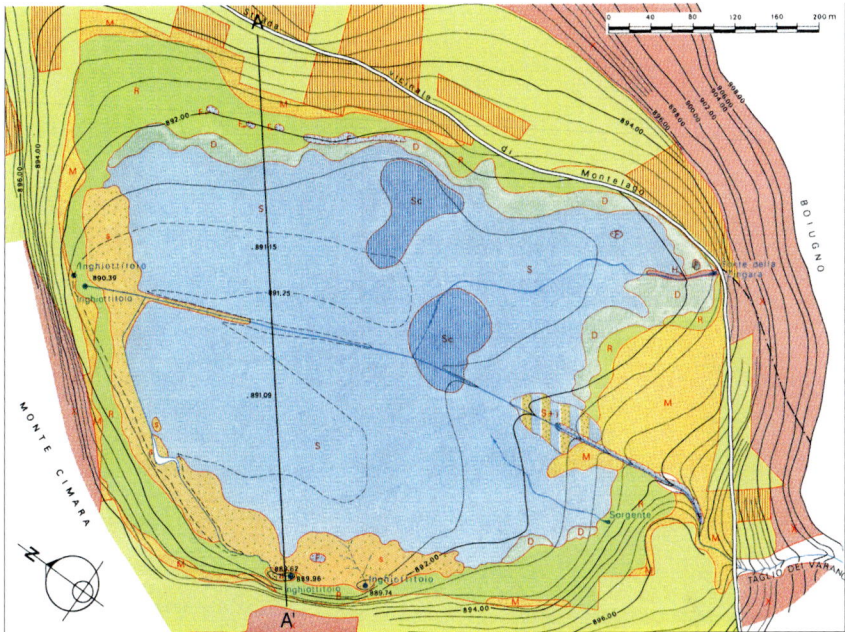

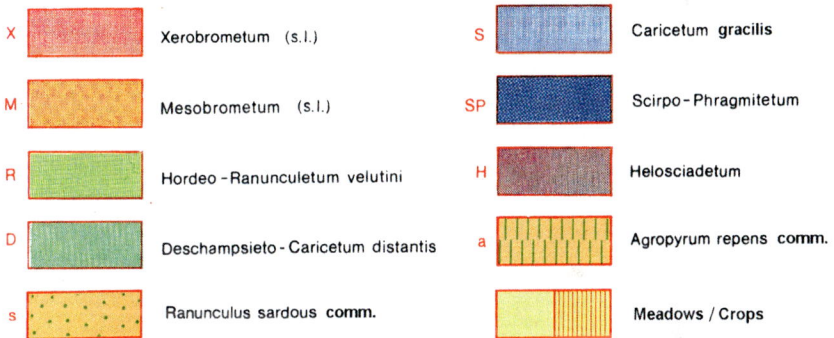

Fig. 7.16 Vegetation map of the lower karst basin of Montelago, Marche Region, central Italy: the transect *A-AI* is shown in Fig. 7.17 (From Pedrotti 1967)

Wetlands, Mires, Lagoons and Rivers

Wetlands are defined as transitional lands between terrestrial and aquatic systems where the water table is at or near the surface or the land is covered by shallow water (Holland et al. 1990). The main consideration for vegetation mapping in wetlands is analysis of the ecological-vegetational gradient (ecotone) on the banks of lakes, lagoons, swamps, peatbogs, etc., through sampling along transects and profiles. In these cases the vegetation forms a distinct ecotone in concentric bands, each of which is characterized by a certain association (Fig. 7.18). This kind of

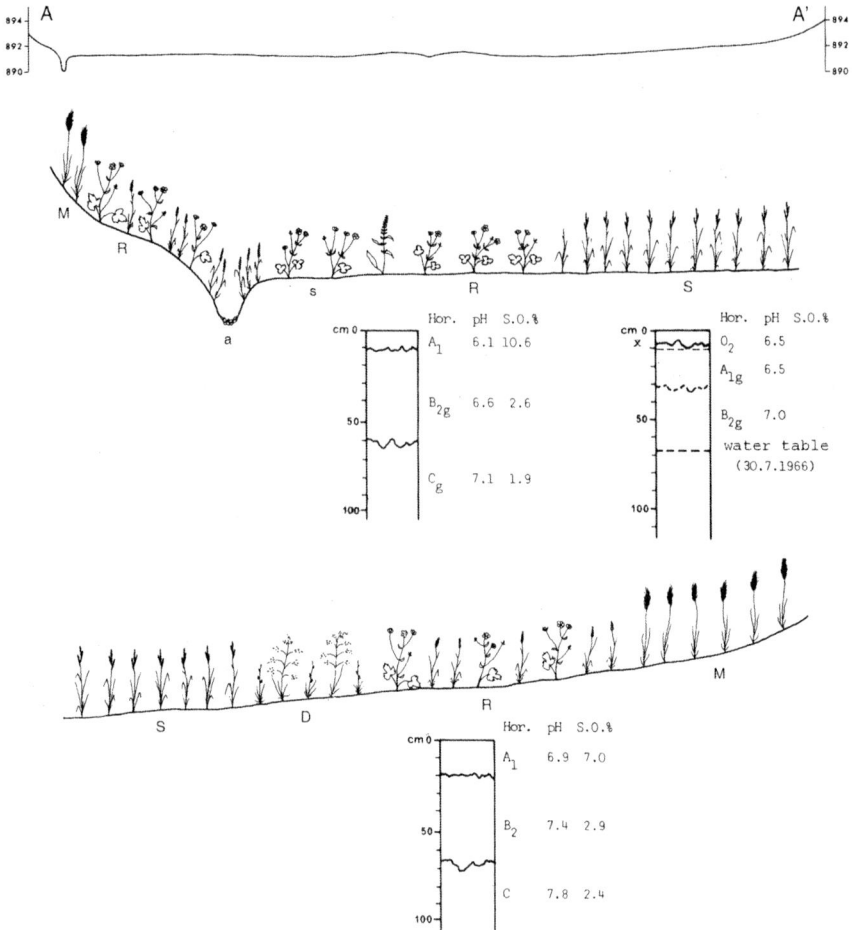

Fig. 7.17 A-A1 geomorphological profile of the lower Piano di Montelago (karst basin), and vegetation transects: M – *Mesobrometum*; R – *Hordeo-Ranunculetum velutini*; 3 – *Agropyrum repens* community; s – areas near sinkhole, with a *Ranunculus sardous* community a – sinkhole, with an *Agropyrum repens* community; S – *Caricetum gracilis*; and D – *Deschampsio-Caricetum distantis*. The soils of the *Caricetum gracilis* are hydromorphic, the soils of the *Hordeo-Ranunculetum velutini* are well drained, and the soils of the *Ranunculus sardous* community are intermediate (From Pedrotti 1982 and Sanesi 1982)

vegetation zonation is found in essentially all wetlands (Pedrotti 2004c). Another problem in wetlands is that the boundaries between aquatic communities are often not well-defined, as for example in floating vegetation; this is the case for associations of the Lemnetea class, such as the Lemnetum trisulcae, Riccietum fluitantis, Ricciocarpetum natantis and Statiotetum aloidis for Europe (Fig. 7.19) and for analogous associations on other continents, such as the Lemnetum minusculae-Lemnetum gibbae and the Lemnetum valdivianae of Lake Titicaca in South America (Kepczynski 1960; Liberman Cruz et al. 1988).

Fig. 7.18 Non-orthogonal aerial photo of Lake Lungo, Rieti, Latium Region, central Italy: the associations *Phragmitetum australis* and *Myriophyllo-Nupharetum* form continuous concentric circles on the lake banks (*Photo by Aldo Labonia, Camerino*)

Fiavé Mire (Trentino, northeastern Italy; Fig. 7.20)

Map of Actual Vegetation, 20 vegetation units (plant associations); **Map of Dynamic Tendency,** 1:8,000; Canullo et al. (1990, 1994); Pedrotti (1997a) (Fig. 7.21).

The map of actual vegetation shows plant associations, especially a *Molinia caerulea* meadow (Gentiano-Molinietum); because of the successive subdivision of

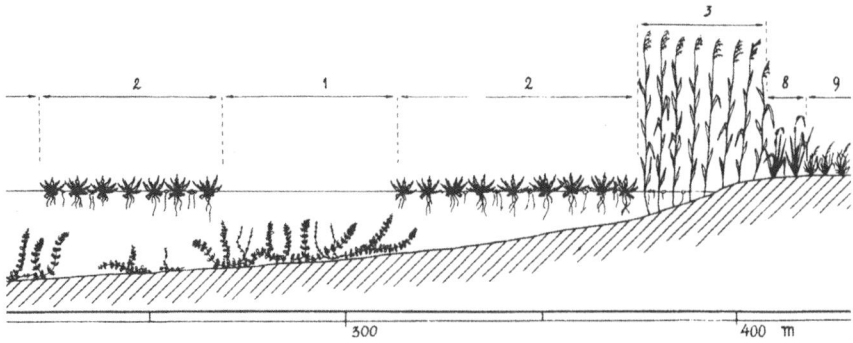

Fig. 7.19 Vegetation profile on the banks of Lake Siete, Poland: *1 – Elodea canadensis* community; *2 – Stratiotetum*; *3 – Phragmitetum australis* (From Kepczynski 1960)

the mire area over many human generations, the mapping units have a form that is always somewhat geometrical, even rectangular. The dynamical tendencies shows the processes at work in the *Molinia* meadows after the cessation of mowing, with the development of a Salicetum cinereae. In the mire there are two vegetation series, of *Salix cineria* and *Salix pentandra*, which correspond to two azonal potential associations, a Salicetum cinereae and a Salicetum pentandrae. The external slopes of the mire are occupied by the zonal potential beech forest (Carici albae-Fagetum).

Vedes Mire (Trentino, northeastern Italy) (Figs. 5.30 and 5.31).

Map of Actual Vegetation, 1:4,000, 7 vegetation units (plant associations), Pedrotti (1980).

This is a phytosociological map of the actual vegetation of a transitional mire with a lake in the center, already divided into two parts by floating meadows of Rhinchosporetum albae and Caricetum limosae; this is surrounded by the vegetation of a raised bog (Sphagnetum magellanici), surrounded in turn by bands of Pino mugo-Sphagnetum and other vegetation types.

Vegetation Map of the Valli di Comacchio (Comacchio valleys, Ferrara, northern Italy), 1:80,000, 4 units (plant associations and groupings), Ferrari et al. (1972) (Fig. 7.22).

This is a phytosociological map of actual vegetation, showing submerged vegetation on the bottom of the Comacchio valleys, with the associations Lamprothamnietum papulosi and Ruppietum spiralis, sampled by boat along transects. Submerged vegetation can be sampled with great precision by using Side-Scan Sonar (SSS), an ecographic device that provides sonar images of the bottom of a water body. These images are called "sonograms" and are very similar to an aerial photograph. The submerged meadows with *Poseidonia oceanica* in Sardinia were mapped with this methodology (Fig. 7.23).

Fig. 7.20 Non-orthogonal aerial photo of the Fiavé peat bog, Trentino-Alto Adige Region, northern Italy. The peat bog is divided into many narrow rectangular stripes, representing the many fragmented private parcels *(Photo Servizio Parchi Foreste Demaniali, Trento 1993)*

Vegetation Map of the Lago Trasimeno (lake, Umbria, central Italy), 1:50,000, 12 units (associations and groupings, only in the lake and on its banks); Pedrotti and Orsomando (1977); Orsomando (1993) (Figs. 7.24, 7.25, and 7.26).

This is a phytosociological map of the actual vegetation of the Lago Trasimeno and its catchment basin. The lakeshore and shallowest part of the lake bottom are

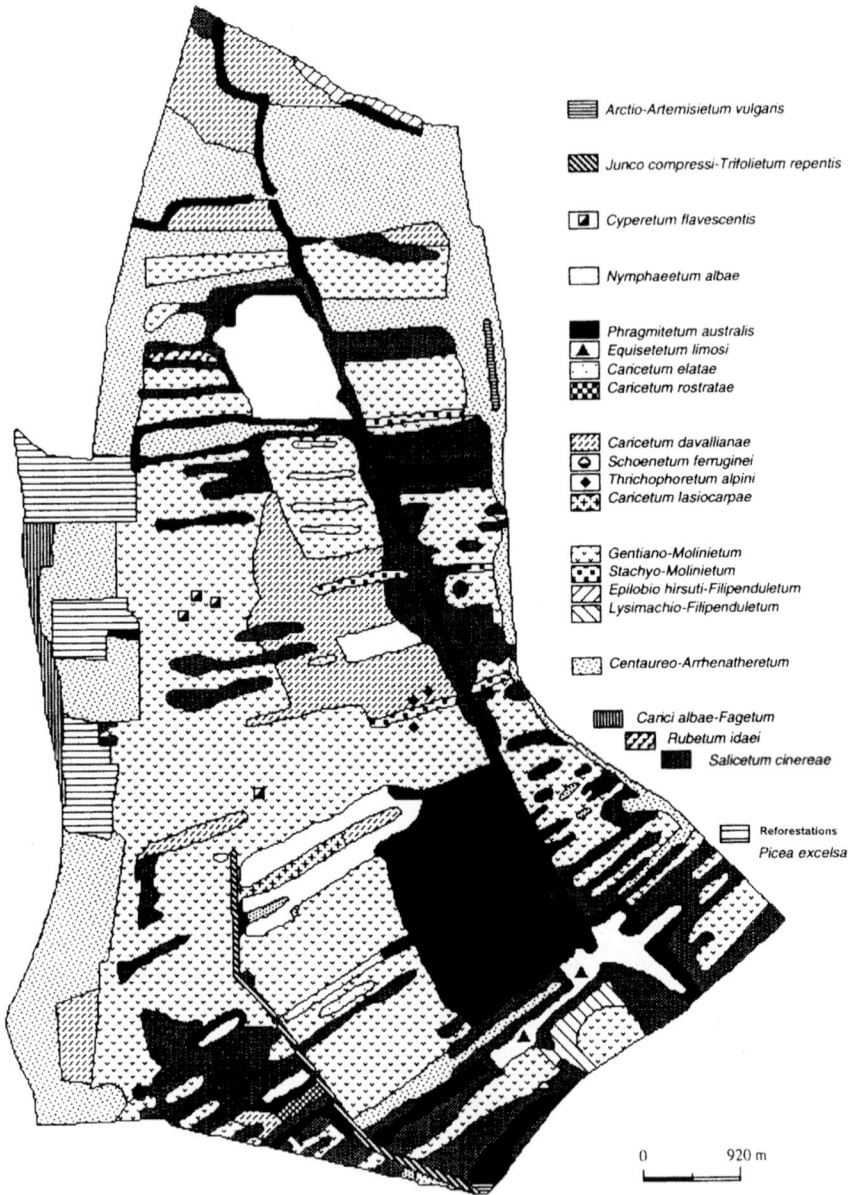

Fig. 7.21 Phytosociological map of the Fiavé peat bog (From Canullo et al. 1994)

occupied by a band of Phragmitetum australis, inside which, in deeper water, there
is an association of floating and submerged hydrophytes of the Potamogetono-
Ceratophylletum demersi. These associations are represented by cartographic units.
The various other associations that occur, in small areas, are indicated by symbols,

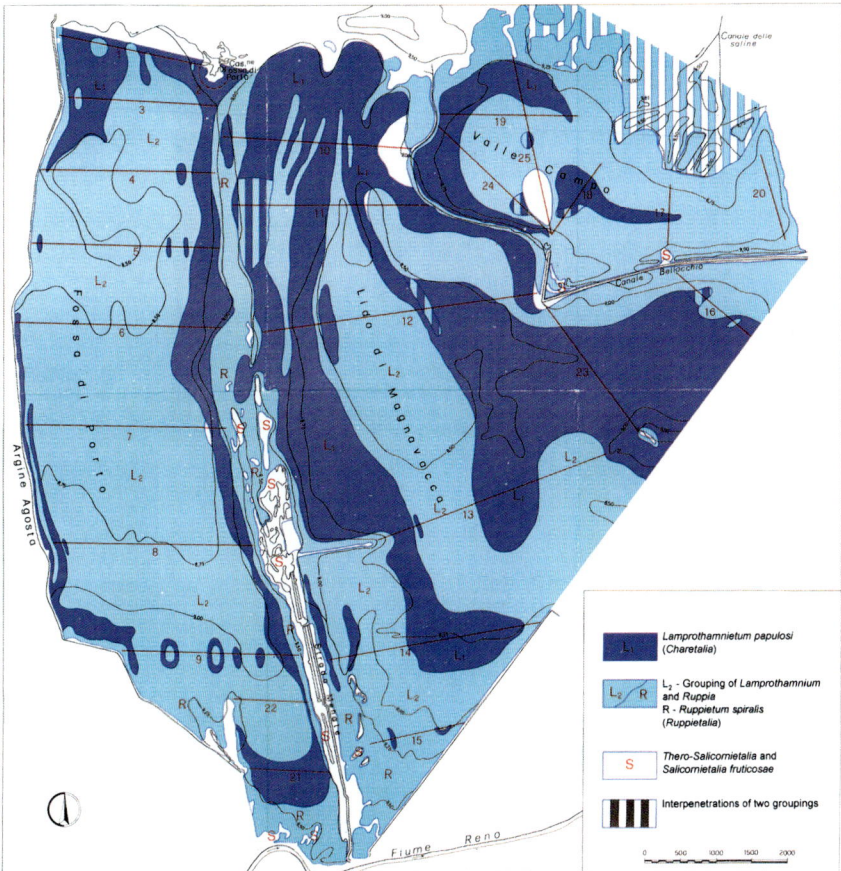

Fig. 7.22 Phytosociological map of the lagoons "Valli di Comacchio", Emilia-Romagna Region, northern Italy (From Ferrari et al. 1972)

such as a Hydrocharitetum, the Potamogetonetum lucentis, Potamogetonetum lucentis nymphaeetosum, Typhetum angustifoliae and Scirpetum lacustris. The submerged vegetation was sampled using a measure of water depth along a transect and by harvesting plant samples using a small dredge.

Vegetation Map of the Chioggiola Wetland (Bologna, north-central Italy), 1:2,000, 4 units (plant associations), Ferrari (1977) (Fig. 7.27).

This is a phytosociological map of actual vegetation, with a water surface at the center with an association with *Potamogeton natans* and *Myriophyllum spicatum*, surrounded by marshes (Caricetum elatae) and cane (Phragmitetum australis).

Vegetation Map of the Punte Alberete and Valle Mandriole (Ravenna, northern Italy), 1:5,000, 20 units (plant associations or higher units), Merloni and Piccoli (2001) (Fig. 7.28).

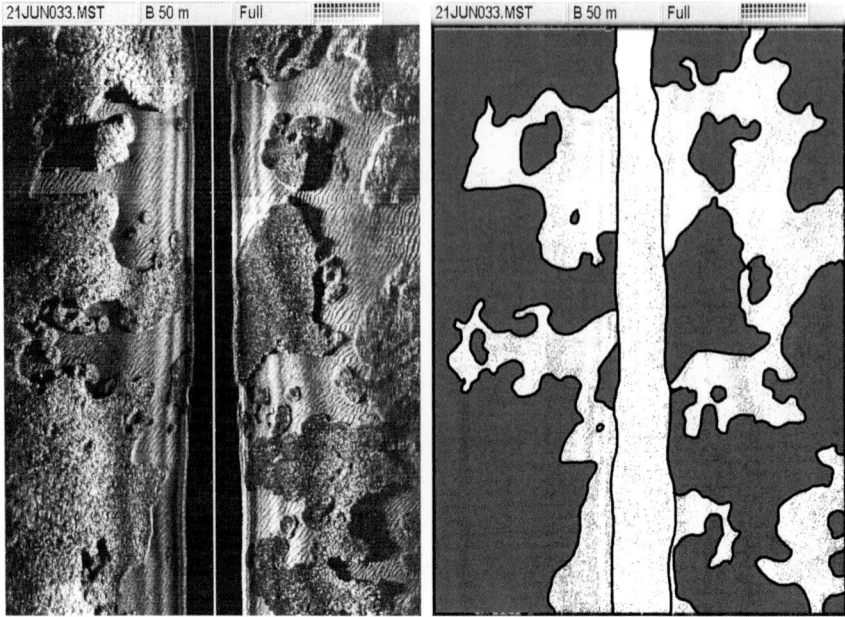

Fig. 7.23 At *left*, a Side Scan Sonar track on which one can see the defined boundaries between ramet groups of a population of *Poseidonia oceanica* ("*matte*") on the sandy seabed, Maddalena National Park, Sardinia; at *right*, a corresponding map of the *Poseidonietum oceanicae* from the previous image. With this methodology it is also possible to monitor damages caused by anchorage and fishing nets (trawls). The *black* strip in the *middle* is the empty track after the passage of the boat (*Survey Giani, Camerino*)

This is a phytosociological map of actual vegetation, with freshwater wetland vegetation formed mainly by cane and related associations (Phragmitetalia), interrupted by water surfaces with floating duckweed (Lemnetalia minoris); where the wetland is filling in there are patches of Salicetum cinereae and of Cladio-Fraxinetum oxycarpae.

Streams

Vegetation Map of the Bosco dell'Incoronata (forest) (Foggia, southern Italy), 1:10,000, 11 units (associations, groupings and physiognomic units), Pedrotti and Venanzoni (1994b) (Fig. 7.29).
This is a phytosociological map of the actual vegetation of the Incoronata Forest along the river Cervaro in Foggia (southern Italy). The riparian vegetation is composed of willow stands (Salicetum triandrae, Saponario-Salicetum purpureae and Salicetum albae) and stands of poplar, ash, and elm (Populetum albae, Carici-Fraxinetum

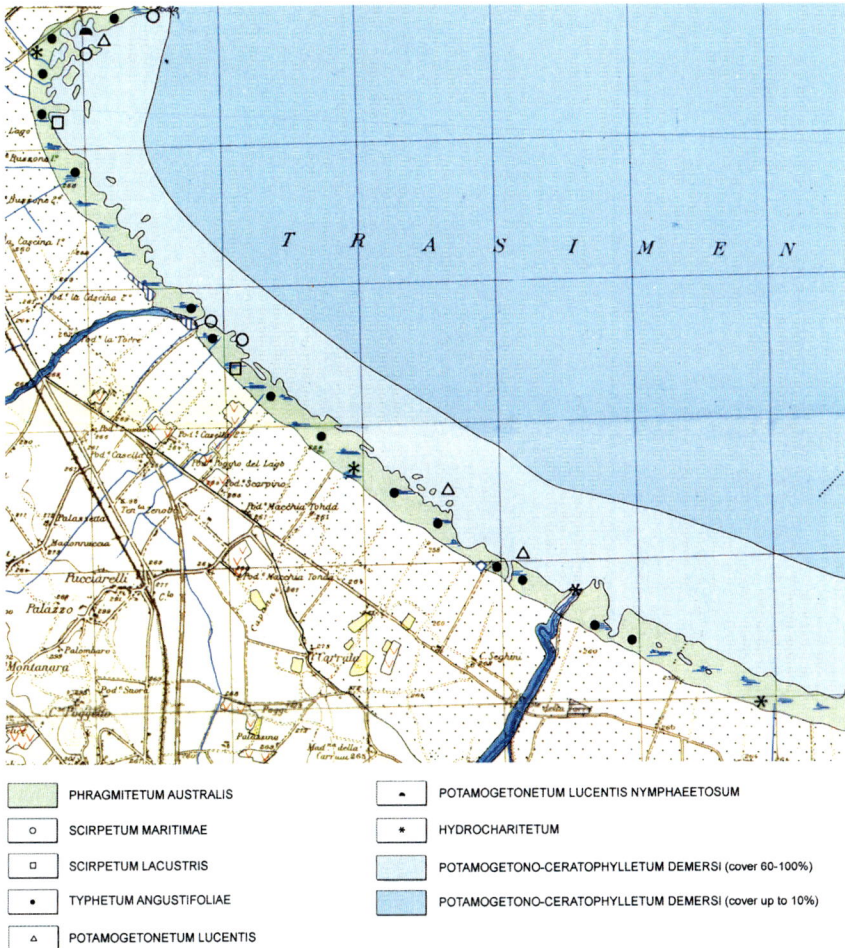

	PHRAGMITETUM AUSTRALIS		▲	POTAMOGETONETUM LUCENTIS NYMPHAEETOSUM
o	SCIRPETUM MARITIMAE		*	HYDROCHARITETUM
□	SCIRPETUM LACUSTRIS			POTAMOGETONO-CERATOPHYLLETUM DEMERSI (cover 60-100%)
•	TYPHETUM ANGUSTIFOLIAE			POTAMOGETONO-CERATOPHYLLETUM DEMERSI (cover up to 10%)
△	POTAMOGETONETUM LUCENTIS			

Fig. 7.24 Phytosociological map of the Trasimeno Lake, Umbria Region, central Italy: the *Phragmitetum* form a continuous band, in which it is also possible to find some other small associations mapped with symbols (Survey Orsomando and Pedrotti from Orsomando 1993)

oxycarpae, Ranunculo-Fraxinetum oxycarpae and Aro italici-Ulmetum minoris), developed along the river banks, in the river bed, and along old and recent river meanders.

Urban Centers

The vegetation of urban centers is formed of anthropogenic ruderal and nitrophilous associations, generally occurring widely, developed on old [stone] walls, the walls of buildings, monuments, street pavement, street verges, gardens of various kinds,

Fig. 7.25 Vegetation of the Trasimeno Lake: *A* – reeds (*Phragmitetum australis*), B – floating hydrophytes (*Potamogetonetum lucentis nymphaeetosum*), and *C* – submerged hydrophytes (*Potamogetono-Ceratophylletum demersi*) (Photo Ettore Orsomando)

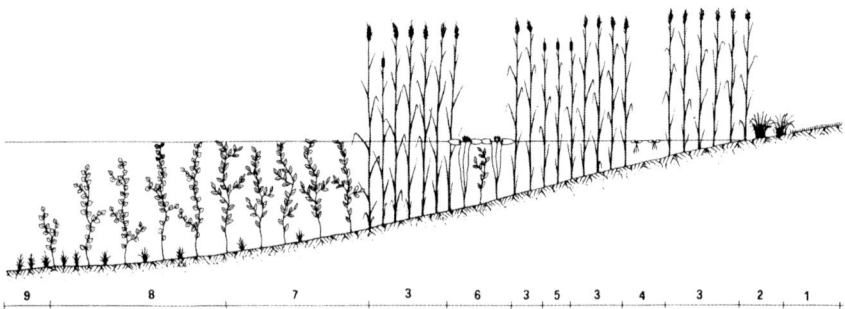

Fig. 7.26 Profile of the vegetation of Trasimeno Lake: *1* – *Holoschoenetum* and *Juncus articulatus* communities, *2* – *Caricetum ripariae*, *3* – *Phragmitetum australis*, *4* – *Hydrocharitetum morsus-ranae*, *5* – *Typhetum angustifoliae*, *6* – *Potamogetonetum lucentis nymphaeetosum*, *7* – *Potamogetono-Ceratophylletum demersi*, *8* – submerged *Potamogetono-Ceratophylletum demersi*, and *9* – *Vallisneria spiralis* community (From Pedrotti and Orsomando 1977)

urban parks, and uncultivated areas. The study and mapping of urban vegetation has a genuinely botanical character, but the results obtained constitute a useful reference in urban ecological research (Hruska 1991).

Vegetation Map of the city of Camerino (central Italy), 1:10,000, 12 units (plant associations), Hruska (1998) (Fig. 7.30).

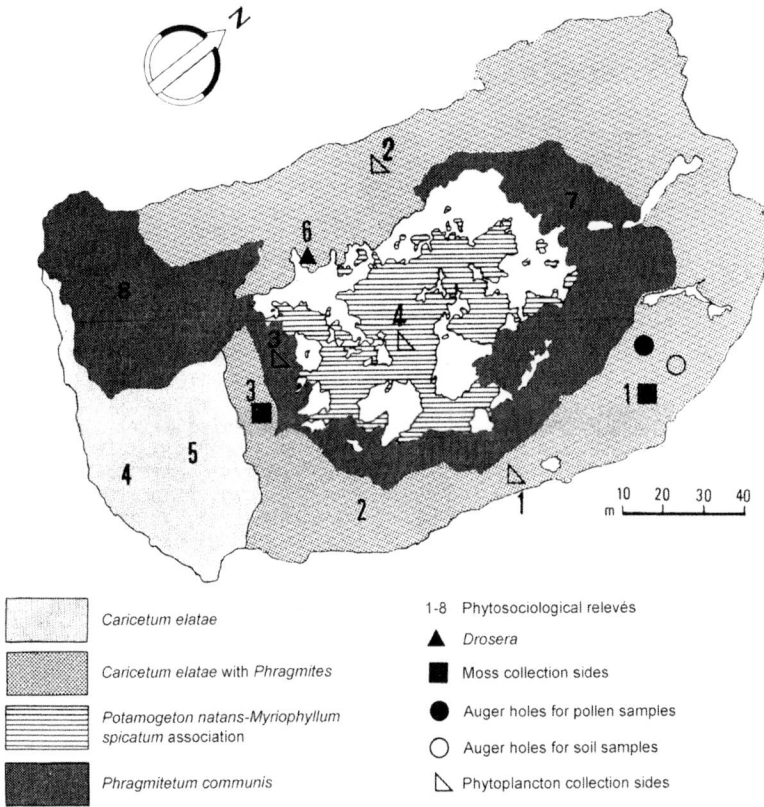

Fig. 7.27 Vegetation map of the Chioggiola wetland, Emilia-Romagna Region, northern Apennines (From Ferrari 1977)

This is a phytosociological map of actual vegetation; the anthropogenic vegetation of the historic center of Camerino, an old hilltop town in east-central Italy, is formed of (trampled) street associations, namely the Poo infirmae-Polycarpetum tetraphylli (occurring everywhere), the Bryo-Saginetum procumbentis and the Parietario diffusae-Amaranthetum deflexi (limited to some streets). On ancient walls there are five associations, including the Capparidetum inermis; the other associations, such as the Rumici-Carduetum pycnocephali and Hordeetum murini, are present only on the periphery of the city. The Lunularietum cruciatae, an association formed exclusively of bryophytes (hepatics and mosses), typical of moist, shady lanes, has been found in a few locations (Cortini Pedrotti 1989; Aleffi 2010).

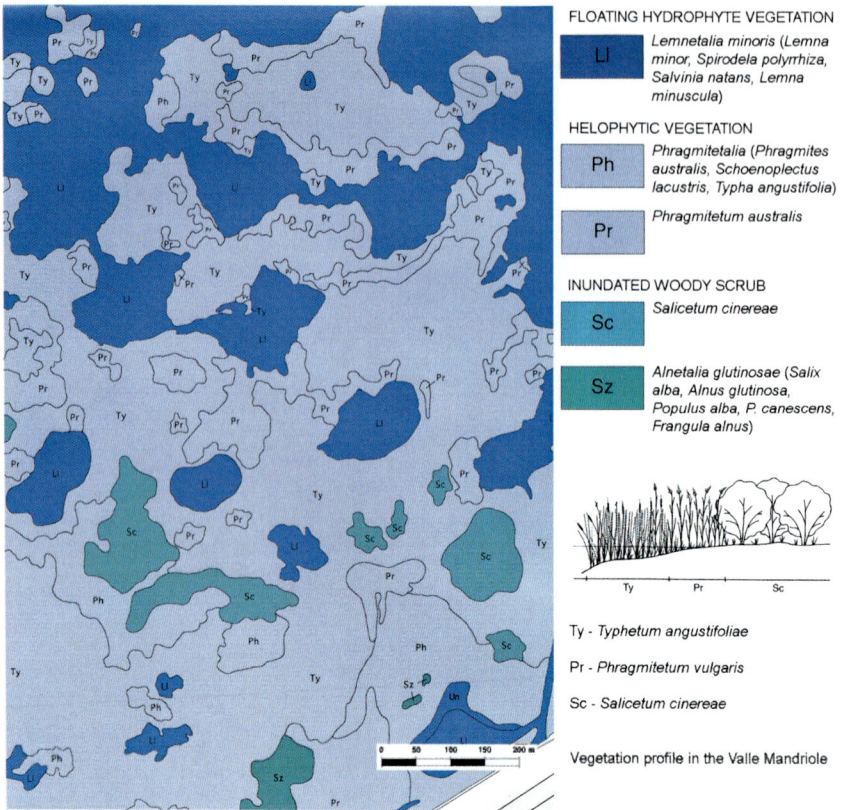

FLOATING HYDROPHYTE VEGETATION

Ll — *Lemnetalia minoris (Lemna minor, Spirodela polyrrhiza, Salvinia natans, Lemna minuscula)*

HELOPHYTIC VEGETATION

Ph — *Phragmitetalia (Phragmites australis, Schoenoplectus lacustris, Typha angustifolia)*

Pr — *Phragmitetum australis*

INUNDATED WOODY SCRUB

Sc — *Salicetum cinereae*

Sz — *Alnetalia glutinosae (Salix alba, Alnus glutinosa, Populus alba, P. canescens, Frangula alnus)*

Ty - *Typhetum angustifoliae*

Pr - *Phragmitetum vulgaris*

Sc - *Salicetum cinereae*

Vegetation profile in the Valle Mandriole

Fig. 7.28 Vegetation map of the Valle Mandriole wetland, Emilia-Romagna Region, northern Italy. The vegetation profile shows a *Salicetum cinereae* (Sc), a *Phragmitetum vulgaris* (Pr), and a *Typhetum angustifoliae* (Ty) (From Merloni and Piccoli 2001)

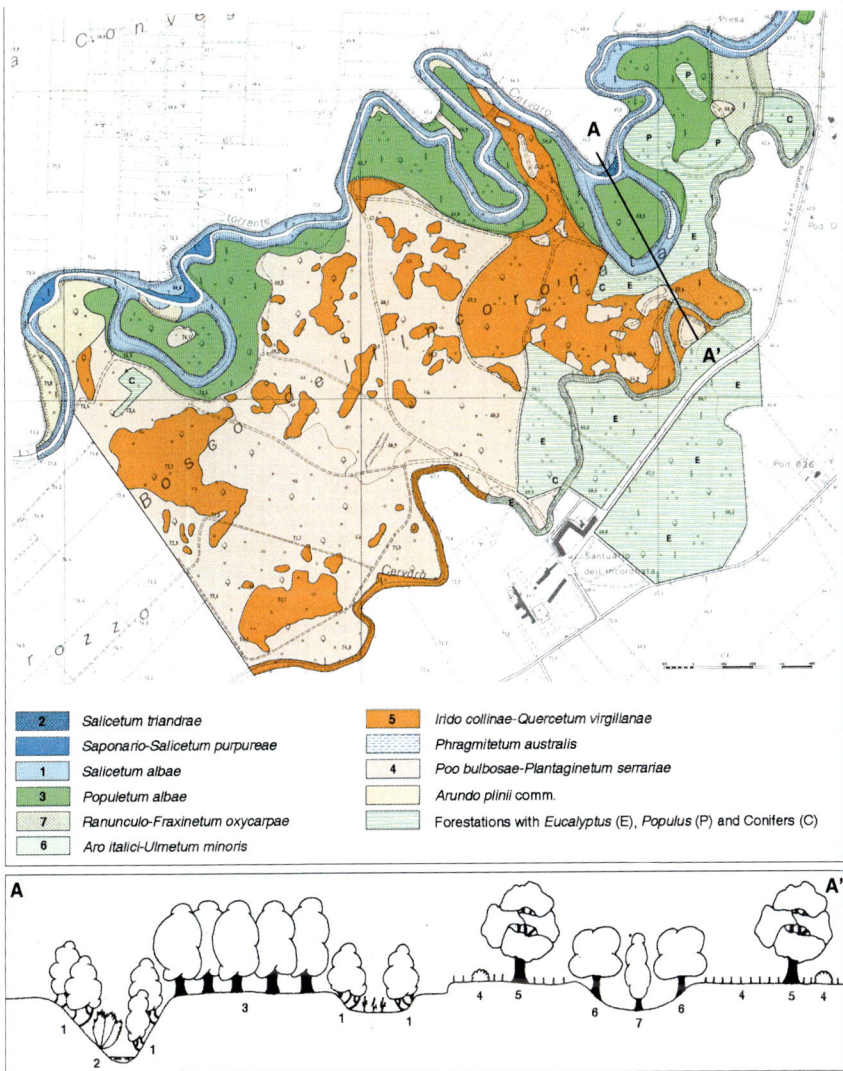

Fig. 7.29 Vegetation map of the Incoronata forest, Foggia, Apulia Region, southern Italy, showing a transect A-A′ through the Incoronata forest (From Pedrotti and Venanzoni 1994b; Pedrotti and Gafta 1996)

Fig. 7.30 Vegetation map of Camerino, an old hilltop town in the Marche Region, central Italy (From Hruska 1998)

A sigmetum, as described in Chap. 6, is a local assemblage of all the communities that have the potential to develop toward a particular single climax. A geosigmetum is the system of multiple sigmeta along a gradient, which in a landscape may correspond to a catena. Synphytosociology is the phytosociology of sigmeta, i.e. the analytical classification, by phytosociological principles, of these related communities (sigmeta). Geo-synphytosociology is the phytosociology of geosigmeta. There is no simple vocabulary for these concepts in English, other than these rather long, Greek-based terms from phytosociology. The term catenal vegetation cartography, though, or catenal vegetation mapping, may suggest what is involved.

Catenal cartography, in this sense, is based on the concept of the geosigmetum, or geo-sigma-association or catena of vegetation series as summarized by Géhu (1991a). All of these terms are well defined, albeit in French, in the dictionary by Géhu (2006). A geosigmetum is recognized in relation to geomorphological and climatic features, and is formed of a set of sigmeta. A geosigmetum is thus an ecologically heterogeneous unit, because it is formed of several sigmeta, each of which has its own ecology and thus a particular type of potential vegetation.

Catenal (geo-synphytosociological) maps represent the geosigmeta (geo-sigma-associations) or higher units into which they can be grouped (geosigmion, geosigmetalia, geosigmetea); these units are also called vegetation geoseries, macro-geoseries, mega-geoseries and hyper-geoseries (Rivas Martínez 2005a, 2007–2011). These are true phytosociological maps of the vegetated landscape and correspond to level VI (see Chap. 1). Many authors have made "maps of the vegetated landscape" showing plant associations. In the phytosociological sense, though, these are not maps of the vegetated landscape but rather maps of plant associations, i.e. vegetation maps. Undoubtedly, the set of associations mapped constitutes the vegetated landscape, but the map was not the result of phytosociological methodology. Only if there is a representation based on sigmeta, as in the case of geo-synphytosociology, can one speak of a true phytosociological map of the vegetated landscape.

In studying vegetated landscapes one can distinguish three analytical phases, serial, catenal and chorological, each of which represents a different level of vegetation complexity. The serial phase studies the plant associations in a given vegetation series (i.e. that have the same potential): the sigmetum (vegetation series) or the perma-sigmetum (perma-series of vegetation). The catenal phase groups the sequences of series occurring along a gradient (catena), i.e. the geosigmetum (vegetation geoseries) or the geo-perma-sigmetum (geo-perma-series of vegetation). The chorological phase considers distinct vegetation units based on geological homogeneity (in districts), climatic homogeneity (in sectors), and floristic uniqueness (in provinces); these are the so-called macro-geoseries, mega-geoseries and hyper-geoseries (Rivas Martínez 2005a) and correspond to the classical phytogeographical subdivisions of district, sector and province, to which the region could be added.

With catenal phytosociology it is thus possible to come to the definition of phytogeographical subdivisions that in the past had been defined only by climate, floristics, or other criteria. Still it must be noted that a map showing units higher than a geosigmetum is no longer a vegetation map but rather a phytogeographic map that subdivides the plant cover of an area based on both species and vegetation types. Such maps are treated further in Chap. 9.

One of the first examples of a catenal (geo-synphytosociological) map is that by Theurillat (1992) of the region of Aletsch in Switzerland, showing 17 geosigmeta.

The *Carta geosinfitosociologica del Trentino-Alto Adige* (Geo-synphytological Map of Trentino-Alto Adige) at scale 1:250,000, by Pedrotti and Gafta (2003), represents the first attempt to distinguish catenal mega-units (mega-geoseries) in Italy. These mega-geoseries were distinguished by districts, which were recognized by superimposing climatic continentality and limits for different parent materials and soils. A district has a typical composition in terms of vegetation series, although in some cases a sigmion has been used, in order to include more than one sigmetum, as in the case of the Erico-Pinion mugo and the Caricion curvulae. Each mega-geoseries groups the complete altitudinal sequence of vegetation series that characterize a given district.

The region Trentino-Alto Adige has three climatic subdivisions, the pre-Alpine, Alpine and Inner Alpine. Each of these possesses its own floristic uniqueness, evaluated by endemic and differential species. The region also includes areas of four different geologic substrates, namely the calcareous-dolomitic, metamorphic with intrusive magma or with effusive magma, and alluvial. The combination of climatic and lithological data led to the distinction of 9 mega-geoseries and one riparian geoseries; some mega-geoseries were in turn subdivided into sub-mega-geoseries.

By way of example, a brief description is given for the mega-geoseries Fraxino orni-Ostryeto carpinifoliae – Cardamino pentaphylli – Fageto sylvaticae geosigmetum, that belongs to the pre-Alpine climatic sector and the calcareous-dolomitic district; it can also be subdivided into three sub-mega-geoseries (Table 8.1). Figure 8.1 shows a sub-mega-geoseries characterized by series of *Quercus ilex*, in particular the Celtidi astralis-Querceto ilicis sigmetum that is found in southern Trentino. Figure 8.2 shows a fragment of the geo-synphytosociological map of Trentino-Alto Adige, which shows the three sub-mega-geoseries of the mega-geoseries Fraxino orni-Ostryeto carpinifoliae – Cardamino pentaphylli-Fageto sylvaticae geosigmetum.

Table 8.1 Serial composition of megageoseries *Fraxino orni-Ostryeto carpinifoliae – Cardamino pentaphylli-Fageto sylvaticae* geosigmetum (I.A.1.) distinguished in submegageoseries *Celtidi australis-Querceto ilicis* (I.A.1.a), submegageoseries typica (I.A.1.b) and submegageoseries *Adenostylo glabrae-Piceeto abietis* (I.A.1.c.) (From Pedrotti and Gafta 2003)

Submegageoseries	I.A.1.a	I.A.1.b	I.A.1.c
Celtidi australi-Qureceto ilicis	+	.	.
Fraxino orni-Ostryeto carpinifoliae	+	+	+
Fraxineto excelsioris l.s.	.	+	+
Chamaecytiso-Pinetum sylvestris	.	.	+
Carici albae-Fageto sylvaticae	+	+	+
Cardamino pent.-Fageto sylvaticae	+	+	+
Adenostylo glabrae-Abieteto albae	.	+	+
Adenostyo glabrae-Piceeto abietis	.	.	+
Erico-Pinion mugo	+	+	+
Seslerion albicantis	+	+	+

Fig. 8.1 Eastern slopes of Monte Bondone, Trento Province, northern Italy, from the top (2,176 m) to the valley bottom (245 m). The picture shows the sub-mega-geoseries *Celtidi australis-Querceto ilicis*, composed of: *1* a *Seslerion albicantis* sigmion, *2* an *Erico-Pinion mugo* sigmion, *3* a *Cardamino pentaphylli-Fageto sylvaticae* sigmetum, *4* a *Cardamino pentaphylli-Abieteto* sigmetum, *5* a *Carici albae-Fageto sylvaticae* sigmetum, *6* a *Fraxino orni-Ostryeto carpinifoliae* sigmetum, *7* a *Celtidi australis-Querceto ilicis* sigmetum (Photo Giovanni Carotti, Camerino)

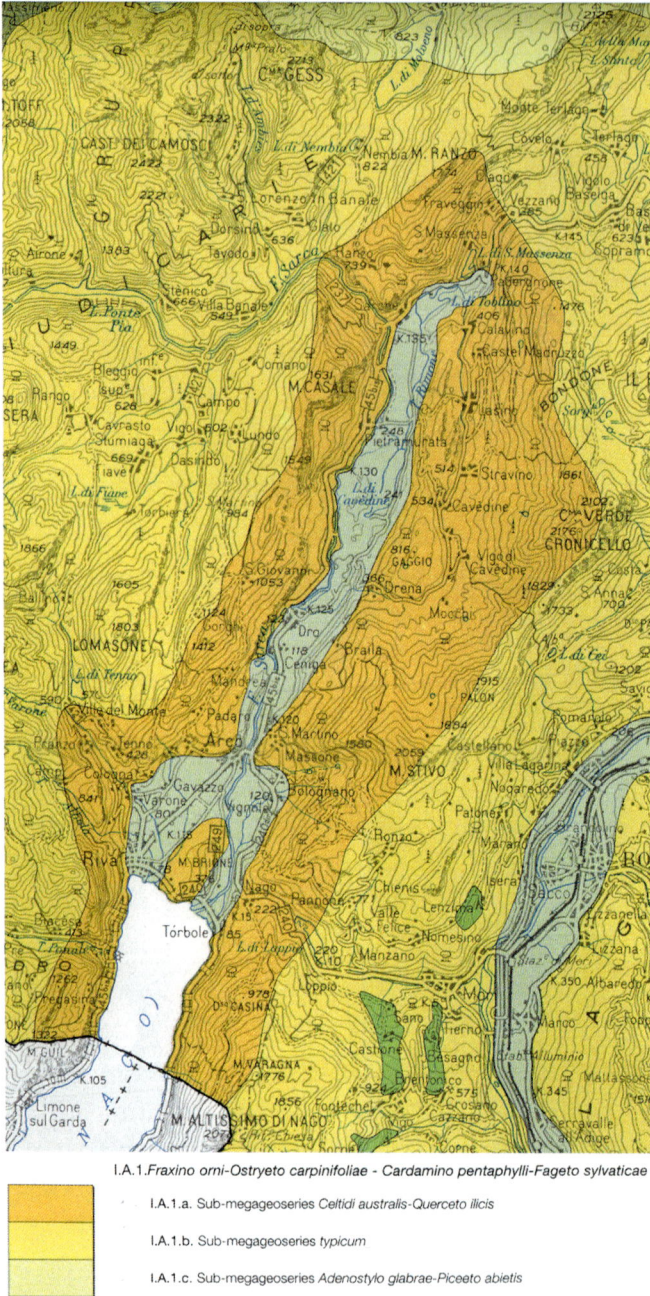

I.A.1.*Fraxino orni-Ostryeto carpinifoliae - Cardamino pentaphylli-Fageto sylvaticae*

I.A.1.a. Sub-megageoseries *Celtidi australis-Querceto ilicis*

I.A.1.b. Sub-megageoseries *typicum*

I.A.1.c. Sub-megageoseries *Adenostylo glabrae-Piceeto abietis*

Fig. 8.2 Map of the mega-geoseries of the Trentino-Alto Adige Region, Sarca Valley sector, at scale 1:250,000 (From Pedrotti and Gafta 2003)

As can be seen in Table 8.1, the sub-mega-geoseries have many sigmeta in common (the sigmeta with *Fagus sylvatica* and those of the Alpine belt, the Erico-Pinion mugo sigmion and Seslerion albicantis sigmion). Two sub-mega-geoseries have an exclusive sigmetum (the Celtidi australis-Querceto ilicis sigmetum or Adenostylo glabrae-Piceeto sigmetum); other differences are noted in Table 8.1.

Phytogeographical Mapping

9

The purpose of phytogeographical mapping is to subdivide land areas of the globe into plant-geographic units (or subdivisions)[1] based on plant-systematic categories, i.e. orders, families, genera and species, especially endemic species and their respective areas, biogenetic centers and geological age. A phytogeographic unit, in this sense, thus represents a territory that possesses a specific, homogeneous flora, with endemic families, genera and species, and that resulted from a common evolutionary history under conditions of relative geographic isolation. Pignatti (1988a) notes that a phytogeographical zone is a unique area with a significant number of species that occur only there.

As noted in Chap. 1, one can distinguish lower phytogeographic units, called districts, sectors and provinces or dominions (level VII) and higher phytogeographic units, namely regions and kingdoms (level VIII).

Analogously, one can also make *zoogeographic* maps, for the distribution of animal species, and *biogeographic* maps, that consider the chorology of plant or animal species. Other maps may consider the whole biota, i.e. the whole set of plant and animal species in a geographic area (Bănărescu and Boşcaiu 1978; Belov et al. 2002; Zunino and Zullini 2004; Ogureeva et al. 2010; Ladle and Whittaker 2011).

The largest phytogeographic subdivisions are the floristic kingdoms, recognized already by Drude (1890), Diels (1929) and others. These include the Holarctic, Paleotropical, Neotropical, Cape, Australian and Antarctic (Fig. 9.1); each kingdom

---

[1] There are several bases for botanical subdivision of the earth, including floristic, geobotanical and botanical-geographic. According to Karamysheva and Rachkovskaya (1975), floristic subdivision is based on systematic, geographic and genetic analysis of the flora and takes into account its qualitative and quantitative composition; geobotanical subdivision is based on data and geographical laws for plant distribution in relation to environmental conditions; and botanical-geographic subdivision combines floristic features and vegetation composition and structure. Thus, the phytogeographic maps described in this chapter are floristic maps (first type); the maps described in Chap. 10 (climatic-vegetational maps) are geobotanical maps (second type); and the maps described in Chap. 8 (catenal, or geo-synphytosociological maps) are botanical-geographic maps (third type).

F. Pedrotti, *Plant and Vegetation Mapping*, Geobotany Studies,
DOI 10.1007/978-3-642-30235-0_9, © Springer-Verlag Berlin Heidelberg 2013

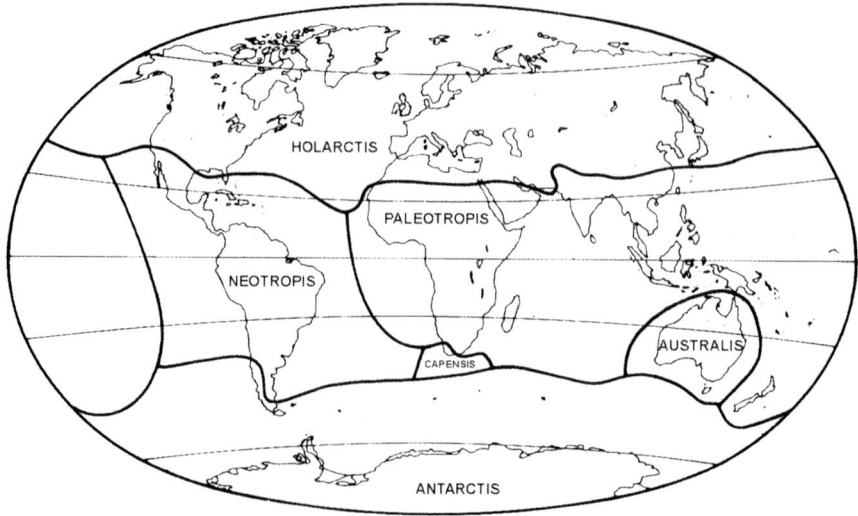

Fig. 9.1 Foristic kingdoms of the world, according to the classical conception

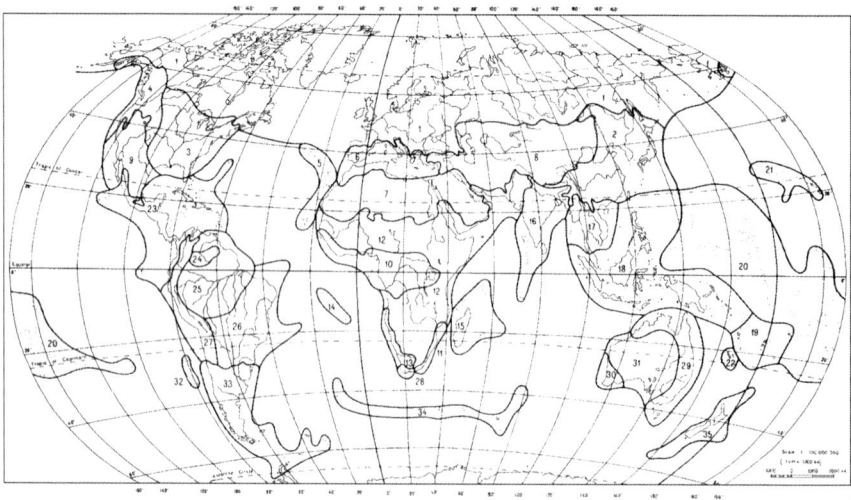

Fig. 9.2 Floristic regions of the world: *1* Circumboreal, *2* East Asiatic, *3* Atlantic North America, *4* Rocky Mountains, *5* Macaronesian, *6* Mediterranean, *7* Saharo-Arabian, *8* Irano-Turanian, *9* Madrean, *10* Guineo-Congolian, *11* Uzambara-Zululand, *12* Sudano-Zambesian, *13* Karoo-Namib, *14* S. Elena and Ascension, *15* Madagascar, *16* Indian, *17* Indochinese, *18* Malesian, *19* Fijian, *20* Polynesian, *21* Hawaiian, *22* New Caledonian, *23* Caribbean, *24* Guayana Highlands, *25* Amazonian, *26* Brazilian, *27* Andean, *28* Cape (of Good Hope), *29* Northeast Australian, *30* Southwest Australian, *31* Central Australian or Eremaean, *32* Fernandezian, *33* Chile-Patagonian, *34* Subantarctic Islands, *35* New Zealand (From Takhtajan 1986)

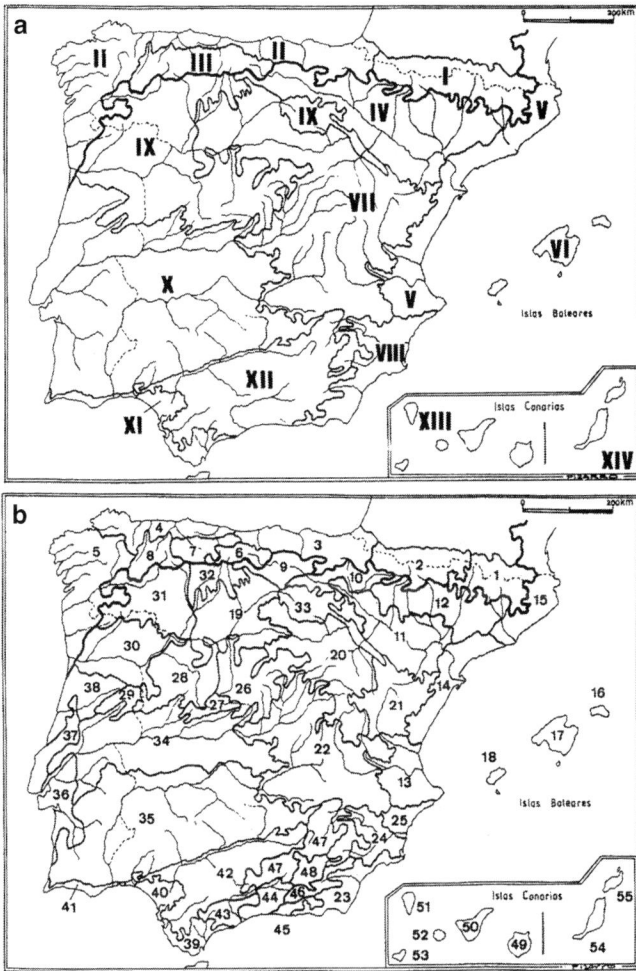

Fig. 9.3 Biogeopraphic subdivisions of Spain and Portugal. (**a**) Regions and provinces: *Eurosiberian region*: *I* Pyrenees, *II* Cantabro-Atlantic, *III* Orocantabrian. *Mediterranean region*: *IV* Aragonesian, *V* Catatalonia-Valencia-Provence, *VI* Balearic, *VII* Castile-Maestrazgo-Manchega, *VIII* Murcia-Almerian, *IX* Carpetania-Iberia-Leon, *X* Luso-Estramadura, *XI* Gaditano-Onubo-Algarve, *XII* Betic. *Macaronesian region*: *XIII* west Canarian, *XIV* east Canarian. (**b**) Sectors: the provinces are subdivided in 55 sectors (From Rivas-Martínez 1987)

can be subdivided into regions and these in turn into provinces (called dominions by some authors), sectors and districts, albeit with differences among various authors. Takhtajan (1986) distinguished the Holarctic, Paleotropical, Neotropical, Cape, Australian and Holantarctic kingdoms, but on his map of floristic regions (Fig. 9.2), only the regions are shown. A recent general biogeographic classification of the world by Rivas Martínez (2008) distinguishes the Holarctic, Paleotropical, Neotropical-Austroamerican, and Neozeland-Australian kingdoms. Compared with

Fig. 9.4 Distribution of some vegetation orders in Italy: *Vaccinio-Piceetalia* (Alps plus northern and central Apennines), *Fagetalia sylvaticae* (everywhere in Italian mountains except Sardinia), *Quercetalia ilicis* (coastal Italy, Apennines and Pre-Alps) and *Pistacio lentisci-Rhamnetalia alaterni* (Southern Italy) (From Pedrotti 1996a, modified)

preceding systems, the differences are the following: the Cape kingdom is included in the Paleotropical as the Cape region; the Antarctic kingdom is included in the Neotropical-Austroamerican; the Australian kingdom is called the New Zealand-Australian, combining Australia and New Zealand into a single kingdom; and various modifications are also made to the individual regions.

For the phytogeographic subdivision of Spain (Fig. 9.3), Rivas Martínez (1987) used the following rankings: kingdom, subkingdom, region, subregion, superprovince (group of provinces), province (dominion), subprovince, sector, subsector, district, subdistrict, landscape cell (group of *teselas*) and *tesela*. A recently published phyto-geographic map of Portugal, at scale 1:1,000,000, used the same criteria as for the map of Spain (Costa et al. 1998). The geo-synphytosociological cartography of Rivas Martínez (2005a, b) represents the most complete synthesis up to now, both floristically (phytogeographically) and vegetationally (phytosociologically), with the vegetation examined by sigmeta, each then assigned to a particular floristic subdivision.

The territory of Italy belongs to the Holarctic kingdom and to the two regions Euro-Siberia (also called circumboreal) and Mediterranean. It is not easy, though, to fix the boundaries in Italy between the two contiguous regions, and in fact the interpretations given by past authors have been quite different (Fiori 1908; Adamović 1933; Marchesoni 1958; Giacomini and Fenaroli 1958; Moggi 1969; Pignatti 1959, 1976, 1988a, 1994; Arrigoni 1983; Blasi 1994; Pedrotti 1996a; Blasi and Michetti 2005; etc.). Partly this is because the Mediterranean phytogeographic region has been confused with the Mediterranean bioclimatic zone, which has a different basis (see Chap. 10).

The division between the two phytogeographic regions may be made based on the distribution of vegetation orders formed by evergreen sclerophylls, namely the Pistacio-Rhamnetalia alaterni and Quercetalia ilicis, which predominate along the

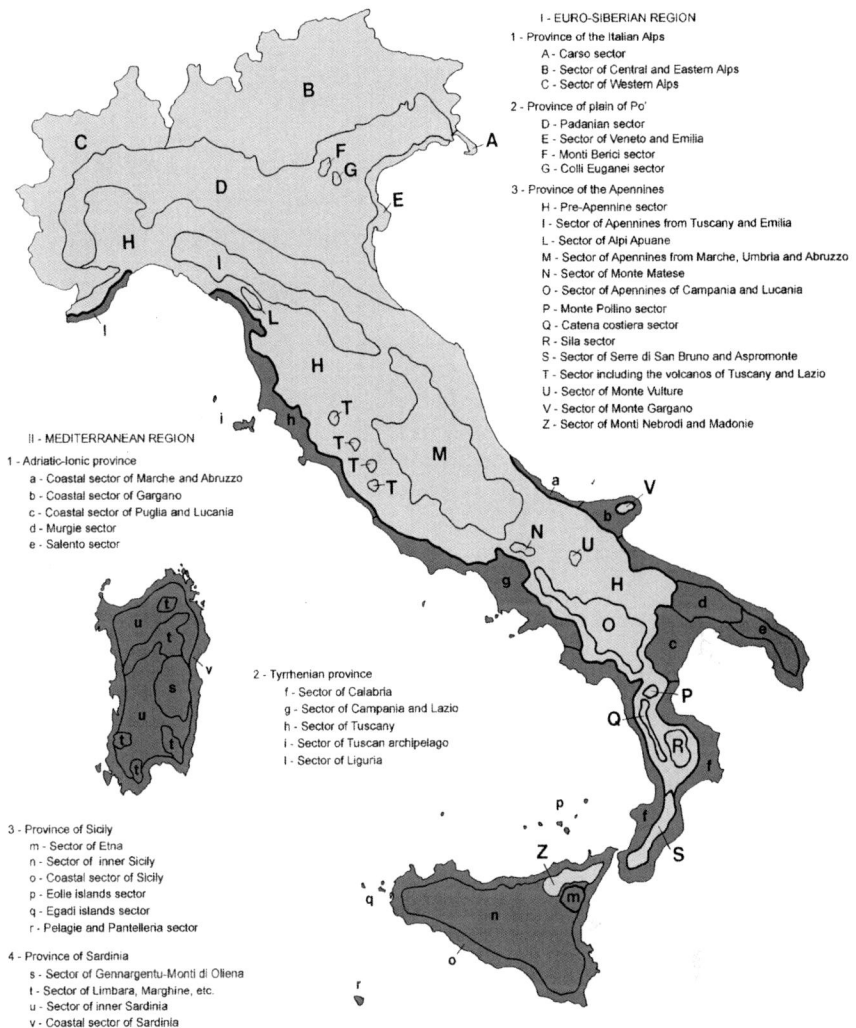

Fig. 9.5 Phytogeographical subdivision of Italy (From Pedrotti 1996a, modified)

coast but advance continuously inland only in Puglia, Lucania, Sicily and Sardinia, and in parts of Campania, Lazio and Tuscany (Fig. 9.4).

The further subdivision of Italy into provinces and sectors is shown in Fig. 9.5, on which districts have been identified only for some regions.

Mapping Vegetation Zones and Belts

10

General maps of the plant cover, understood to be a complex of different phytocoenoses, are also called maps of vegetation zonation or maps of vegetation regionalization. These are based on qualitative and quantitative characteristics of the plant (vegetation) formations, using quite different criteria (e.g. Lavrenko 1950, 1964; Doniţă 1979; Ivan 1979; Rivas Martínez 1987; Ozenda 1994; Ogureeva 1999).

Vegetation zonation is the process of delimiting zonal units of different rank, identified by bioclimate and roughly following latitude (*vegetation zones*) or altitude (*vegetation belts*).

A vegetation zone is a broad latitudinal band with a large east–west extent, characterized by particular morph-ecological plant types (plant life forms), a particular structural-functional type of phytocoenosis constructed by particular main plant species (e.g. edificators, or main structural types), a particular type of zonal soil, and a particular type of natural landscape. The zone is situated between particular latitudes and is characterized by a particular range of mean temperature and precipitation, to which the plants that populate the zone are adapted. The specific character of the zone is expressed under the conditions prevailing at low elevation, in general up to 300–500 m above sea level.

A vegetation belt is an altitudinal band in a colline-montane massif that possesses specific pedoclimatic and vegetation characteristics. Vegetation belts depend on temperature, which decreases more or less linearly with increasing elevation, at a rate of about 0.55 °C per 100 m in most land areas (Ozenda 1982). Vegetation belts become evident only above about 300–500 m, above the corresponding vegetation zone. The belts are recognized as different when the vegetation changes: alpine (above tree line), subalpine (just below tree line), montane (mid-mountain, but different from the more stressed subalpine vegetation and from the foothill vegetation), and colline (on the foothills, if that vegetation is different from that of the lowland zone).

Maps of vegetation zones are more useful over vast territories, such as continents. The map of the vegetation zones of European Russia, at scale 1:20,000,000, shows

F. Pedrotti, *Plant and Vegetation Mapping*, Geobotany Studies,
DOI 10.1007/978-3-642-30235-0_10, © Springer-Verlag Berlin Heidelberg 2013

zones for the polar desert, tundra, taiga (boreal coniferous forest), deciduous forests, steppe, and shrub steppe (Fig. 10.1).

Maps of vegetation belts on mountain ranges are of particular interest (Ozenda 1985), since the vegetation appears in successively higher bands following eleva-tion and the geomorphology of the mountain system; this can be seen on the vegetation map of Trentino-Alto Adige (south-central Alps), all of which is moun-tainous territory (Fig. 10.2).

In Italy the first maps of altitudinal belts were made by the Phytographic School of Firenze (Florence; Negri 1934, 1947, 1951, 1954) for various locations in the Apennines and Alps (Gavioli 1936; Zenari 1941; etc.). Subsequently, standard altitudinal belts were defined by the Phytosociological School of Pavia, and these are now used throughout Italy.

The climatic-physiognomic map of world vegetation by Tomaselli (1970b, 1977, 1981) is an example of the type of map treated in this chapter. This map shows 72 distinct ecological-structural types based on vegetation physiognomy and climate, each of which characterizes definite zones of the earth's surface. Four types are shown for Italy: (a) vegetation of the montane (subalpine) belt of the Eurasian mountain ranges; (b) mixed forests (broad-leaved deciduous plus evergreen coni-fer) of the humid cool-temperate zone or belt; (c) broad-leaved deciduous forests of the humid temperate zone; and (d) forests, maquis and garrigue of Mediterranean-type evergreen sclerophyll broadleaf taxa (locally with evergreen conifers).

The map of European bioclimates by Rivas-Martínez (1987) corresponds rela-tively well to the biomes or groups of biomes that follow the individual bioclimates: mediterranean (sclerophyll forests), temperate (deciduous forests and steppe), boreal (conifer forests) and polar (tundra).

Broad-scale maps can also involve vegetation types defined entirely by pheno-physiognomy (structure plus its seasonality), which are thus readily "visible" to satellite-based remote sensors. The map in Fig. 10.3 shows 14 world pheno-physiognomic vegetation types obtained by grouping 50 types predicted by climatic envelopes (Box 1995). Each of the 50 types was defined by structure and seasonal-ity, which formed the basis for grouping like types. The 50 types, called potential dominant vegetation (dominant over significant areas, e.g. zonal vegetation), were identified as necessary to cover the variation in world vegetation by means of a sort of "geographic regression" that added, deleted and grouped types as necessary until all areas seemed to be covered as well as possible.

Finally, vegetation regionalization consists of delimiting regional vegetation units, within a zonal unit, and identifying their areas by specific climatic indices, substrate, or longitude; Ivan (1979) called these units zones, regions and provinces.

Zonation of the Vegetation of Italy

Italy extends from 36 °N to 47 °N latitude and belongs to two vegetation zones, the Mediterranean (sometimes misleadingly called subtropical) and the Central Euro-pean (or nemoral). The Mediterranean zone is characterized by evergreen sclerophyll

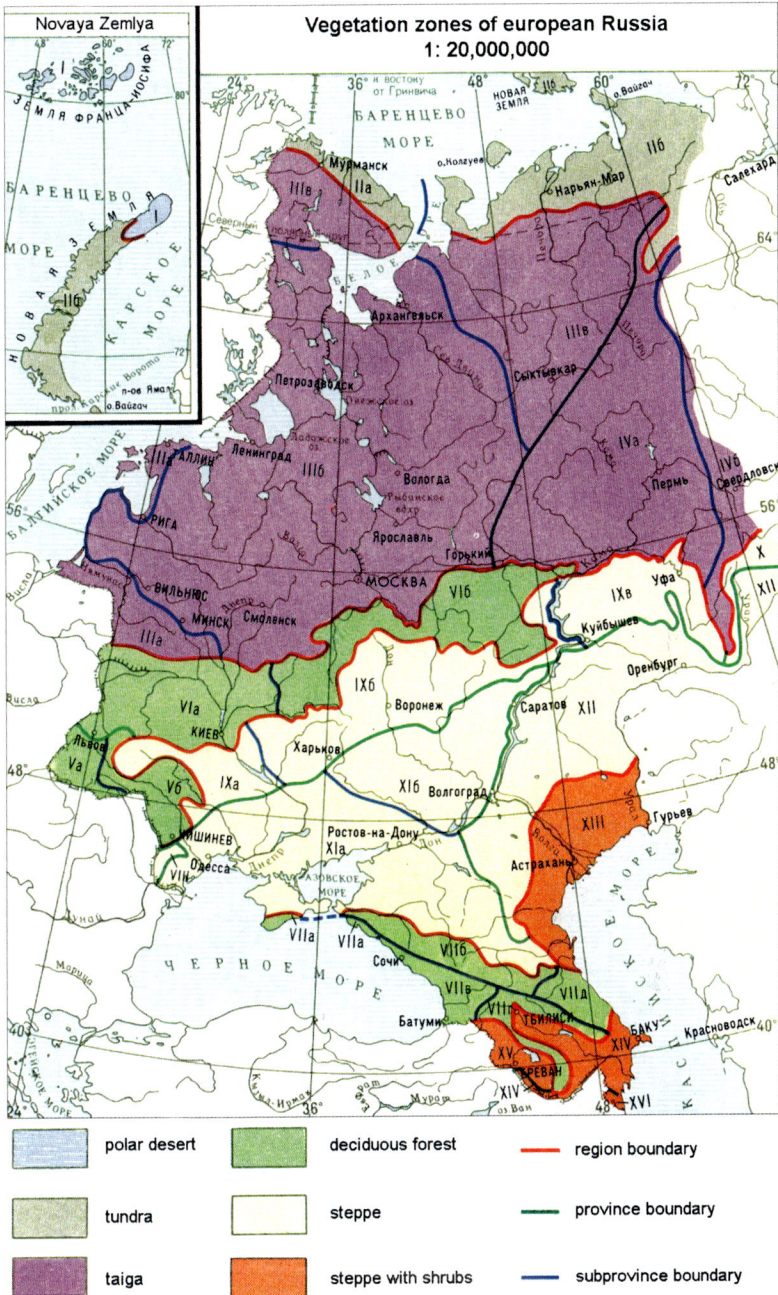

Fig. 10.1 Vegetation zones of European Russia: the map also shows the phytogeographical subdivision into regions, provinces and subprovinces (From Isachenko and Lavrenko 1979)

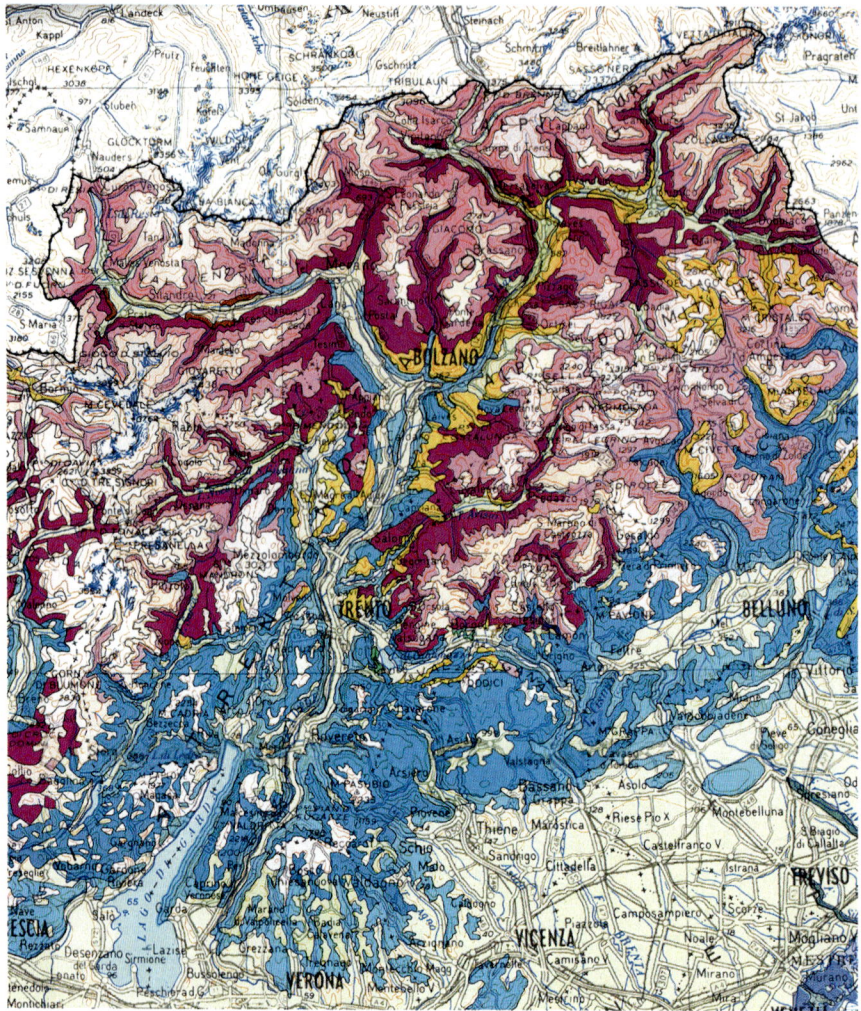

Fig. 10.2 Vegetation belts of the central Italian Alps, between the Alpine crests (continental divide) and the Po River plain: nival belt (pionieer vegetation), alpine belt (*Caricetalia curvulae*), subalpine belt (*Vaccinio-Piceetalia, Vaccinio-Piceion*), montaine belt (*Fagetalia sylvaticae, Eu-Fagion; Vaccinio-Piceetalia, Vaccinio-Abietion; Pulsatillo-Pinetalia, Ononido-Pinion; Erico-Pinetalia, Erico-Pinion*); and foothill belt (*Quercetalia pubescenti-petraeae; Quercetalia robori-petraeae; Fagetalia sylvaticae, Carpinion*); 1:1,000,000 (From Pedrotti 1992)

formations of thermo-mediterranean Rhamno-Pistacietalia and meso-mediterranean Quercetalia ilicis. The Central European zone comprises deciduous broadleaf formations of orders Quercetalia pubescentis, Quercetalia roboris and Fagetalia sylvaticae, plus the needle-leaved orders Vaccinio-Piceetalia and Erico-Pinetalia.

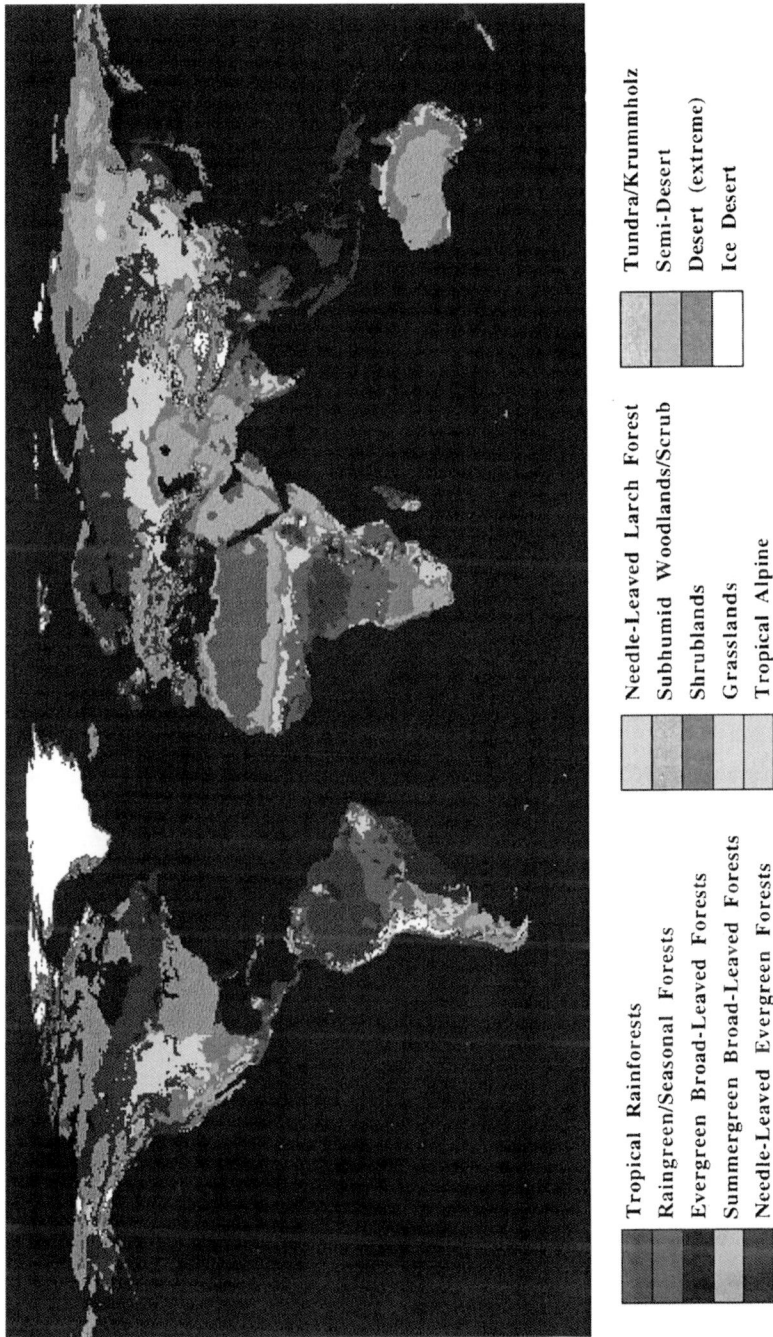

Fig. 10.3 World pheno-physiognomic vegetation pattern predicted from climate. The map shows 14 pheno-physiognomic vegetation types, obtained by grouping 50 types predicted by climatic envelopes (From Box 1995)

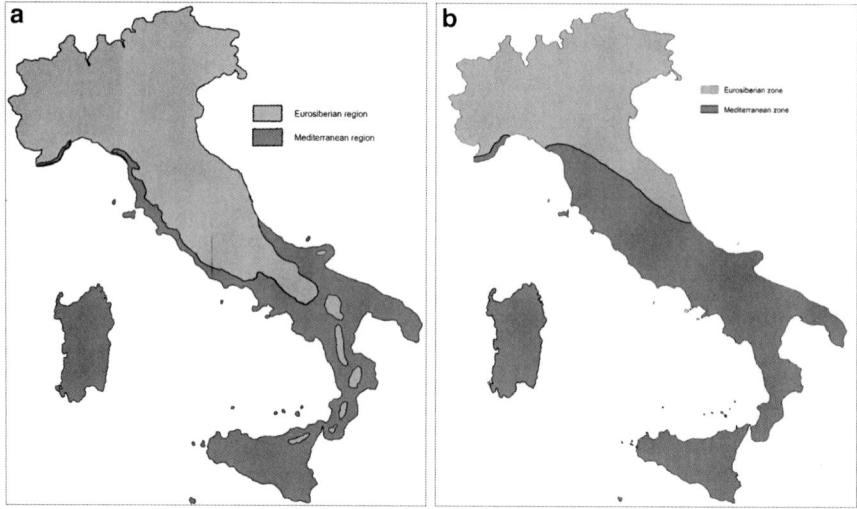

Fig. 10.4 (a) Phytogeographical map of Italy with subdivision into regions; the Eurosiberian region in southern Italy occur only in mountains; (b) phytoclimatical map of Italy, with subdivision into zones (From Pedrotti 1996a, modified)

The Central European zone includes the Alps, with elevation-related formations of evergreen needle-leaved trees of the orders Piceetalia excelsae, Athyrio-Piceetalia, Erico-Pinetalia and Pulsatillo-Pinetalia. The Apennine range, on the other hand, includes formations of deciduous broad-leaved trees of the orders Fagetalia sylvaticae and Quercetalia pubescentis. Sometimes there are also conifers, such as *Abies alba* in the montane belt and *Pinus halepensis* in the foothill belt. The order Vaccinio-Piceetalia is very localized, only in North Apennine.

Recognition of the boundary between these two zones in Italy is complicated by the long, narrow shape of the peninsula, with many individual mountain ranges. Also, the Alps are oriented mainly east–west but the Apennines mainly north–south. Because of the Apennines, the Mediterranean zone occurs mainly along the coast of the peninsula, though it may extend into the interior in some places. To the north, the Mediterranean zone reaches 42° 30′ N on the Adriatic side and about 44 °N on the Tyrrhenian side (Fig. 10.4).

Thus, the subdivisions are climatic-vegetational in character and can only be recognized theoretically, following the mountainous relief; these can be imagined as projections in a band that extends from sea level up to about 300 m. In fact, if we consider also the relief, the Euro-Siberian region reaches all the way to Aspromonte in southern peninsular Italy, with some outliers on mountains even in Sicily; this, though, is a phytogeographic rather than a climatic-vegetational subdivision. Figure 10.4 attempts to clarify the relation between climatic-vegetational zones and phytogeographic zones in Italy.

The altitudinal belts in the Central European zone are: a foothill belt [mesotemperat] (forests of deciduous trees such as *Carpinus betulus, Quercus robur, Quercus petraea, Quercus pubescens, Castanea sativa, Ostrya carpinifolia* and *Fraxinus ornus*); a montane belt [supratemperat] (forests of deciduous *Fagus sylvatica* and needle-leaved *Abies alba, Pinus sylvestris, Pinus nigra* and *Picea abies*); a subalpine belt [orotemperat] (forests of needle-leaved *Picea abies, Larix decidua, Pinus cembra* and shrubs, including *Pinus mugo, Pinus uncinata, Rhododendron ferrugineum, R. hirsutum, Empetrum nigrum, Vaccinium gaultherioides* and *Loiseleuria procumbens*); an alpine belt [cryo-orotemperat] (meadows with *Festuca halleri* and *Carex curvula* on siliceous substrates or with *Sesleria albicans, Carex firma* and *Dryas octopetala* on calcareous-dolomitic substrates, plus alpine tundra with *Polytrichum norvegicum, Salix herbacea, Salix retusa* and *Salix reticulata*) and finally a nival desert with pioneer vegetation involving *Andreaea nivalis, Andreaea rupestris*, etc. An example of altitudinal zoning of the Alps chain is provided in Fig. 10.4, which refers to the Trentino-Alto Adige Region.

The altitudinal belts in the Mediterranean zone (sensu Rivas MartÍnez 1996; see also Biondi and Baldoni 1995 and Blasi and Michetti 2005) are: infra-mediterranean (*Periploca laevigata* ssp. *angustifolia* and *Lycium intricatum*), present only in the Canale di Sicilia, which separates Europe from Africa; thermo-mediterranean (*Olea sylvestris, Pistacia lentiscus, Rhmanus alaternus, Ceratonia siliqua, Chamaerops humilis*), distributed along the coast in southern Italy (Puglia, Lucania, Calabria, Campania, Sicily, Sardinia); meso-mediterranean (*Quercus ilex, Quercus suber, Q. virgiliana, Q. trojana*), present along the coast from the Adriatic See (from Monte Conero southwards), and along the Tyrrhenian coasts; in Sardinia and Sicily this altitudinal belt concerns most of the internal sectors; supra-mediterranean (*Quercus ilex, Q. pubescens*), in some internal parts of Basilicata and Puglia (Gargano), Sicily and Sardinia; oro-mediterranean (*Abies nebrodensis*, etc.) the mountains of Sicily and Sardinia; cryo-oro-mediterranean (*Astragalus siculus*, etc.) on Mount Etna.

Mapping Plant Biodiversity   11

Plant biodiversity is defined by the variety in all the plant-taxonomic categories (species, genera, families, orders, etc.) and by the variety of ecosystems and plant communities to which they belong (Wilson 1992). One may distinguish the following levels of biodiversity, according to a hierarchy of increasing spatial extent: $\alpha - diversity$ (number of species present in a single community, or local species richness); $\beta - diversity$ (variation in the number of species along an environmental or geographic gradient); $\gamma - diversity$ (number of species present in a relatively wide area, or regional species richness); and $\delta - diversity$ (number of species present on a continent). The study of biodiversity today constitutes a principal theme in conservation biology (Primack 2002; Blasi 2005; Poldini 2009) and is at the center of a grand theoretical and practical discussion about environmental implications and the survival of species. In this chapter we introduce briefly some aspects relative to the possibility of mapping biodiversity.

Estimation of plant biodiversity is done first by counting all the plant species in the area of interest. Such areas may be predefined areal units subdivided by a grid or units representing some classification of vegetation (phytocoenoses, etc.) or environments (environmental units, etc.). Barthlott et al. (1999) speak of *autodiversity* when referring only to autochthonous species and *allodiversity* when also including allochthonous species introduced by man. This distinction is important because the number of species can increase significantly going from a natural environment to one affected by man (or other frequent disturbance). In this latter case one can speak of *negative* biodiversity. Knowledge of data on allodiversity (negative biodiversity) can be used for scrutiny of anthropogenic changes in the vegetation produced over the years.

Quantitative evaluation of biodiversity may be done iteratively by means of synthetic indices, of which many have been proposed (e.g. Magurran 1988; Ferrari 2001; Wilson 1992); the most complete manual of sampling methods for "measuring" the various levels of biodiversity is that by Stohlgren (2007), which also outlines the classical phytosociological method, albeit generally.

Floristic sampling by phytosociological relevés, always done for the smallest areas that capture all the vegetation characteristics, permits comparing the biodiversity of

F. Pedrotti, *Plant and Vegetation Mapping*, Geobotany Studies,
DOI 10.1007/978-3-642-30235-0_11, © Springer-Verlag Berlin Heidelberg 2013

233

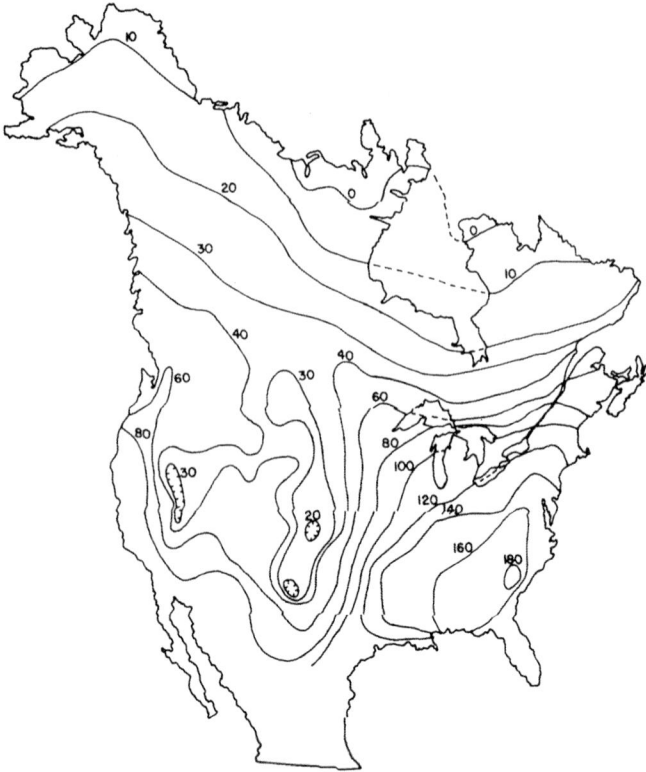

Fig. 11.1 Species richness of trees in Canada and the United States: the contours connect points with the same approximate number of species per quadrat (From Currie 1991)

very different plant formations (forests, woods, scrub, grasslands, etc.) by means of the specific richness and relative abundance (frequency) of the species.

Now comes the problem of the methodology to follow for mapping the data obtained. There are two main ways, by isolines or by symbols or colors for the cells of a pre-established grid. Representation by isolines provides a spatially continuous distribution of biodiversity levels, since each isoline represents a precise value. This is shown very well on the map of tree species richness of Canada and the USA (Fig. 11.1) (Currie 1991) and on maps of North and South America that show the numbers of species or genera of Cactaceae (Fig. 11.2) (Mutke and Barthlott 2005).

Mapping by symbols and colors, on the other hand, produces a spatially discontinuous distribution of biodiversity levels, since the grid cells are discrete and each color or symbol corresponds to a range of values over the area of its cell.

For the province of Palermo (northern Sicily), the biodiversity of the vascular flora has been evaluated in pre-established areas (Fig. 11.3). The territory was subdivided into a grid, in each cell of which the endemics and all plant species present were counted. Biodiversity was mapped by assigning each cell to one of eight classes, each representing a range in number of taxa, increasing from fewer

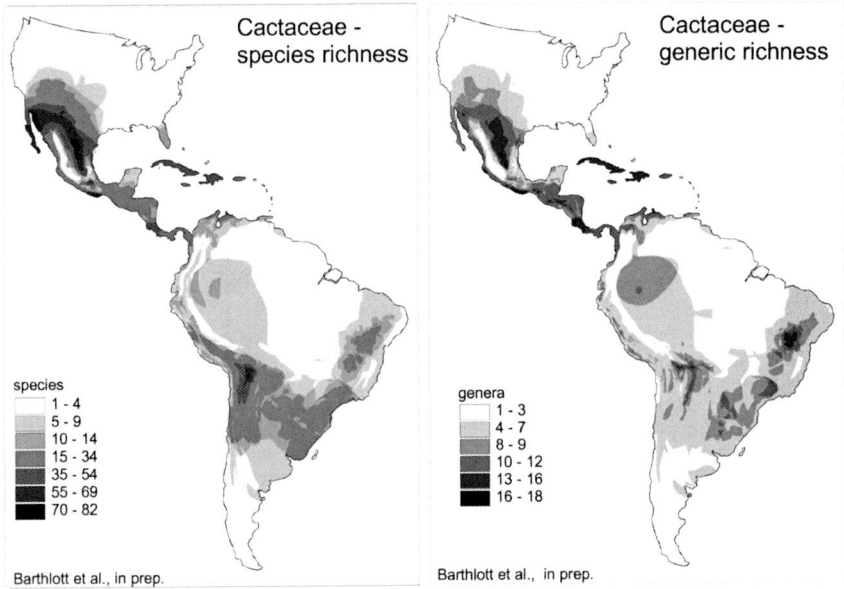

Fig. 11.2 Species richness (*left*) and genus richness (*right*) of *Cactaceae* in the new world (where *Cactaceae* are native) (From Mutke and Barthlott 2005)

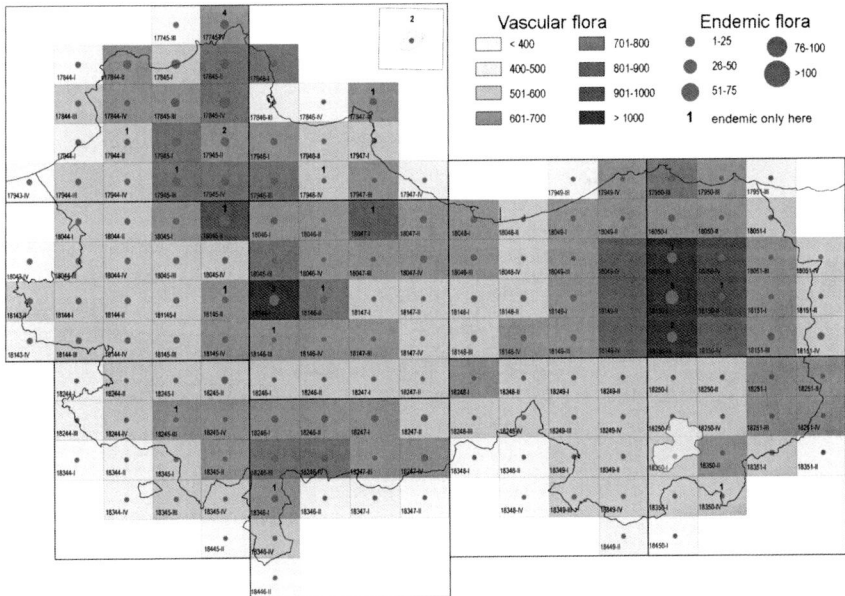

Fig. 11.3 Numbers of endemic and other vascular species per quadrat, by classes, in Palermo province of Sicily (From Raimondo 2000)

Observed species richness

Number of species

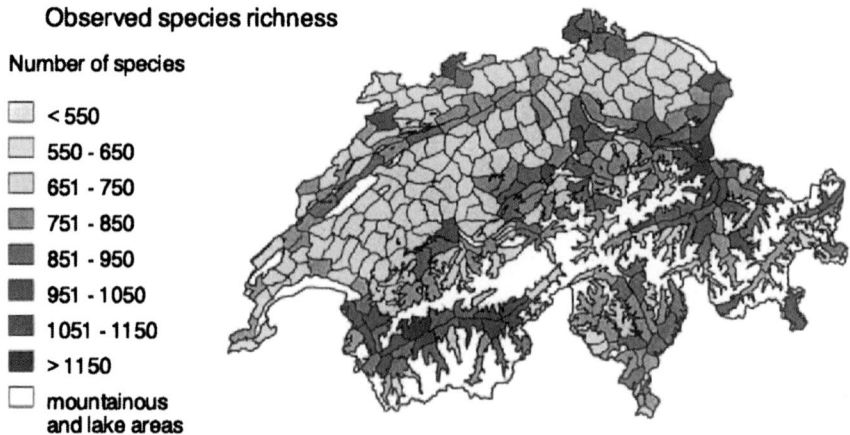

- ☐ < 550
- ☐ 550 - 650
- ☐ 651 - 750
- ☐ 751 - 850
- ■ 851 - 950
- ■ 951 - 1050
- ■ 1051 - 1150
- ■ > 1150
- ☐ mountainous
  and lake areas

Fig. 11.4 Observed species richness in Switzerland, based on mapping (1967–1979) and two additional supplements (From Wohlgemuth 1998)

than 400 to more than 1,000 (Raimondo 2000). The starting point for creating the map is then constituted by chorological maps (see Chap. 4).

Many maps of floristic richness have been made in different parts of the world, including the species richness of Switzerland (Fig. 11.4) by Wohlgemuth (1998), using data taken from the atlas of the Swiss flora by Welten and Sutter (1982), as illustrated in Chap. 4 by Fig. 4.17; the richness of Spermatophytes and Pteridophytes of part of the Białowieza forest (Poland), subdivided into quadrats of 100 m on a side, and of the forest associations (Fig. 11.5) (Faliński and Mulenko 1995); the species richness of Spain by Lobo et al. (2001), based on 254 UTM grid cells (Fig. 11.6); the richness of vascular plants in California (Fig. 11.7) (Williams et al. 2004) the distribution of vascular plants of Austria by Moser et al. (2005) evaluated in sampling areas of 5 × 3 arc minutes; and the species richness of South Africa by Thuiller et al. (2006) based on a mesh of 25 × 25 km.

The data for the maps just cited can also be used further. Pignatti (2004), for example, starting from a map of the floristic richness of northeastern Italy, calculated the relation between the total frequency of certain kinds of species (Eurasian, endemic, etc.) and all species counted in each cell. The fewest Eurasian species were found in an area of continental climate, while the highest percentage of endemic species was found on nunataks or other areas of refuge during the glaciations (Fig. 11.8).

Pignatti (2004), however, also criticized the use of quantitative measures based on number of species, noting that biodiversity results substantially from relations between organisms; thus, to have a valid quantitative measure, it would be necessary to count these relationships too. Applying very sophisticated methodologies like the "tree alpha-diversity" (Fishers' alpha), based on local spatial regression (Venables and Ripley 1997; Kaluzny et al. 1998; see Stropp et al. 2009), one can obtain good results, such as for tree species diversity in Amazonia (Stropp 2011) (Fig. 11.9).

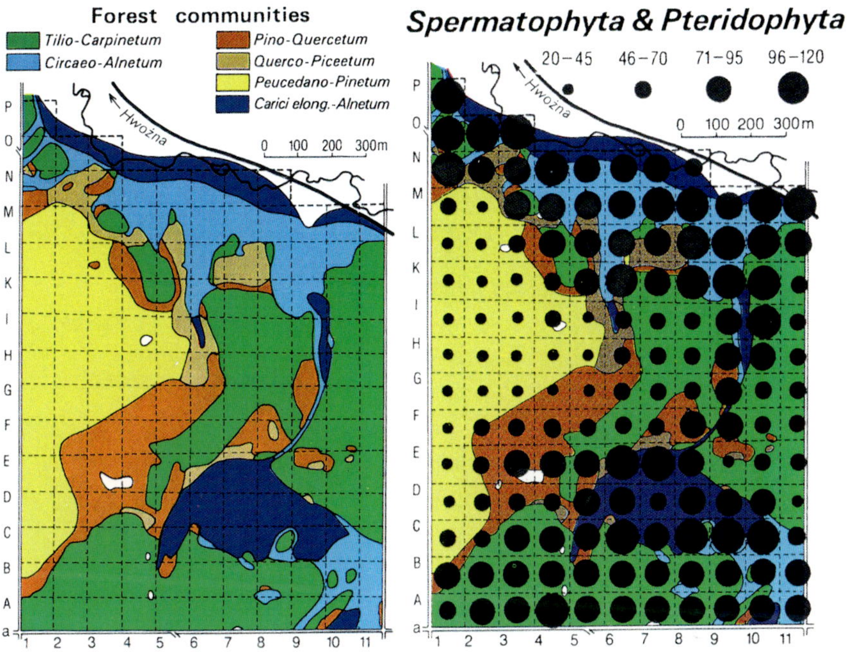

Forest communities

- Tilio-Carpinetum
- Circaeo-Alnetum
- Pino-Quercetum
- Querco-Piceetum
- Peucedano-Pinetum
- Carici elong.-Alnetum

0 100 200 300m

Spermatophyta & Pteridophyta

20–45 46–70 71–95 96–120

0 100 200 300m

Fig. 11.5 At *left*, the forest communities in part of the Białowieza primeval forest, Poland; at *right*, the number of species in each square (From Faliński and Mulenko 1995)

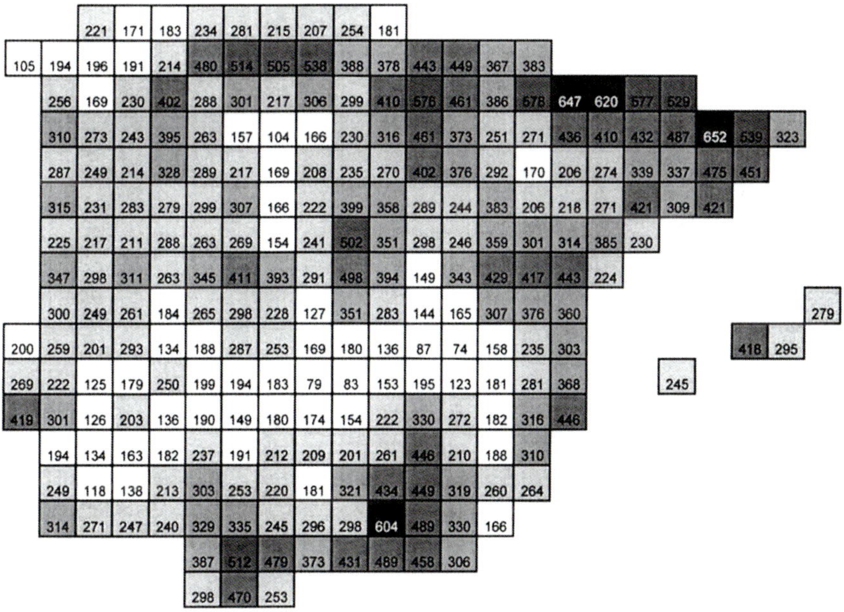

		221	171	183	234	281	215	207	254	181												
105	194	196	191	214	480	514	505	538	388	378	443	449	367	383								
	256	169	230	402	288	301	217	306	299	410	578	461	386	678	647	620	577	529				
	310	273	243	395	263	157	104	166	230	316	461	373	251	271	436	410	432	487	652	539	323	
	287	249	214	328	289	217	169	208	235	270	402	376	292	170	206	274	339	337	475	451		
	315	231	283	279	299	307	166	222	399	358	289	244	383	206	218	271	421	309	421			
	225	217	211	288	263	269	154	241	502	351	298	246	359	301	314	385	230					
	347	298	311	263	345	411	393	291	498	394	149	343	429	417	443	224						
	300	249	261	184	265	298	228	127	351	283	144	165	307	376	360							279
200	259	201	293	134	188	287	253	169	180	136	87	74	158	235	303					418	295	
269	222	125	179	250	199	194	183	79	83	153	195	123	181	281	368			245				
419	301	126	203	136	190	149	180	174	154	222	330	272	182	316	446							
	194	134	163	182	237	191	212	209	201	261	446	210	188	310								
	249	118	138	213	303	253	220	181	321	434	449	319	260	264								
	314	271	247	240	329	335	245	296	298	604	489	330	166									
				387	512	479	373	431	489	458	306											
				298	470	253																

Fig. 11.6 Number of vascular plant species recorded in each of the 254 UTM grid cells of the Iberian Peninsula. The gray scale changes at 200 species (From Lobo et al. 2001)

Vascular plant richness

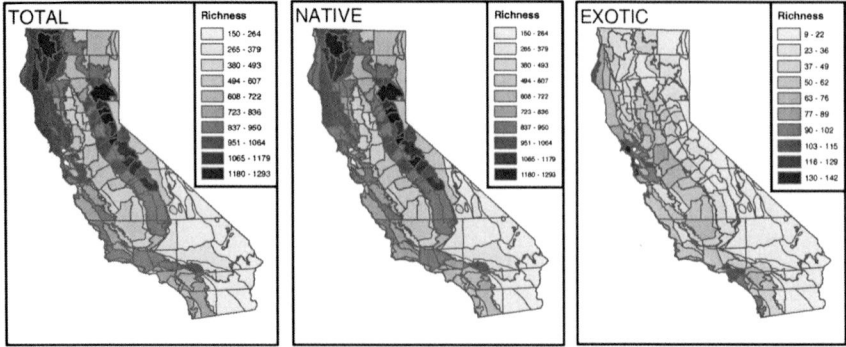

Fig. 11.7 Vascular plant species richness of California for (*left*) native plus exotic species, (*middle*) native species only, and (*right*) exotic species (From Williams et al. 2004)

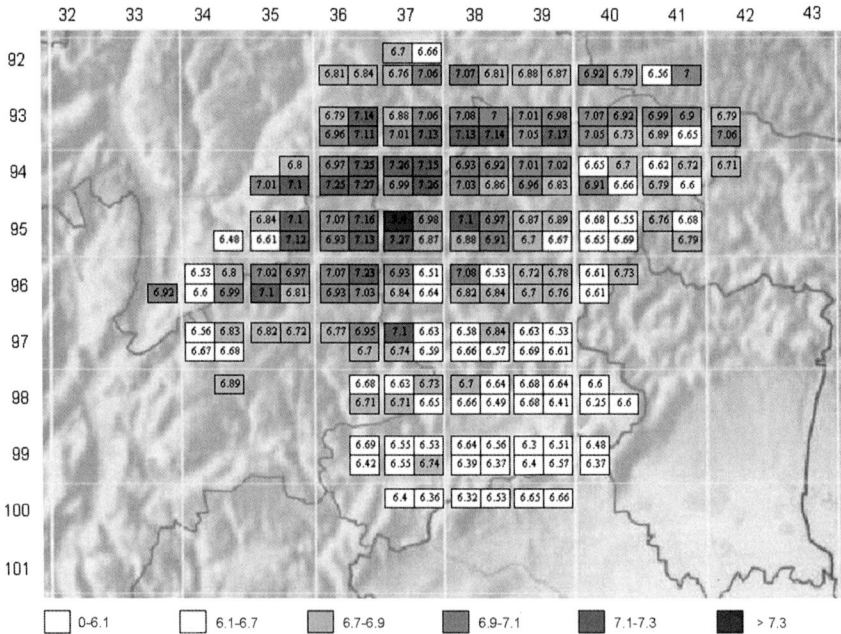

Fig. 11.8 Frequency of Eurasian species compared to the total number of species listed in each OGU in the Dolomiti Mountains of northern Italy (From Pignatti 2004)

Returning briefly to the map of the Palermo province, we note that the number of endemic species present is also given for every quadrat (Fig. 11.3). Two areas were indicated as possessing the highest diversity of plant species, namely the Madonie and Rocca Busambra mountain areas.

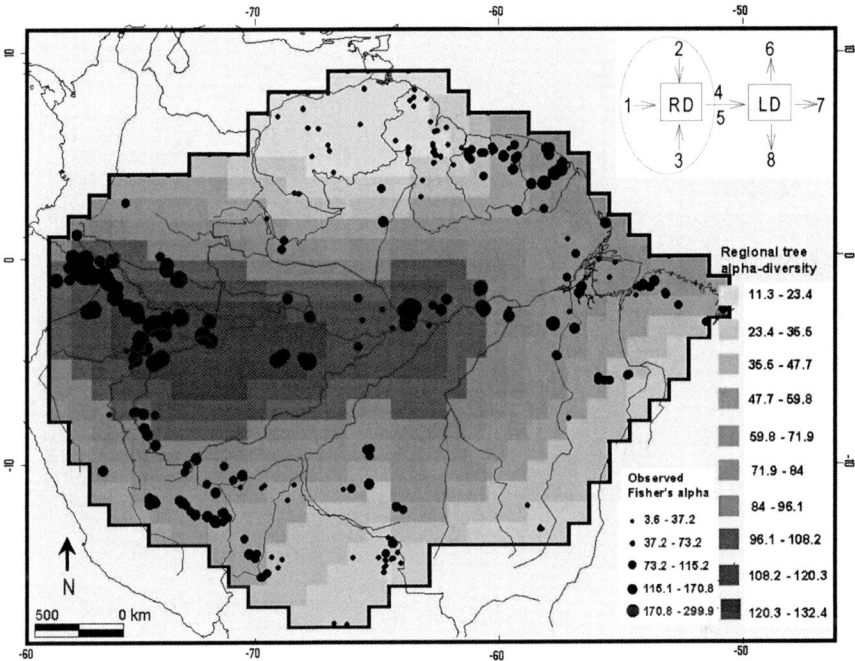

Fig. 11.9 Regional variation of tree alpha diversity (Fishers' alpha) based on a spatial regression. The tree alpha diversity of 752 one hectare inventory plots was modeled as a function of latitude and longitude, and mapped on a one-degree grid Amazonia (From Stropp 2011)

It is thus possible to identify, for areas of increasing size, the areas that have the highest biodiversity. An example for Italy is the map of areas richest in endemic species, produced by Pignatti (1982b) (Fig. 11.10).

At global scale, "biodiversity hot spots" have been identified, i.e. the land areas in the world most important for their biodiversity but gravely threatened (Mittermeier et al. 1999; Myers et al. 2000). This evaluation was made based on the number of endemic vascular species, which in 25 very rich areas of the world account for percentages up to 81.1 % (New Zealand), 80.9 % (Madagascar and Indian Ocean), and 76.5 % (Philippines).

It remains to examine diversity at the level of plant communities and ecosystems; also for this the problem can be addressed by counting the plant communities present in pre-established cells of a grid. Synchorological and integrated phytosociological mapping (mapping of vegetation series) are particularly useful as a starting point.

Trentino-Alto Adige (central Alps of northern Italy) has 95 vegetation series present, to which can be added the vegetation series of wet environments (lakes, mires, etc.), which are not considered here. This large number of series is the result of several factors: long north–south extent of the territory, parent bedrock types, geomorphological conditions, and phytoclimate. The complexity and great phytogeographic diversity of the territory are related also to the very steep gradient

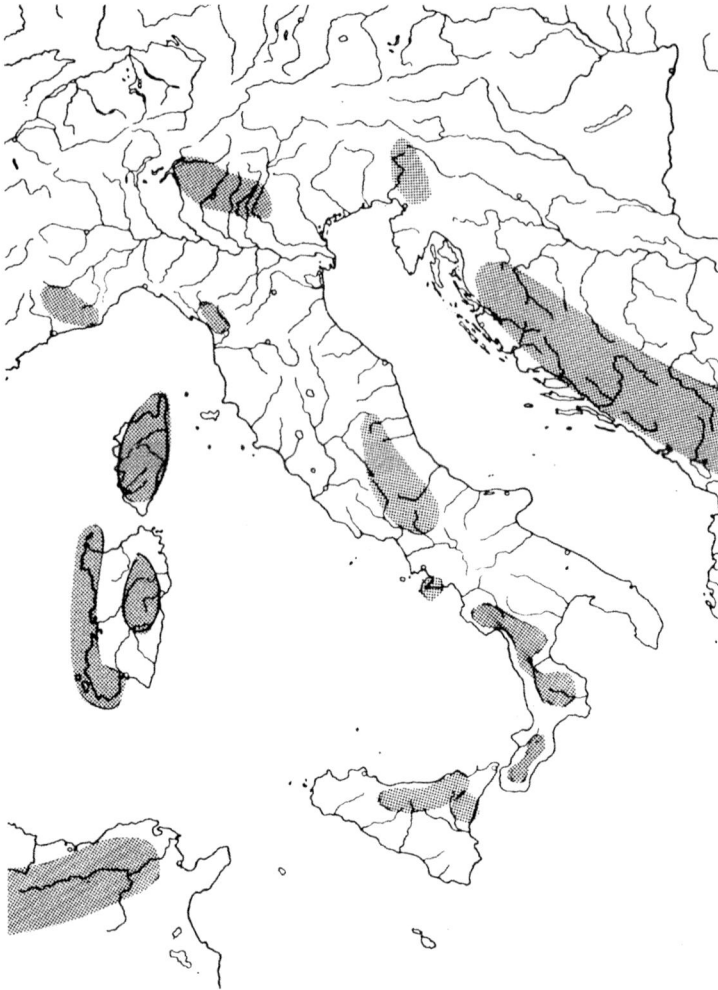

Fig. 11.10 Areas of Italy rich in endemic species (From Pignatti 1982b)

from pre-Alpine to Inner-Alpine sectors, yielding a flora with high endemism and many vegetation series and mega-geoseries. The series occur as follow in the various altitudinal belts: nival belt 2, alpine belt 11, subalpine belt 15, montane belt 26, and colline belt 12, to which can be added 23 azonal series. The largest number of series (26) is found in the montane belt, where the altitudinal range is wider (1,000–1,800 m) and where the greatest number of environments is found, which favors the development of more intrazonal series. The species that occurs in the most series is *Pinus sylvestris*, with ten series, of which one is zonal, one is extrazonal and the others are intrazonal or azonal. Another species with a wide ecological adaptation, albeit limited to north-facing slopes, is *Abies alba*, appearing in eight vegetation series. *Fagus sylvatica* occurs in six series. On the other hand, the two typical conifers of the Alpine and Inner Alpine sectors, *Picea abies* and *Pinus cembra*, only enter into five and three series respectively; this confirms the ecological and phytogeographical specialization especially of *Pinus cembra*. The same can be said for *Larix decidua*, which appears in only three series. In the colline belt, *Carpinus betulus* and *Quercus petraea* occur in only one vegetation series each. *Alnus glutinosa* occurs in the largest number of azonal series (6), all riparian or wetland series (Pedrotti 2010).

These data have been mapped on a square-grid map like those employed for distribution maps of the associations of *Pinus sylvestris* (Fig. 6.65).

Applied Geobotanical Mapping

12

Geobotanical maps are being used more and more for applied problems such as environmental management, land-use planning, nature conservation, management of protected areas, agriculture, silviculture, evaluation of environmental impacts, and so on.

Vegetation also performs important functions within ecosystems as well as for human benefit, albeit generally at more local scales. Control, recycling and production are just a few obviously important ecosystem services that must be replaced after extinctions or other ecosystem degradation (Ehrlich and Mooney 1983). Functions of vegetation in built and other modified landscapes include providing green space for wildlife, human recreation and conservation education in urban areas; green screens for localizing industrial effects or buffering schools or residential areas against disturbances; provision of escape routes and refuge during earthquakes; and even serving as barriers to the spread of large urban fires (cf the "environmental protection forests" of Miyawaki et al. 1987). Functions of vegetation in both natural and modified landscapes in Japan have been presented on maps by Fujiwara et al. (1991).

Experience has shown that the simultaneous use of phytosociological maps of actual vegetation and dynamic tendencies of the vegetation, of integrated phytosociological maps, and of phytosociological maps of potential vegetation is especially useful (see Chap. 6).

Of these four map types, and adding other specific data, it is also possible to derive (Pedrotti 1983):

(a) Inventory maps (or maps of actual land use)
(b) Predictive maps (or maps of potential land use)
(c) Planning maps (or maps of recommended land use).

These three map types have eminent practical value for problems of environmental management. In most cases, though, such maps are not published but remain in the institutions that requested them. Also, both the classifications on these maps and the mapping methods are quite variable, since precise codification and conventions are lacking.

F. Pedrotti, *Plant and Vegetation Mapping*, Geobotany Studies,
DOI 10.1007/978-3-642-30235-0_12, © Springer-Verlag Berlin Heidelberg 2013

Inventory Mapping

For that part of a land or other resource inventory that involves vegetation, *inventory maps*, also called *diagnostic* maps, can be derived from phytosociological maps of actual vegetation. These are essentially maps of land use and illustrate the "actual use" of a given territory. With these it is possible to evaluate the current situation of the vegetation and its conservation status, for example, the distribution of wooded areas, percentage of land covered by natural vegetation, or the continuity or fragmentation of the vegetation.

Cadastral maps constitute an older example of land-use map and may show, for individual land holdings (parcels), the form of cultivation, such as woods, pasture, or cultivated fields. Often these are represented by different colors, as in the cadastral maps at scale 1:2,880 of Trentino appearing in the second half of the 1700 at the time of Maria Teresa (Fig. 12.1). The cadastral maps of the Papal State are also in colors, with the addition of symbols of trees and of the various forms of agricultural cultivation, as on the maps of Abbadia di Fiastra in 1722, those of connected "possessions" (Fig. 12.2), and of the Castello di Lanciano near Camerino at the beginning of the 1800s (Aleffi 1987, 1993, 1994). The *Carta delle pinete ravennati* of Ginanni (1774) has characteristics similar to those of cadastral maps; all these documents, and many others of various parts of Italy that could be cited, have preceded the production of true geobotanical maps.

Figure 12.3 shows a fragment of the *Carta di uso del suolo della Regione Abruzzo* (map of land use of the Abruzzo region) produced at scale 1:25,000, using map sheets (*tavolette*) from the Italian *Istituto Geografico Militare* plus photointerpretation (Regione Abruzzo 2000). A similar land-use map of Lazio was produced using identical criteria (Regione Lazio 2003).

The *Carta della montagna of Italy* at scale 1:500,000 is a very generic land-use map that shows only woods, pastures and permanent meadows, cultivated fields and valley meadows, unproductive uncultivated areas, surface rocks and glaciers, and main urban and industrial areas (Ministero Agricoltura Foreste 1976). More detailed land-use maps of the various *Regioni* have also been made but are not coordinated with each other as can be seen if maps of contiguous regions are placed side by side.

When necessary, one can also make maps of past land use, relying on cadastral maps that arose at various times and from which it is possible to deduce ancient forms of cultivation. One example is a map of the current Torricchio Nature Reserve (near Camerino), for which it was possible to reconstruct the situation at the time of the 1833 and 1943 cadastres (Fig. 12.4) and to ascertain the changes up to today, such as the abandonment of some cultivated areas and the return of forest (Pedrotti 1978); other examples are the maps of Monte Cardosa and Monte Letegge, near Camerino (Rosi 2005; Catorci and Gatti 2007).

Forestry maps represent areas occupied by "forest" vegetation, which often, however, also includes open or shorter woods and macchia scrub. Often shown are dominant tree species (sometimes with more detailed classifications), the architecture of the forest (coppice, large old trunks, etc.), the annual growth rate and resulting standing wood in production forests, the potential forest vegetation, etc. These are

Fig. 12.1 Ancient cadastral map from 1780 of Laghestel di Piné, Trentino-Alto Adige Region, at scale 1:2,880 (From Catasto Teresiano, *Archive of "Libro fondiario e Catasto", Trento*)

considered land-use maps and are not rigorously full-floristic maps representing actual vegetation. The *Carta della vegetazione forestale della Toscana* (map of the forest vegetation of Tuscany), at scale 1:250,000, shows 18 types of forest vegetation, distinguished by physiognomy: evergreen oak forests (*Quercus ilex*), beech forests

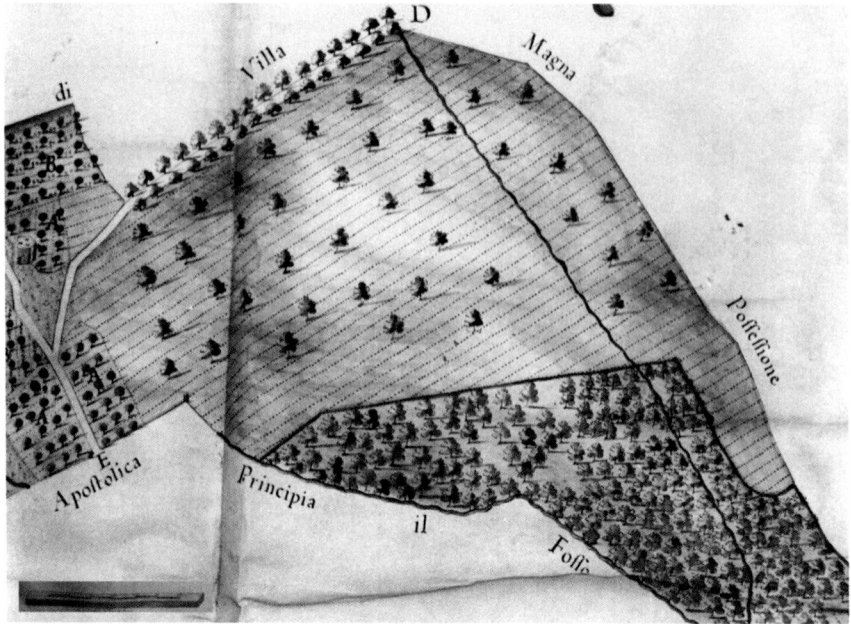

Fig. 12.2 Ancient cadastral map from 1722 of the San Pietro, Marche Region, central Italy, showing different land uses: forest of *Quercus pubescens*, vineyards, cultures with isolated trees of *Quercus pubescens*, and road with *Quercus pubescens* (From Catasto Gregoriano, *Archive of Fiastra Abbey*)

(*Fagus sylvatica*), locust stands (*Robinia pseudacacia*, an aggressively invasive tree from North America), pure pine forests and those mixed with indigenous species, etc.; to this is added a *Carta della vegetazione forestale potenziale* (map of the potential forest vegetation) at the same scale (Regione Toscana 1998a, b).

Maps of green spaces in urban areas, or of individual parks and gardens, can also be seen as maps of land use; for example, the *Carta della vegetazione del Parco della Favorita* (vegetation map of the Favorita Park, in Palermo), shows the flower beds with then-current cultivations, as well as orchards, gardens, ornamental plantings, and cypress or conifer groves, etc., plus remaining strips of natural vegetation such as garrigue, macchia, and existing evergreen oakwoods on some rocky areas inside the park (Buffa et al. 1986).

Finally, we note that in 1985 the Council of the European Community approved the program COR.IN.E (Coordination de l'Information pour l'Environnement), which envisions mapping land use for all countries of the European Community using satellite imagery. Italy is in the process of making land-use maps at scales 1:100,000 and 1:250,000 using Landsat TM imagery.

The COR.IN.E Biotopes program, on the other hand, intends to inventory the sites of greatest importance for nature conservation in the European Community (Commission European Communities 1991), but so far the nomenclature and classification have not been checked with complete satisfaction.

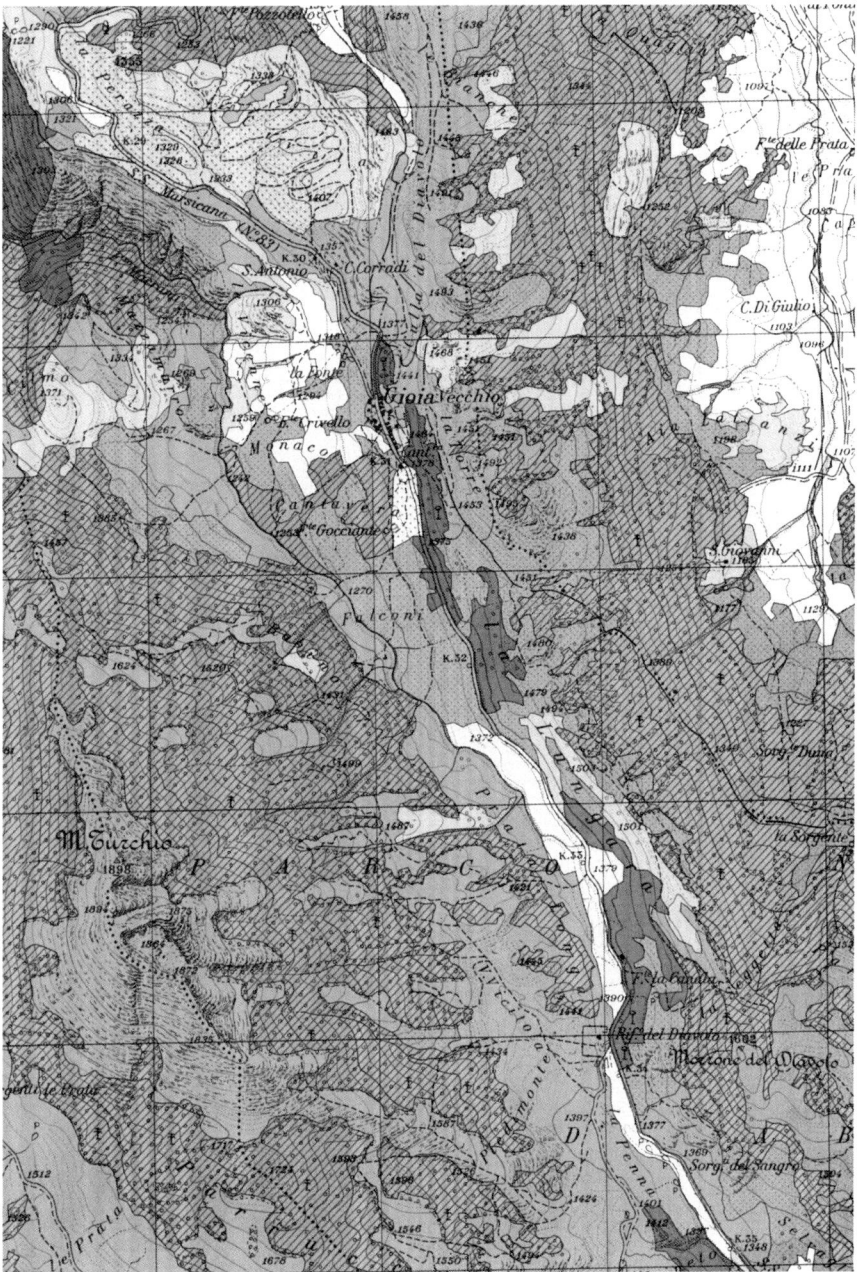

Fig. 12.3 Land use map of the Abruzzo Region, sector of Sangro Valley (From Regione Abruzzo 2000)

Fig. 12.4 Land use in the
Torricchio Nature Reserve,
Marche Region, central Italy,
at three times in the past:
1833 (Catasto Salimbeni),
1943 (Catasto Italiano) and
1973 (Catasto Italiano) (From
Pedrotti 1978)

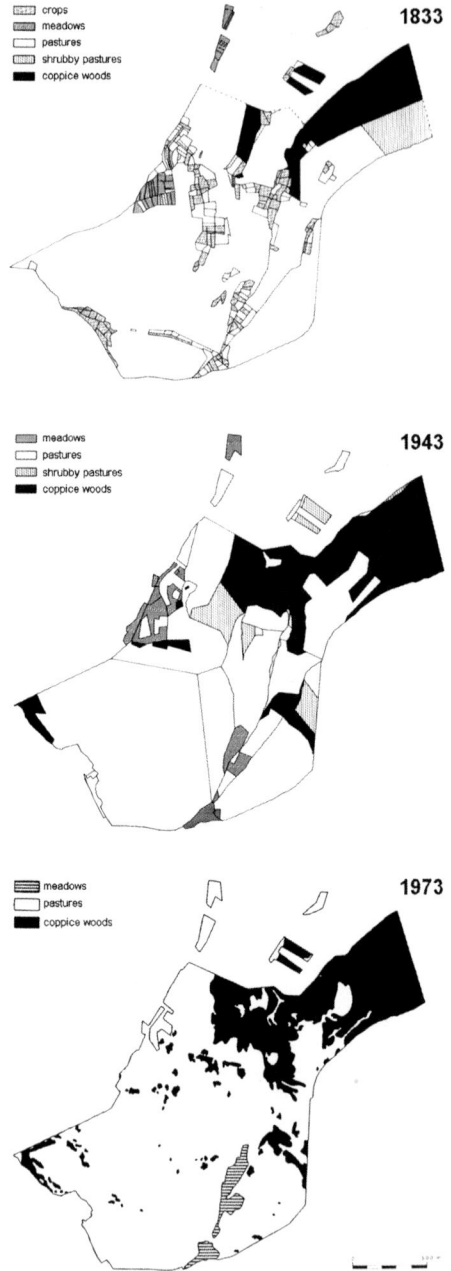

Predictive Mapping

Predictive maps for land management are derived from maps of potential vegetation and illustrate the "potential land use" that would be possible in a given territory; with such maps it is possible to formulate hypotheses on the agricultural and forestry potentials of a given zone and on the productivity of these ecosystems. Two examples of predictive map are shown. The Map of the Potential Natural Vegetation of West Germany, at scale 1:200,000, includes several inset maps that show the potential cultivation zones for some forest species of economic importance, the potential cultivation area of barley, and yield probabilities for some field and meadow crops, all deduced from the map of potential vegetation (Trautmann 1966). An example of a predictive map made in Italy is that of the Val Pellice, a river valley in the Piemonte southwest of Torino, showing *utilizzazioni prevedibili* (foreseeable uses) at scale 1:25,000, with the following legend items: classical reforestations, grazed meadows, cultures of fast-growing conifers, and field crops. Limitations on the use of the area, due to topography or urbanism, are also shown. A map of Chiomonte was produced, using analogous criteria, at scale 1:10,000; and a map of the whole Valle di Susa (near the French border, including Chiomonte) was produced at scale 1:50,000, on which, however, the legend is much more developed and the protected natural areas are also shown (Giordano et al. 1970, 1972, 1974).

Mapping for Land-Use Planning

Land planning should consider all environmental aspects (geology, geobotany, zoology, ecology, human geography, agronomy, silviculture, urbanism, etc.), integrated with the socio-economic and historical reality, and should thus propose land-use trajectories deemed most adapted for the various parts of the territory considered, according to environmental suitability and the diverse interests involved.

It is good to note, though, that planning and zoning of any territory, including protected areas, are the result not only of the physical, biological and demographic features that characterize an area but follow also from a broader cultural, socio-economical and political context that conditions the choices and decisions.

Zoning maps result from collaboration among specialists in the various disciplines listed above. The potential contribution of geobotany can be seen, from time to time, in some of the types of geobotanical map already described, such as maps of real or potential vegetation, dynamic tendencies, vegetation series, synchorology of associations, degree of naturalness or adaptation to human impact ("synanthropization"), or the conservation status of the vegetation. Conservation status depends on the intensity of past disturbance, to both vegetation and physical environment. Geobotanical maps summarize and document not only the effects of human intervention on the natural environment and the different plant formations, over the centuries, but also the ongoing processes. It suffices, perhaps, to mention the secondary meadows that have been created over centuries and that now, due to changed socio-economic conditions, have been abandoned and are undergoing secondary succession, with progressive development of shrub and tree species, that will lead to the transformation of the meadow first into scrub and then into forest. Geobotanical maps are thus not only useful but also indispensable in research on relations between man and vegetation, even if Gambino (1991), coming perhaps from a "social theory" attitude that disdains science, is strongly critical that these maps do not represent "social processes". Of course they do not, but the maps surely do represent the *results* of social processes, i.e. different types of human activities according to different socio-economic conditions that have taken place over the centuries.

The contribution of geobotany is also useful in environmental cartography, to produce maps of environmental units (see Chap. 14). These constitute the base of reference for the successive production of zoning maps.

Zoning maps are one tool used in planning and thus also in management (Fig. 12.5). These maps are made for particular, administratively defined territories such as a commune, a region or a protected area, that are subdivided into zones based on environmental characteristics, the socio-economic situation, and political choices. Zoning maps show the cartographic delimitation of identified zones, for each of which a particular law is proposed. This corresponds, then, to a form of "recommended land use" that aims to lead to an optimal occupation of the territory.

Planning consists of choosing destined uses for the various parts of a territory, in a proposal for an optimal spatial organization of the territory that designs management options deemed adapted to the various zones identified, considering the natural aspects of the territory.

If the land-use proposals are approved by the proper public authorities (e.g. regional government or a park service), then the proposed regulations become law.

For national parks, for example, according to the law for protected areas of Italy (law no. 394, of 6 December 1991), the territory is subdivided, based on environmental characteristics and on the political choices that inspired the conservation, into four zones: A for strict protection, B for general protection, C for agricultural use, and D for inhabited space. Management proposals are made for each according to the destined use.

The first proposal for zoning the territory of national parks in Italy was made by Renzo Videsott, who proposed the "adaptation of various zones within the park for nature study, visitors and mountaineering". These were defined as a scientific zone (strictly protected); a zone for orderly tourism (under general protection); and a pre-park zone, destined mainly for the necessities of the park inhabitants and adjacent areas (Videsott 1955).

The first park zoning maps made in Italy pre-dated this general law for protected areas; these include maps of the Abruzzo and Stelvio National Parks, and that proposed for Mt. Etna (Bruno and Bazzichelli 1966; Pratesi 1968; Pedrotti et al. 1969; Poli 1973). These were planning proposals with promotional aims inspired by the park philosophy of the time and by legislative projects for parks, very much discussed in Italy at the time.

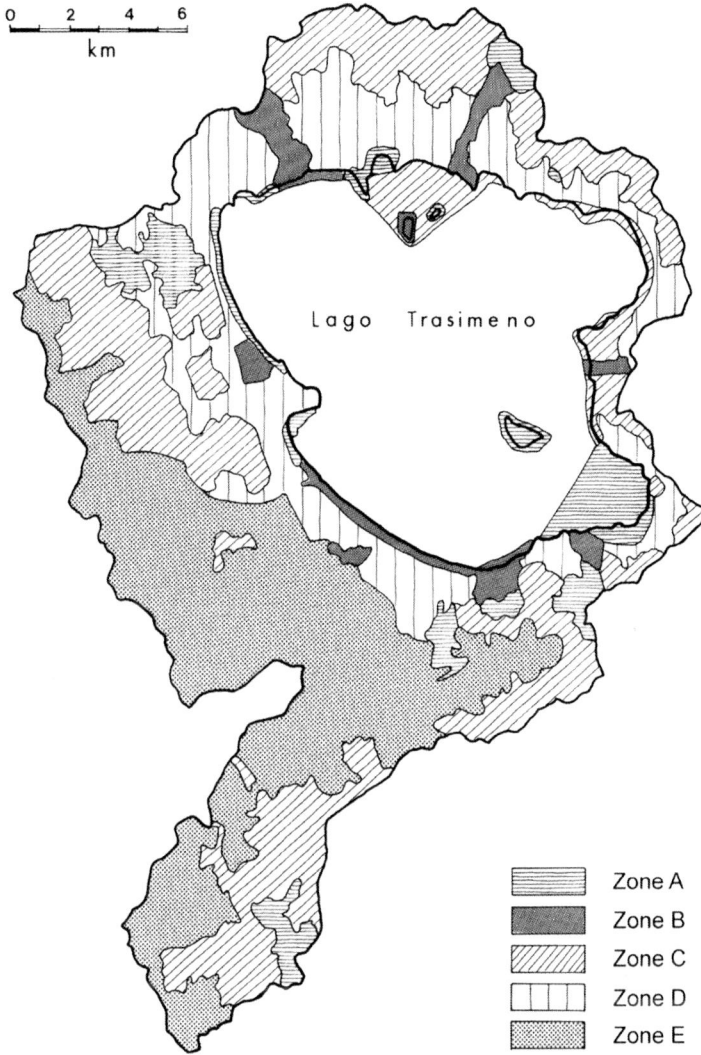

Fig. 12.5 Proposal for zoning of the Trasimeno Lake basin, Umbria Region, central Italy: *A–E* are proposed conservation zones, from strict to general (From Pedrotti 1983)

In the following years, and especially after approval of the general legislation of 1991 on protected areas, the problem of zoning the national parks was examined widely from two rather different but tightly intertwined viewpoints: the aims and objectives of the national parks, which could change significantly under different socio-economic and environmental conditions (Tassi 1983; Cassola 1985; Ceruti 1993; etc.); and the methodology for putting the zoning into effect, which is a complex operation involving participation by specialists from various disciplines from time to time.

Mapping Habitats

This kind of cartography involves the mapping of areas occupied by one or more habitats listed in the "Habitat" directive of the European Community (Directive 92/43/CEE); this directive defines as a *habitat of community interest* an environment of limited extent, containing significant biological diversity, designated for the conservation of animal and plant species in risk of disappearance or that have a distributional range reduced by regression or that present characteristics typical of the biogeographic region to which they belong. These thus constitute important specimens for the conservation of biodiversity of a biogeographic region. Habitats of community interest are subdivided into priority and non-priority habitats. The presence of a priority habitat at a place is motive enough for the proposal and definition of a Site of Community Importance. Examples of priority habitats include the "grassy, dry, semi-natural formations on calcareous substrate (Festuco-Brometalia) with the presence of orchids"; the "alluvial forests of *Alnus glutinosa* and *Fraxinus excelsior* (Alno-Padion, Alnion incanae, Salicion albae)"; the "natural eutrophic lakes with vegetation of Magnopotamion or Hydrocharition"; and the "beechwoods of the Apennines with *Taxus baccata* and *Ilex aquifolium*". All these habitats of community interest are listed in the Appendix I of Directive 97/62/CE (which updated Directive 92/43/CEE), which includes 198 habitats of which 64 are priority habitats.

Habitats can be mapped based on the principles of vegetation mapping, involving a map derived from classical phytosociological maps, with some integrations. Habitat mapping, however, may result in maps with mixed characteristics, since definition of a given habitat may vary quite significantly, based in some cases on phytosociological units (associations, alliances and orders) or in other cases on plant formations or environments with certain types of vegetation (see earlier examples).

For the habitats of the Marche Region (as mandated by the Habitat directive of the European Community), it was decided to make a phytosociological map, an integrated phytosociological map (called a map of the plant landscape by the authors), and a map of habitats (Biondi et al. 2007). The Monte Conero sheet shows 10 types of habitat plus 10 vegetation series (sigmeta). The methodology was the same as already used for sampling the biotopes of Trentino (Pedrotti 2001, 2004a), for which a map of the dynamic tendencies of the vegetation was also always made. In effect, this map contains information about the vegetation changes currently underway, which is very useful for the agencies and other entities that manage the protected areas.

Mapping for Landscape Planning

Vegetation mapping constitutes the indispensable basis for holistic planning and landscape protection; for this the European Community requires a landscape plan and the so-called VIA-VAS (environmental impact evaluation and strategic

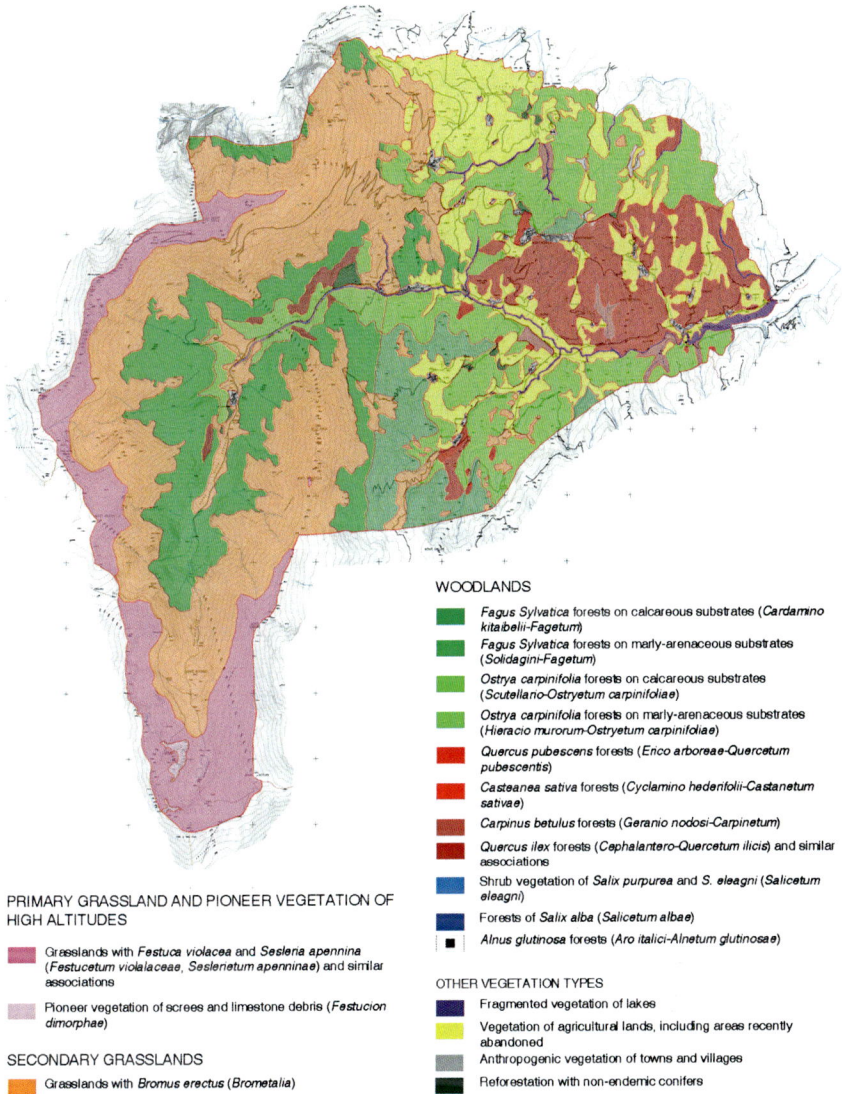

WOODLANDS

- Fagus Sylvatica forests on calcareous substrates (*Cardamino kitaibelii-Fagetum*)
- Fagus Sylvatica forests on marly-arenaceous substrates (*Solidagini-Fagetum*)
- Ostrya carpinifolia forests on calcareous substrates (*Scutellario-Ostryetum carpinifoliae*)
- Ostrya carpinifolia forests on marly-arenaceous substrates (*Hieracio murorum-Ostryetum carpinifoliae*)
- Quercus pubescens forests (*Erico arboreae-Quercetum pubescentis*)
- Casteanea sativa forests (*Cyclamino hederifolii-Castanetum sativae*)
- Carpinus betulus forests (*Geranio nodosi-Carpinetum*)
- Quercus ilex forests (*Cephalantero-Quercetum ilicis*) and similar associations
- Shrub vegetation of Salix purpurea and S. eleagni (*Salicetum eleagni*)
- Forests of Salix alba (*Salicetum albae*)
- Alnus glutinosa forests (*Aro italici-Alnetum glutinosae*)

PRIMARY GRASSLAND AND PIONEER VEGETATION OF HIGH ALTITUDES

- Grasslands with *Festuca violacea* and *Sesleria apennina* (*Festucetum violalaceae, Seslerietum apenninae*) and similar associations
- Pioneer vegetation of screes and limestone debris (*Festucion dimorphae*)

SECONDARY GRASSLANDS

- Grasslands with *Bromus erectus* (*Brometalia*)

OTHER VEGETATION TYPES

- Fragmented vegetation of lakes
- Vegetation of agricultural lands, including areas recently abandoned
- Anthropogenic vegetation of towns and villages
- Reforestation with non-endemic conifers

Fig. 12.6 Map of the actual vegetation of Montemonaco municipality, Marche Region, central Italy (*Surveyed by Cianfaglione and Pedrotti*)

environmental evaluation). The landscape plan is illustrated by two maps made for Montemonaco (Marche, Adriatic central Italy), namely a vegetation map (Fig. 12.6) and a map of individual main elements of the landscape (Fig. 12.7), both at scale 1:10,000. The first is a map of actual (including natural) vegetation, showing 18 vegetation units; mainly these are associations but in one case an order

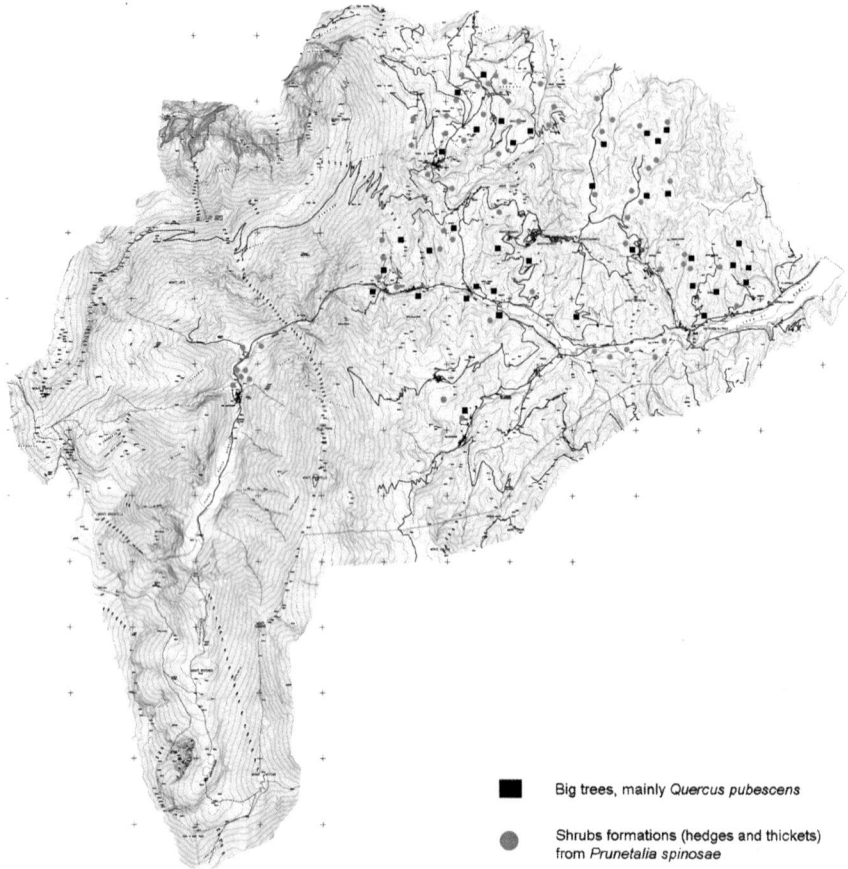

Big trees, mainly *Quercus pubescens*

Shrubs formations (hedges and thickets)
from *Prunetalia spinosae*

Fig. 12.7 Map of plant landscape elements of Montemonaco municipality (*Surveyed by Cianfaglione and Pedrotti*)

(Brometalia), plus some generic groupings, including anthropogenic vegetation of human settlements, vegetation of crop fields including the abandoned areas, lake vegetation, and reforestations with exotic conifers.

Historical Overview

The first geobotanical map in Italy was made in 1863, namely the *Carta agrologica della Provincia di Reggio Calabria Ultra* (toe of southern Italy) at scale 1:280,000, by Pasquale (1863). It is a color map showing the following very general units: citrus orchards, olive groves, forests, areas of *sulla* (*Hedysarum coronarium*) or wheat cultivation, and bare rocks. It is thus a physiognomic map, but given the type of legend, it could also be considered a map of land use. The first map that covered all of Italy, the *Carta botanica d'Italia* (botanical map of Italy), at scale 1:5,000,000, by Fiori (1908), is shown in the next section; it was also a physiognomic map.

Among the first chorological maps is the monograph by Pampanini (1903), which analyzed and mapped the areas of 159 species of the Alps, grouped into categories by chorological affinity. Chorological cartography in Italy developed widely in the following decades, as described by Pignatti (1988b) in a historical review.

Among vegetation maps (physiognomic at first), one can also cite those by Zenari (1925) at scale 1:250,000 for the Val Cellina (Friulia, NE Italy); by Gavioli (1936) at scale 1:132,500 for the Gruppo del Pollino (Lucania, southern peninsula), showing both the main forest formations and the altitudinal belts and horizons (Fig. 13.1); and by Zangheri (1936) of the pine forest of Ravenna (northern Adriatic coast) at scale 1:14,000. In the same years some detailed maps of small biotopes also appeared, such as the *Carta della vegetazione del Lago Baccio*, a high-altitude lake in the Apennines of Emilia-Romagna, made at scale 1:1,540 by Provasi (1926); and the *Carta della vegetazione del Lago e del Padule di Sibolla* (a lake and wetland area in Tuscany), at scale 1:6,000, by Francini (1936). All these maps were black and white, and were rather schematic (Fig. 13.2).

The first color maps were published in the following years, with individual cartographic units defined by a boundary and shown in different colors, such as the *Carta della vegetazione della Val Sangone* (a small side valley of the Valle di Susa in western Piemonte), at scale 1:50,000 by Sappa and Charrier (1949).

F. Pedrotti, *Plant and Vegetation Mapping*, Geobotany Studies,
DOI 10.1007/978-3-642-30235-0_13, © Springer-Verlag Berlin Heidelberg 2013

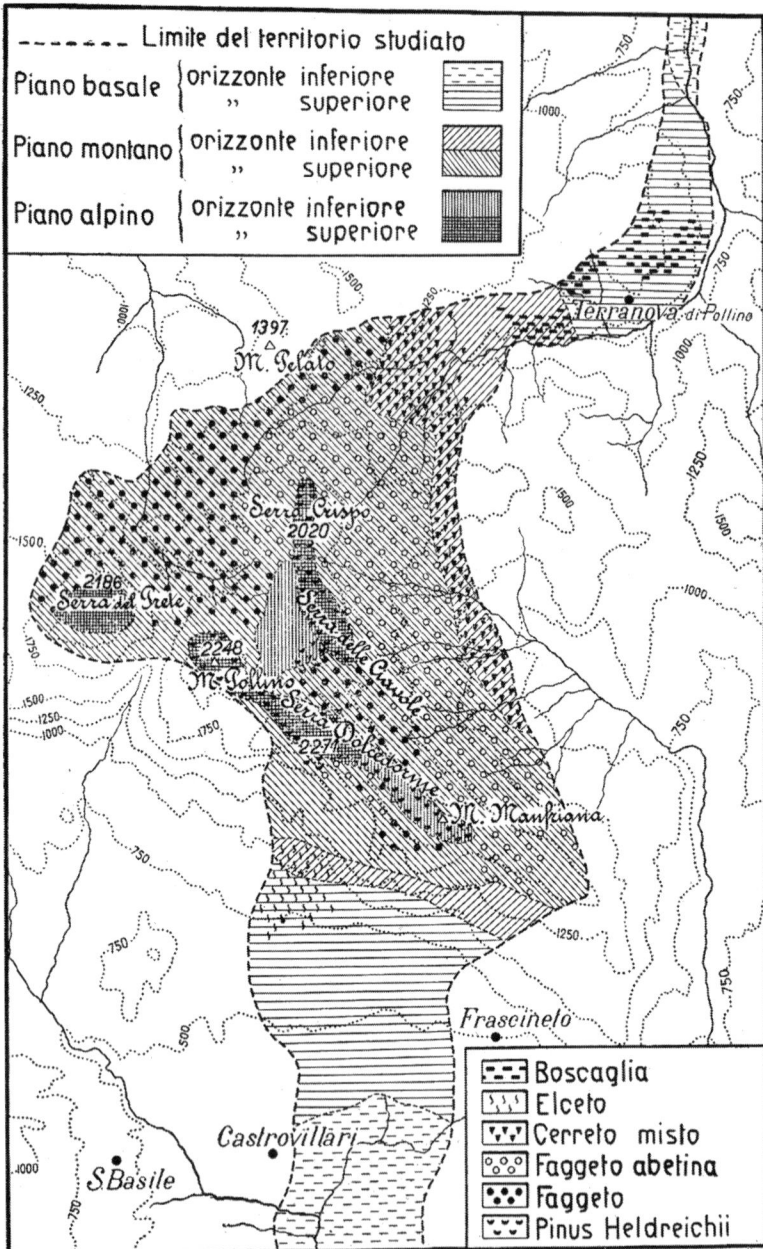

Fig. 13.1 Ancient vegetation map of Monte Pollino, Calabria Region, southern Italy. *Limite del territorio studiato* limits of the territory surveyed, *piano basale, montano, alpino* lowland, montane and alpine belts, *orizzonte inferiore/superiore* lower/upper layers, *boscaglia* thicket, *Elceto* forest of *Quercus ilex*, *Cerreto misto* mixed forest with *Quercus cerris*, *Faggeto abetina* forest of *Fagus sylvatica* and *Abies alba*, *Faggeto* forest of *Fagus sylvatica* (From Gavioli 1936)

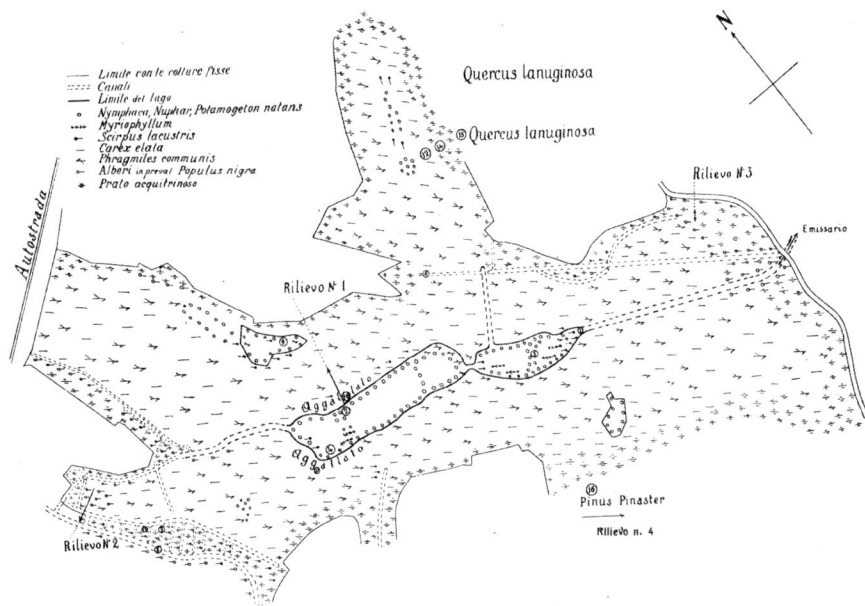

Fig. 13.2 Ancient vegetation map of Sibolla Lake, Tuscany Region, central Italy (From Francini 1936)

The range of map types broadened, and maps followed that showed vegetation bands (sensu Schmid 1961), as on the map of the Val Sangone just mentioned, by Sappa and Charrier (1949); the *Carta della vegetazione forestale delle Langhe* (southern Piemonte, southeast of Torino), at scale 1:50,000, by Sappa (1955); the *Carta della vegetazione dell'Appennino Lucano centrale* (southern peninsula), at scale 1:50,000, by Famiglietti and Schmid (1969); and some others.

The first phytosociological maps to see the light of day in Italy were both for mountain areas of northernmost Lombardia, in the Italian Alps and foothills. One map was by Giacomini (1954) for a pasture area of the Valtellina valley, at scale 1:8,300 (in black and white), and another was by Tomaselli (1956b) for the Val di Scalve, at scale 1:7,500 (in two colors). These were schematic maps but were followed fairly quickly by the *Carta fitosociologica dei pascoli dell'Alta Valle del Braulio* (phytosociological map of the pastures of the upper Braulio valley, also northern Lombardia), at scale 1:12,500, by Giacomini and Pignatti (1955), a color map showing 24 plant associations, indicated according to phytosociological alliances, orders and classes (and with vegetation boudaries in red). The production of phytosociological maps over the years was inspired by this model from Italy, and numerous such maps were made. Among these were maps made under the project "Promotion of Environmental Quality" completed by the Consiglio Nazionale Ricerche (National Research Council), which stimulated the production of many maps at various scales; the complete list is given by the Consiglio Nazionale Ricerche (1982). Annotated lists of vegetation maps made in Italy have been provided by Giacomini (1966), Bruno et al. (1976),

Pirola (1988), Ferrari and Rossi (1990), Pirola and Vianello (1992), and Pedrotti (1988b, 1990, 1993), to which can be added the most recent and complete, computerized, list of Bruno et al. (2003).

General Vegetation Maps of Italy

Listed here (in chronological order) and commented on briefly are the general vegetation maps of Italy made up to now. Each map is identified by name, scale, author, year printed, map type, and is described briefly. Also listed are some general maps of Europe or the Mediterranean Basin that include Italy. The list demonstrates the development of geobotanical research in Italy over the last 100 years, as reflected in the maps produced.

- Botanical map of Italy, 1:5,000,000, Fiori (1908), vegetation physiognomy, showing five vegetation regions indicated by colors and hatching: alpine (small herbaceous plants in the upper zone, low shrubs in the lower); montane (conifer forests, beech forests, chestnut forests, oak forests of two kinds, *Quercus petraea* and *Q. cerris*); the Po valley (cultivated plains, without characteristic plants); Mediterranean (pinewoods of *Pinus pinea*, *P. pinaster*, or *P. halepensis*; evergreen trees and shrubs of olive, myrtle, *Quercus ilex*, *Q. suber*, etc.); and submerged (algae, Najadaceae, Lemnaceae, Hydrocharitaceae, etc.).
- Map of vegetation zones, 1:2,000,000, Béguinot (1933), vegetation physiognomy. The following regions and zones are distinguished: a high-mountain region; conifer, beech and chestnut-oak zones; and a region of Mediterranean macchia and olives, in the midst of which are shown individual pinewoods.
- Map of Italian floristic zones, 1:7,500,000, Adamović (1933), floristic subdivisions; the following zones are distinguished: Sardinia-Corsica, Liguria, Veneto, pelagic Tyrrhenian, circum-Adriatic, and western Apennine.
- Regional map of the vegetation of the Apennines, about 1:7,000,000, Lüdi (1935 and 1946), climax vegetation. Four types of climax vegetation are shown: Quercion ilicis, Quercion pubescentis, Fagion sylvaticae and alpine vegetation.
- Map of forest-climatic zones of Italy, 1:2,500,000, de Philippis (1937). The following zones are distinguished: Lauretum, Castanetum, Fagetum and Piceetum.
- Map of plant formations, 1:2,500,000, Fiori (1939), vegetation physiognomy. The following formations are distinguished: submerged, marine and lacustrine zone; Mediterranean zone; Po valley; and submontane, montane, and alpine belts.
- Map of the vegetation of Italy, 1:6,000,000, Giacomini and Fenaroli (1958), potential vegetation, distinguishing five types of climax: Quercion ilicis, Quercion pubescenti-petraeae, Fagion sylvaticae, Piceion excelsae and a hypsophilic climax, plus 17 floristic symbols.
- Map of regional botanical subdivisions, 1:6,000,000, Giacomini and Fenaroli (1958), botanical subdivisions (regions, provinces, districts, sectors).
- Bioclimatic map of the Mediterranean zone, 1:5,000,000, UNESCO-FAO (1963). The following climate types are distinguished: warm and warm-temperate

climates (accentuated meso-mediterranean, mitigated meso-mediterranean, submediterranean, non-xeric temperate with a subhumid period); cool and cool-temperate climates (cool non-xeric).

– Map of floristic kingdoms and regions of the world, 1:7,500,000, Engler and Gilg (1924); this map shows the boundary between the Mediterranean and the central-European vegetation in Italy.

– Map of the vegetation of the Mediterranean region, 1:5,000,000 (UNESCO-FAO 1970). This is a map of potential vegetation that shows various grand formations, such as the lowland formation of evergreen Mediterranean oaks, foothill formations, montane formations, etc.

– Map of the actual vegetation of Italy, 1:1,000,000, Fenaroli (1970), vegetation physiognomy, with eight grand physiognomic types and some tree species by means of symbols.

– Map of the potential natural vegetation of Italy, 1:1,000,000, Tomaselli (1970a), potential vegetation. This map shows the potential natural vegetation in more detail than did preceding maps; 18 types of climax vegetation are shown, grouped according to altitudinal belts. Each type is mapped with the corresponding phytosociological units (associations and alliances) and with some characteristic species. The map shows only zonal vegetation. This map has been the object of two successive editions, at scales 1:2,500,000 and 1:2,000,000 (Tomaselli 1973a, b).

– Map of forest vegetation of Italy, 1:2,000,000, Tomaselli (1973a), physiognomy of forest vegetation. On this map are shown seven forest formations characterized by as many tree species.

– Map of the vegetation of the member states of the Council of Europe, 1:3,000,000, Ozenda (1979), potential vegetation. This map shows the types of zonal climax vegetation.

– Map of the natural vegetation of the member states of the European Community and Council of Europe, 1:3,000,000, Noirfalise (1987), potential vegetation. This map shows types of zonal and azonal climax vegetation.

– Map of the potential natural vegetation of Italy, 1:5,000,000, Pedrotti (1989, 1991), potential vegetation. This map has 42 vegetation units and shows, for the first time, also the azonal and extrazonal vegetation and some new vegetation types not considered before. Each vegetation unit is indicated with one or more diagnostic (separating) species and corresponds to a group of plant associations that belong to the same alliance or order. The 42 types are represented on the map by different colors according to their phytosociological and systematic affinity, using warm colors (red and orange) for Mediterranean vegetation and cool colors (blue and violet) for montane and subalpine vegetation. From this map were derived the four synchorological maps that indicate the distribution in Italy of the following vegetation orders (Fig. 9.4): Vaccinio-Piceetalia (Alps), Fagetalia sylvaticae (Alps and Apennines), Quercetalia pubescenti-petraeae and Quercetalia robori-petraeae (Alps and Apennines), and Quercetalia ilicis and Rhamno-Pistacietalia (Mediterranean region). From this map was also derived a later Map of the potential vegetation of Italy, 1:6,200,000 (Pedrotti 1996a).

– Map of the actual natural vegetation of Italy, 1:1,000,000, Pedrotti (1992), actual vegetation. This is a phytosociological map showing the actual vegetation, or actual natural vegetation, distinguished as zonal vegetation (belonging to the Euro-Siberian and Mediterranean regions) and azonal vegetation (dependent on particular edaphic conditions). This map shows 53 vegetation types, each of which corresponds to an order or an alliance and only in a few cases to individual associations. This map also shows the distribution of some rare and localized tree and shrub species of great phytogeographic interest.

– Map of the landscape systems in Italy, about 1:7,000,000, Pignatti (1994). This map shows 31 landscape systems, understood by the author as "the sum of interacting elements" interpreted holistically (i.e. all together) rather than by a reductionist approach (i.e. one at a time). The systems are determined by the substrate, by the vegetation and by man.

– Map of the geobotanical subdivisions of Italy, 1:6,200,000, Pedrotti (1996a), phytogeographical subdivisions. Italy is seen as belonging to the Euro-Siberian and Mediterranean regions, and each region is divided into provinces and districts (Fig. 10.4).

– Map of the vegetation naturalness in Italy, 1:6,450,000, Pedrotti (1996a), vegetation naturalness. Naturalness is expressed in six levels, identified on the basis of plant communities of primary or secondary origin, semi-natural vegetation, and "synanthropic" vegetation.

– Map of the natural vegetation of Europe, 1:2,500,000, Bohn et al. (2000a, b, 2003), potential vegetation (Fig. 6.17); a short description of this map is given in Chap. 6. In the sector shown in Fig. 6.18 (south-central Italy) one can see the distribution of the Mediterranean vegetation (Quercetalia ilicis) along the coast, submediterranean (Quercetalia pubescenti-petraeae) on the foothills, montane (Fagetalia sylvaticae) on the Apennine topography, and alpine (Seslerietalia apenninae) in the Apennines above 1,800 m. Some vegetation types have very limited distributions, such as the oakwoods of *Quercus trojana* in Puglia (southeastern Italy), the pinewoods of *Pinus nigra* in the Sangro Valley in Abruzzo (central Italy), the pinewoods of *Pinus leucodermis* on Mt. Pollino (southern Apennines), and the pinewoods of *Pinus laricio* in Sila and Aspromonte (southern Italy).

– *Carta dei sistemi di paesaggio,* Blasi (2005), landscape systems distinguished by geomorphological characteristics.

– *Carta del fitoclima d'Italia,* Blasi and Michetti (2005), 28 types of phytoclimate.

– *Regioni fitoclimatiche,* Blasi and Michetti (2005), distribution in Italy of the Mediterranean, temperate, transitional Mediterranean, and transitional temperate phytoclimatic regions.

– *Carta delle serie di vegetazione dell'Italia,* 1:500,000, Blasi (2010), vegetation series; comprehensive work on the vegetation of Italy described in terms of vegetation series (total of 279); the map, which combines all the preceding considerations, was published in three sheets for northern, central and southern Italy (the last including Sardinia and Sicily); the text contains a monograph on the vegetation of each of the 20 regions of Italy.

Mapping Environments

14

Definition and Trends in Environmental Mapping

Environmental mapping is a kind of synthetic cartography that attempts to consider the total of all possible ecological factors (physical, biological and anthropogenic) and to obtain a synthetic and unique representation of the environment (Ozenda 1974, 1975, 1986).

Nowadays environmental cartography is producing documents of many different kinds, due to the diverse range of content that such maps can have: environmental units, environmental risks, maps evaluating human impacts or environmental naturalness, maps related to environmental restoration, etc.

Here only basic environmental mapping is discussed, i.e. the general interpretation of the environment and its cartographic representation by various methodologies, from which environmental units are recognized and in turn arranged into higher categories or environmental systems.

Environmental mapping, even if it is not yet systematized like geobotanical or geological mapping, is today of great interest for its range of possible uses in studying applied environmental problems.

Approaches and Study Methods

Environmental maps can be made by four main methods (Ingegnoli 2002), briefly described here.

The first, which by now can be considered classical (Naveh and Lieberman 1984), starts from the existence of seven major types of landscape (the first four "open" and the other three "constructed"), listed here in decreasing order of naturalness:

- Natural landscapes (including only primary ecosystems);
- Semi-natural landscapes (that also include secondary ecosystems, such as pastures with native species);

F. Pedrotti, *Plant and Vegetation Mapping*, Geobotany Studies,
DOI 10.1007/978-3-642-30235-0_14, © Springer-Verlag Berlin Heidelberg 2013

- Semi-agricultural landscapes (that include agricultural ecosystems in addition to the preceding);
- Agricultural landscapes (consisting entirely of secondary and rural ecosystems);
- Rural landscapes (composed only of rural ecosystems, i.e. those associated with agriculture or other farming);
- Suburban landscapes (composed of rural and techno-urban ecosystems); and
- Urban landscapes (composed only of techno-urban ecosystems).

More recently, Haber (1990) proposed a more synthetic classification that has five landscape types, the first four constituted by bio-anthropogenic meta-ecosystems and the last by techno-meta-ecosystems:

- Natural landscapes (not influenced by man and totally self-regulating);
- Quasi-natural landscapes (influenced only slightly by man and still capable of self-regulation);
- Semi-natural landscapes (composed of natural and semi-natural ecosystems that require some management for their maintenance or regulation);
- Biotic-anthropogenic (or cultural) landscapes (including man-made ecosystems that depend totally on management by man); and
- Urban-industrial (or technical) landscapes (dominated by human artifacts due to industrial, agricultural and socio-cultural activities).

The second method is based on concepts of (geo)synphytosociology, i.e. the *tesela* and catena of vegetation (see Chaps. 6 and 8). In the case of "microheterogeneous" landscapes (wetlands, coastal dunes, dolines, etc.), the landscape units can be described beginning from the catenas, while in "macroheterogeneous" landscapes (steppe, desert, high plateaus, etc.) this can be done based on the vegetation series.

The third method, presented by Forman and Godron (1986), uses five levels of ecological determinants as the criteria for classifying landscapes:

- Zonal climates, i.e. macrobioclimates, *sensu* Walter 1977 (cf Troll 1964; Walter and Box 1976), such as the equatorial climate or the boreal climate;
- Climatic regions, i.e. based on mesoclimates, for example the Mediterranean region;
- Bioclimatic units, for example the smaller phytoclimatic units of Forman and Godron (1986);
- Pedo-geomorphological units, for example the floodplains of major rivers, with alluvial substrates;
- Human influence through land uses (orchards, introductions by man, etc.).

The fourth method uses as the determining criterion the functional processes of the components of the terrestrial landscape, distinguished by different landscape "apparatuses" involving specific landscape functions (Ingegnoli 2002):

- Geological apparatus (e.g. base areas, rocky crests);
- Connective apparatus (linkage elements that permit or enhance flows of materials or energy, such as hedges);
- Stabilizing apparatus (elements with high meta-stability, such as primary ecosystems);

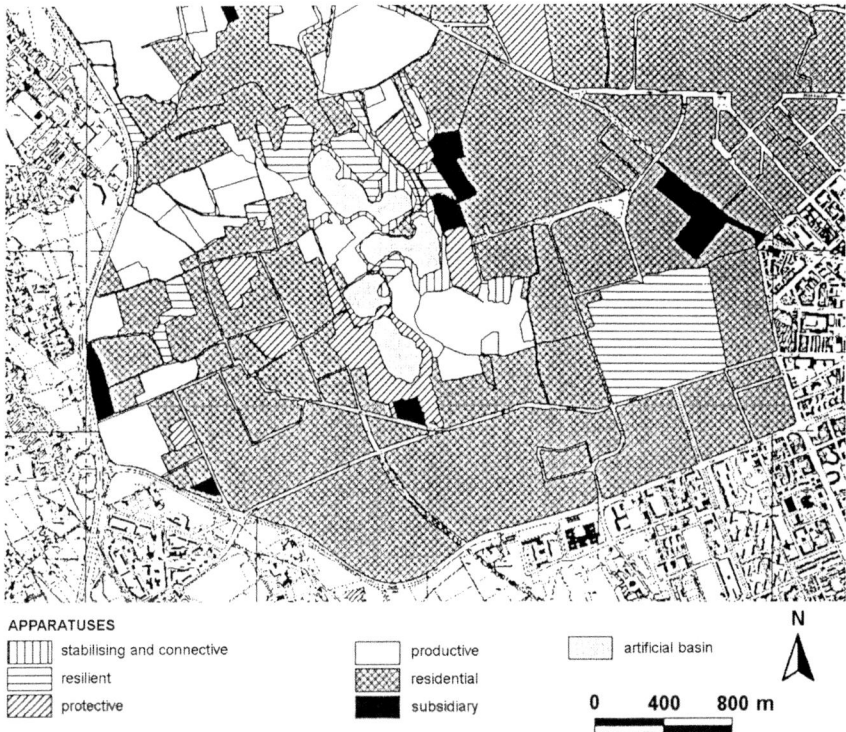

Fig. 14.1 Landscape apparatuses from a part of the Lombardia Region, northern Italy (See main text; from Ingegnoli 2002)

- Resilience apparatus (elements with high capacity for regeneration, for example ruderal ecosystems);
- Storage apparatus (resource centers, for example ecosystems with high biodiversity or rich seedbanks);
- Disturbance apparatus (elements that cause significant disturbances, for example active volcanos);
- Change apparatus (elements that can change readily, for example unstable slopes, mobile dunes, etc.);
- Protective apparatus (elements that perform important protective functions, for example riparian forests for flood protection, windbreaks, etc.);
- Productive apparatus (elements with high biological productivity, such as agricultural fields and crops);
- Subsidiary apparatus (systems for energy and industrial production);
- Residential apparatus (systems related to human habitations and other buildings functioning socio-culturally).

A landscape map made by this last method is shown in Fig. 14.1 and represents a suburban area near Milano.

Environmental Mapping by Geo-Synphytosociology

Application of data from phytosociological vegetation series, and from what was above called catenal mapping, to the study of the plant landscape has permitted going from geobotanical cartography to environmental cartography. According to Ozenda (1974, 1975, 1986), one can do this by adding to the vegetation map ancillary data on the main environmental factors (substrate and climate) and notations on influences by man. From this perspective, what results is a proper vegetation map that can be considered an intermediate but not final document. Subsequent environmental maps made from this can be used for broader objectives, namely as a means of synthetic environmental interpretation.

In practice, the step from (geo)synphytosociological mapping to environmental mapping involves the step from vegetation series and catenas to environmental units. These last are ecologically heterogeneous areas composed of a set of ecotopes that show characteristic structural patterns that repeat themselves spatially and give a landscape a particular identity (Forman and Godron 1986). More synthetically, one can say that an environmental unit is characterized by a specific vegetation mosaic due to primary differentiation or secondary (induced by natural perturbations) and tertiary (induced by anthropogenic perturbations). Thus, homogeneity in the environmental units is found only on a functional plane. Finally, one must remember that the content and amplitude of the environmental units depends on the scale at which the landscape is analyzed.

As an example of synphytosociological units used to distinguish environmental units, one can cite the series of Norway spruce (*Picea abies*) in the continental valleys of Alto Adige, on slopes of siliceous rocks (pegmatitic metamorphic series), on which a humic-ferric podzolic soil has developed. A natural spruce forest belonging to the association Homogyno-Piceetum (Fig. 14.2) has developed here between 1,600 and 2,100 m. The forest is interrupted by clearings with permanent habitations, temporary summer habitations, and huts, with vegetation of mowable meadows, pastures and anthropogenic vegetation (Pedrotti et al. 1997). The vegetation of the forest and the clearings in it can be assigned to the subalpine series of Norway spruce, formed by associations shown in Fig. 14.2. The associations that constitute the series are involved in various ecological processes, especially the processes of regeneration and degeneration (spruce forest), secondary succession (meadows and abandoned pastures) and fluctuation (meadows, still-used pastures, and anthropogenic associations).

According to the type of human settlement in the clearing and the consequent human activities, three types of clearings can be distinguished, each characterized by particular plant associations and with its own physiognomy and landscape functions: clearings with permanent habitations are characterized by cultivated fields with the association Galinsogo-Portulacetum and by mowable meadows with the association Trisetetum flavescentis; clearings with only summer habitations have mowable meadows; and clearings with only huts have pastures with the association Sieversio montanae-Nardetum. A group of associations, though, is common to all three types of clearing: Salicetum capreae, Sambucetum racemosae, Rubetum idaei, etc.

For environmental evaluation, such differences are important and it is necessary to consider these for the synthesis by environmental units. From this it follows that

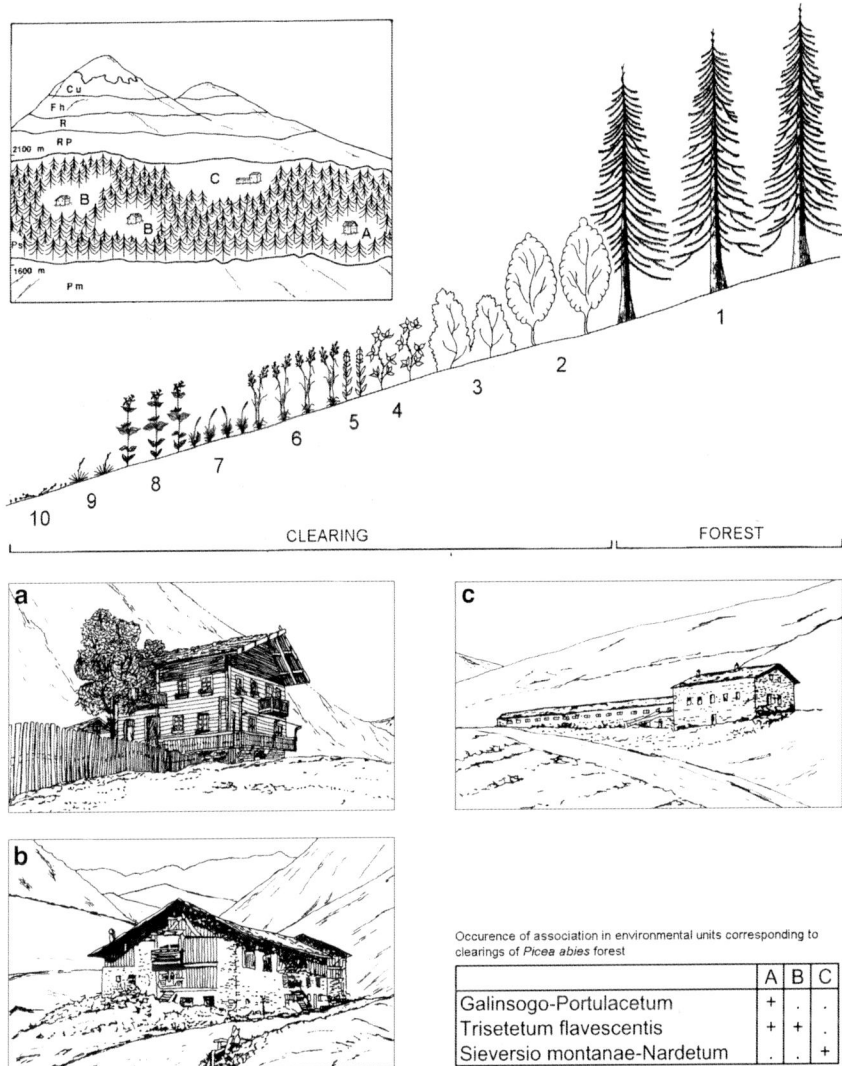

Fig. 14.2 Martello Valley, southern slopes: Pm *Oxali-Piceetum*; Ps *Homogyno-Piceetum*; RP *Rhododendro-Pinetum cembrae*; R *Rhododendretum ferruginei*; Fh *Festucetum halleri*; Cu *Caricetum curvulae*, Acidophilous climax series of *Picea abies* [*Homogyno-Piceeto* sigmetum], composed of the following associations: *1 Homogyno-Pieetum, 2 Salicetum capreae, 3 Sambucetum racemosae, 4 Rubetum idaei, 5 Epilobietum angustifolii, 6 Trisetetum flavescentis, 7 Sieversio montanae-Nardetum, 8 Galinsogo-Portulacetum, 9 Lolietum perennis, 10 Sagino-Bryetum argentei.* Also the following "environmental units": (**a**) "clearing with houses inhabited year-round", (**b**) "clearing with summer houses", and (**c**) "clearing with huts for summer use by herders" (From Pedrotti et al. 1997, modified)

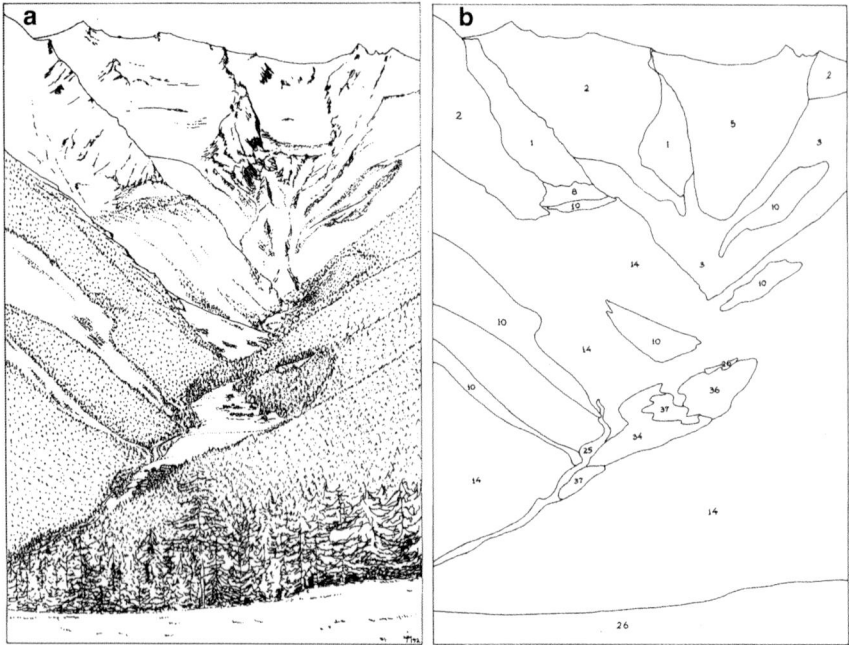

Fig. 14.3 (a) Panorama of the Trafoi Valley (Stelvio National Park, central Italian Alps); (b) environmental units of the Trafoi Valley: *1* high peaks and rocky ridges, *2* permanent glaciers and snowfields, *3* slopes, debris, moraines, *5* glacial amphitheaters, *8* upper slopes with more level areas with primary meadows (*Caricetum firmae* and *Seslerio-Caricetum sempervirentis*), *10* very steep detrital calcareous deposits, partialy with *Mugo-Ericetum* and *Mugo-Rhododendretum hirsuti*, *14* middle and lower slopes with coniferous forests (mainly *Picea abies*), *25* rivers with riparian vegetation, *26* terraces and slopes with secondary meadows (mainly *Sieversio montanae-Nardetum*) and "malghe" which are often associated, *36* slopes and alluvial cones, sometimes quite steep with *Larix decidua* pastures "*pascoli a larice*", *37* villages (From Pedrotti et al. 1997)

the vegetation series of Norway spruce belongs to four different types of environment unit, namely: spruce forest (Fig. 14.2, cf the "head" of the series) and clearings with permanent habitations, with summer habitations, or with huts.

The delimitation of environmental units, by this method, proceeds through the following steps:

– Delimitation of the patch (*tesela*) using the concept of a sigmetum and the relevant potential natural vegetation;
– Recognition and mapping of the actual vegetation of the patch;
– Optionally, identification of the dominant dynamic processes that involve the various plant associations of the series;
– Identification of the human activities that have resulted in specific structural characteristics in the vegetation series; and
– Identification of the environmental units by grouping or subdividing the associations that compose the vegetation series based on landscape functions.

One can then proceed to the mapping of the distinct environmental units thus recognized (Fig. 14.3).

The importance of vegetation maps is thus obvious as a starting point for making environmental maps. In this regard it is worth noting what Pettit et al. (2008) said in their volume *Landscape Analysis and Visualisation*, namely that "vegetation maps are a typical starting point for mapping and modelling landscapes, whether it is the vegetation itself that is of interest or its role as habitat for other native biota" [*sic*], i.e. organisms.

Methods of Representation in Environmental Mapping

This type of cartography is considered relatively complex, and in fact the basis for its systematization is not yet well defined, as opposed to botanical, geological or other thematic cartography.

Two main methods have been proposed for making environmental maps: overlaying of information on the same map and synthetic maps.

The attempt to obtain an exhaustive map, pretending to show on one document all the components of the environment through overlaying, has not produced satisfactory results, for both technical and graphical reasons (Ozenda 1974; Journaux 1975): printing so many symbols and using so many different colors makes the map hard to read.

The method that seems to respond best to these objectives, at the moment, is synthetic cartography. With this method it is possible to recognize synthetic spatial units whose names, based on aggregates of descriptors, vary according to their homogeneity and the scale selected. Various authors have employed terms like ecotopes ("homeo-ecological areas"), eco-patches (*ecotessuti*, or "ecotissues" *sensu* Ingegnoli 2002; environmental units or land units), territorial systems or land systems, etc. (Forman and Godron 1986; Zonneveld 1989; Vos and Stortelder 1992; Martinelli 1999; Martinelli and Pedrotti 2001; etc.).

These two methods have been applied contemporaneously by Martinelli (1990) to the same study area, namely the territory of Camerino; thus it was possible to obtain a map through superposition of various thematics, including some smaller related maps of individual themes (Fig. 14.4) and a map of environmental units (Fig. 14.5), deduced from the preceding. The results confirm the validity of a cartography done by identifying the environmental units.

Recently, quantitative methods have been developed that permit evaluating the degree of spatial agreement among different thematic maps, based on their resemblance and by a non-parametric coefficient of uncertainty, estimated for each pair of maps compared (De Agart et al. 1995; Bernert et al. 1997).

Mapping environmental units, i.e. delimiting them concretely, still remains artificial because the ecosystems and landscapes are open systems (Bertalanffy 1950; Chorley and Kennedy 1971). Still, for practical problems, such as communication with administrators about land management, it is justifiable to use quite specific, "closed" natural and cultural units as basic criteria. These are obtained from integration of the physical (habitat), biotic (biocoenoses) and human environment, as well as their functions (Froment 1987).

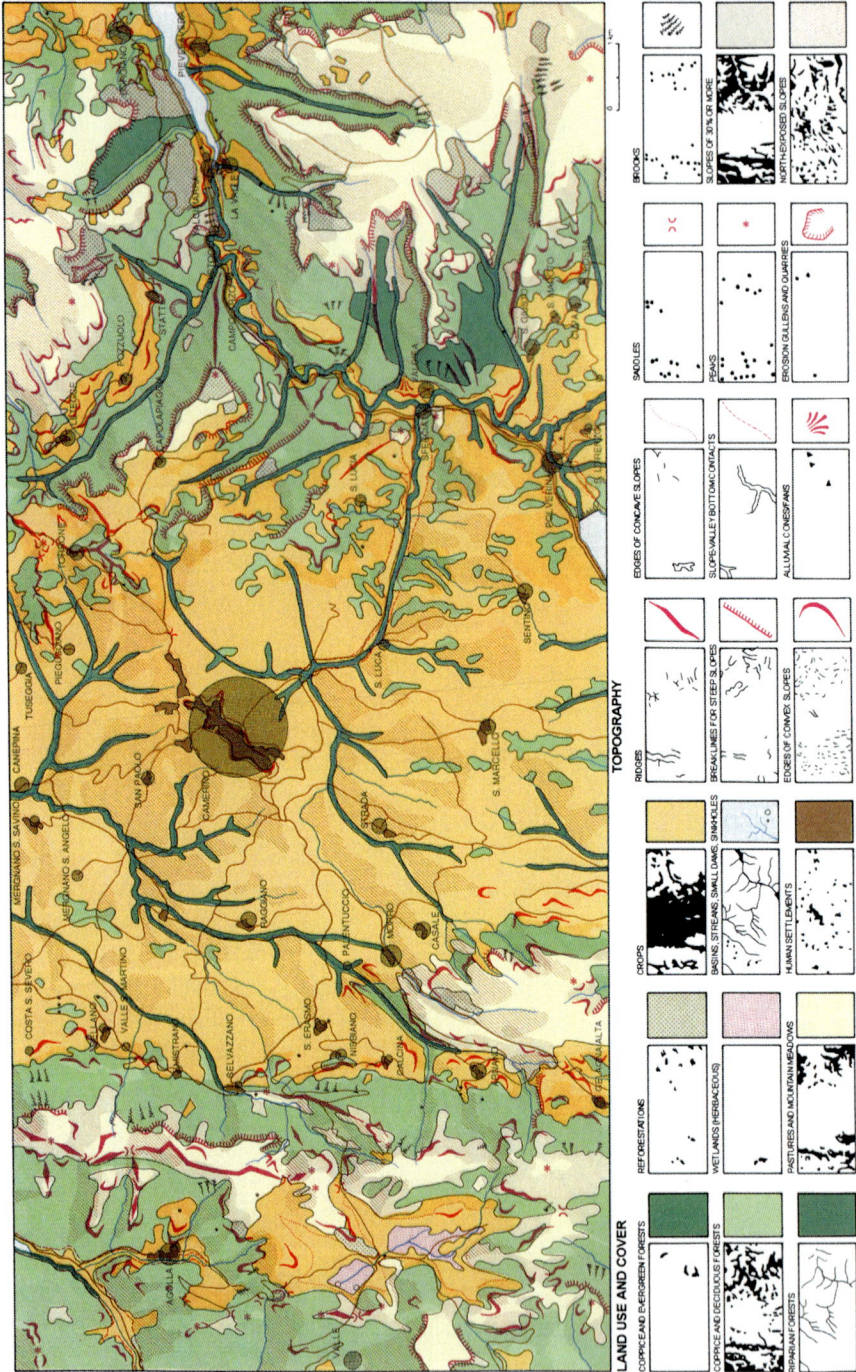

Fig. 14.4 Map of natural features and anthropogenic impacts in Camerino municipality, Marche Region, central Italy. The map was obtained by overlaying 21 theme: original scale 1:50,000, re-sized to 1:100,000 for printing (From Martinelli 1990)

a	Mountain landscapes with steep slopes mainly covered by coppice woods with a few human settlements
b	Valley bottoms flanked by slopes with crops on alluvial soils and with human settlements along the axis of the valley
c	Karst plains with wetland vegetation, in part used for crops on colluvial soils
d	Flat mountaintop landscapes with pastures, without human settlements, showing traces of abandoned camps sometimes with terracing
e	Foothill landscapes mostly cultivated, heavily populated and with dense road networks
f	Artificial bassins
g	Urban landscapes

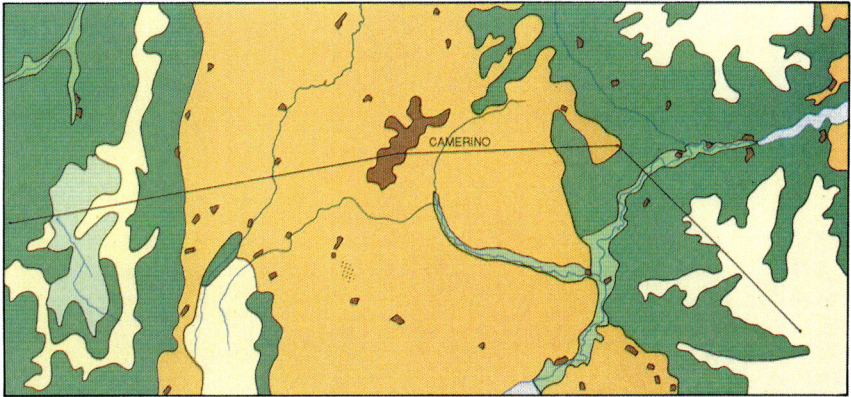

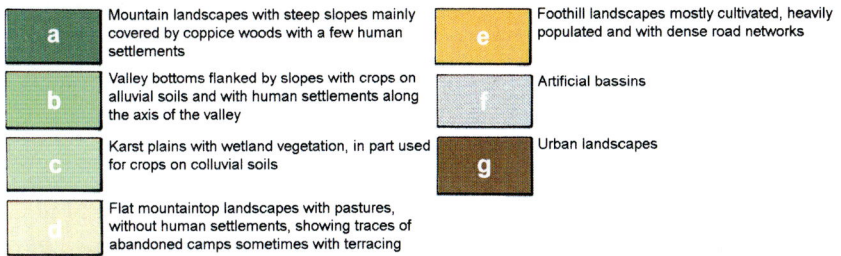

Fig. 14.5 Map of environmental units of Camerino municipality, originally at scale 1:100,000 re-sized to 1:150,000 for printing; *below*, environmental units (From Martinelli 1990, photo F. Pedrotti)

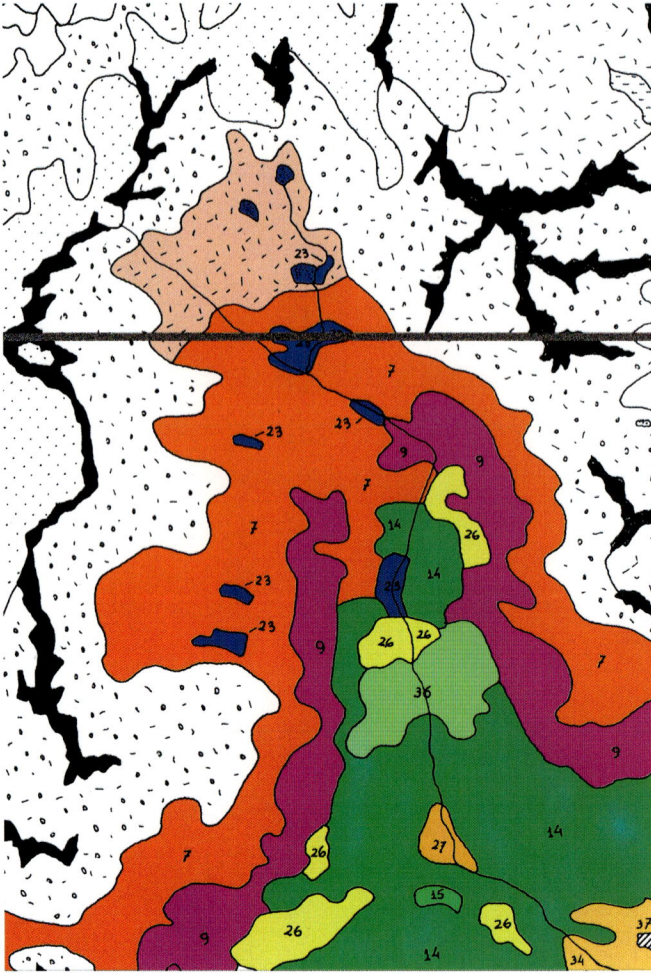

Fig. 14.6 Map of environmental units of Stelvio National Park, Rabbi Valley sector, at scale 1:50,000 (From Pedrotti et al. 1997)

Each distinct unit on the map of environmental units of Ayers Rock and Mount Olga in Australia was described in a standard form that includes geology, topography, soil, vegetation, erosion, stability, fauna and concludes with notes on human influences (Hooper et al. 1973).

Another example of a map with environmental units is the *Carta delle Unità Ambientali del Parco nazionale dello Stelvio* (Map of Environmental Units of the Stelvio National Park, central Italian Alps) at scale 1:50,000 (Pedrotti et al. 1997; Gafta and Pedrotti 1997). This map shows 37 different environmental units, each of which carries a definition (periphrasis) that provides the essential characteristics of the unit, a description and a symbol (Figs. 14.6 and 14.7).

Fig. 14.7 Legend of the previous map (Fig. 14.6) with iconographs for the environmental units: *1* high peaks and rocky ridges, *2* permanent glaciers and snowfields, *3* slopes, debris, moraines, *5* glacial amphitheaters, *7* upper slopes with level areas covered by primary meadows

Maps of the environmental units of the Abruzzo National Park and of the Sibillini Mountains were made by the same methodology. In the latter the environmental units were regrouped into three large, distinct environmental systems based on geomorphologic and lithologic criteria: calcareous massifs and mountain ridgelines, mainly sandstone upland and mountain topography, and alluvial valley bottoms (Pedrotti 1999a).

Environmental units can also be arranged in a hierarchical system. Similar environmental units then come to constitute the landscape units, which in turn constitute parts of a given landscape system; multiple landscape systems form landscape regions. The *Pian Grande di Castelluccio* (great plain of Castelluccio, in Norcia) is a vast karst basin in the calcareous Sibillini Mountains of central Italy. Here it is possible to distinguish two landscape units, one at the bottom of the basin and one on the slopes, with the boundary between the two given by the break-point in the slope (Fig. 14.8). On the basin bottom we can distinguish four environmental units, as indicated in Fig. 14.8. The levels in question are thus: environmental units, landscape units, landscape systems (broad geographical, lithological and physiographic units), and landscape regions (in Italy only two, the Mediterranean and the temperate). It is also possible to distinguish subunits, such as subsystems that correspond to local geomorphological units.

For understanding the landscape in all its dimensions, a hierarchical system is mandatory and fundamental. For applied objectives the basic reference is to units of the lowest hierarchical level, namely the environmental units, which permit a reasonable appreciation of all the variability of the given landscape.

The environments of Camerino, the Stelvio National Park, the Sibillini Mountains and Abruzzo have been described almost exclusively by qualitative terms (geomorphology, vegetation, land use, human settlements, etc.). Still it is possible to go further and consider the spatial configuration resulting from the vegetation patches and corridors, using the textural indices from landscape ecology (Farina 2001), as was done for the map of the Torricchio Nature Reserve in central Italy. The map of vegetation patches and "ecotissue" differentiation by Gafta (2006) distinguishes two main landscape units, a foothill forest-meadow ecotissue and a montane meadow ecotissue (Fig. 14.9). The units of the plant landscape were differentiated and interpreted as ecotissues, i.e. spatial models coinciding with recurring vegetation patches (venation or elementary grain of the landscape

---

Fig. 14.7 (Continued) (*Primulo-Caricetum curvulae* and *Festucetum halleri*), *9* elevated parts of slopes with shallow morainal deposits covered by dwarf shrub woods (mainly *Empetro-Vaccinietum*), *23* glacial cirques with peat bogs, fens and small glacial lakes, *26* terraces and slopes with secondary meadows (mainly *Sieversio montanae-Nardetum*) and "malghe" which are often associated, *27* terrace sequences on mid-slopes with movable grazing areas (*Trisetetum flavescentis* and *Melandrio-Arrhenatheretum*) and summer establishments which are often associated, *34* outskirts of rural centers, more or less urbanized, with grassy surfaces and crops, *36* slopes and alluvial cones, sometimes quite steep with *Larix decidua* pastures "*pascoli a larice*", *37* villages. The icons for 14 (slopes with *Picea abies* forests) and 15 (slopes with *Abies alba*) are not reproduced (From Pedrotti et al. 1997)

LANDSCAPE AND ENVIRONMENTAL UNITS

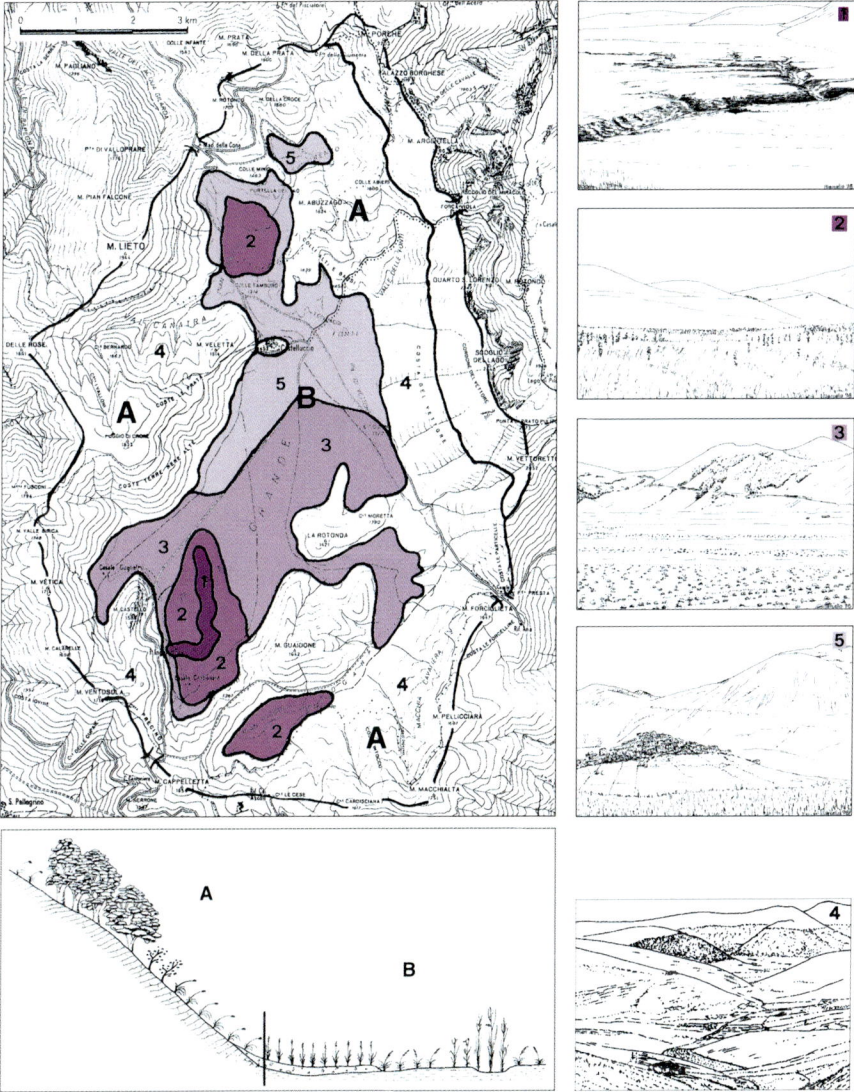

Fig. 14.8 Landscape and environmental units of the Sibillini Mountains: Landscape A shows calcareous mountain slopes, with the following environmental units: *4* slopes with dry meadows (*Brometalia*) and residual forests of *Fagus sylvatica*, *6* human settlements with anthropogenic vegetation; landscape B shows the bottom of karst basin, with the following environmental units: *1* natural channel with sinkhole (*Ranunculus trichophyllus* ssp. *trichophyllus* community and *Potamogeton natans* community) *2* wet plain and dolines with meadows (*Nardetalia* and *Magnocaricetalia* in dolines), *3* wet plain with meadows (*Cynosuro-Trifolietum pratensis*); *5* dry plain with cultivated areas (From Pedrotti 1997a, 1999a modified)

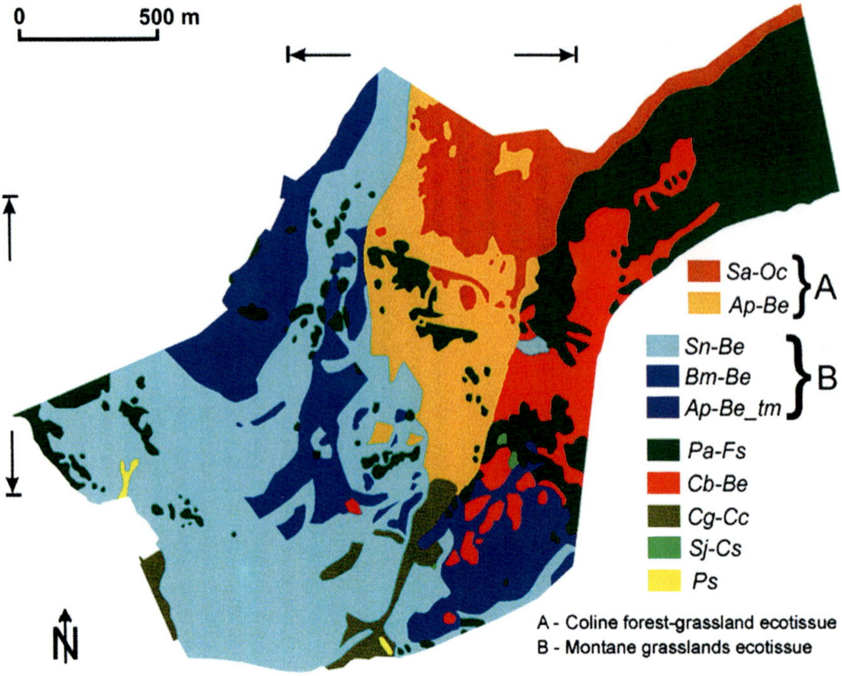

Fig. 14.9 Distribution of vegetation patches and differentiation of ecotissues within the Torricchio Nature Reserve, Marche, central Italy: the *left* and *upper arrows* indicate the position and extension of the square landscape analysed to distinguish ecotissues. Sa-Oc *Scutellario-Ostryetum carpinifoliae*, Ps *Prunetalia spinosae*, Pa-Fs *Polysticho aculeati-Fagetum sylvaticae*, Bm-Be *Brizo mediae-Brometum erecti*, Cg-Cc *Campanulo glomeratae-Cynosuretum cristati*, Ap-Be_tm *Asperulo purpureae-Brometum teucrietosum montani*, Sj-Cs *Spartio juncei-Cytisetum sessilifoii*, Sn-Be *Seslerio nitidae-Brometum erecti*, Cb-Be *Centaureo bracteatae-Brometum erecti*, Ap-Be *Asperulo purpureae-Brometum erecti* (From Gafta 2006)

mosaic) determined by the underlying environmental heterogeneity (distribution of types of ecotopes and coenoses composing the dynamic series). These units of the plant landscape are identified through textural analysis of the digital image, which represents a map of the actual vegetation at a resolution of 1 m^2 on a square area of 100 ha within the reserve. Two kinds of plant-landscape unit were distinguished based on significant spatial association of the vegetation patches: the foothill forest-meadow ecotissue (A) and the montane meadow ecotissue (B).

References

Adamović L (1933) Die pflanzengeographische Stellung und Gliederung Italiens. Fischer, Jena
Aeschimann D, Lauber K, Moser DM, Theurillat J-P (2004) Flora alpina. Zanichelli, Bologna
Aichinger E (1951) Vegetationsentwicklung als Grundlage unserer land-und forstwirtschaftlichen
    Arbeit. Angew Pflanzensoz 1:17–20
Aichinger E (1967) Die Waldentwicklungstypen im Raume von Kirchleerau. Veröff Geobot Inst
    ETH Stiftung Rübel 39:187–270
Aleffi M (1987) La rappresentazione del paesaggio vegetale in un'antica mappa catastale del
    territorio di Lanciano (Camerino). Inform Bot Ital 18(1-2-3):125–138
Aleffi M (1993) Gli antichi documenti come base per la ricostruzione del paeaggio vegetale. Coll
    Phytosoc XXI:303–310
Aleffi M (1994) L'evoluzione del paesaggio vegetale attraverso lo studio delle Piante delle tenute
    dell'Abbazia di S. Maria di Chiaravalle di Fiastra. La Riserva Naturale Abbadia di Fiastra
    2:39–69
Aleffi M (2010) L'associazione Lunularietum cruciatae Giac. 1951 nella città di Camerino
    (Italia centrale). Braun-Blanquetia 46:101–102
Alexander R, Millington AC (eds) (2000) Vegetation mapping. Wiley, Chichester
Amadesi E (1993) Manuale di fotointerpretazione e aerofotogrammetria. Pitagora, Bologna
Arrigoni PV (1981) Le piante endemiche della Sardegna: 84–90. Boll Soc Sarda Sc Nat 20:268
Arrigoni PV (1983) Aspetti corologici della flora sarda. Lav Soc It Biogeogr VIII:83–109
Arrigoni PV, Nardi E (1975) Documenti per la carta della vegetazione del Monte Amiata. Webbia
    29:717–785
Aspinall RJ, Pearson DM (1995) Describing and managing uncertainly of categorical maps in GIS.
    In: Fisher P (ed) Innovations in GIS 2. Taylor & Francis, London, pp 71–83
Austin MP, Smith TM (1989) A new model for the continuum concept. Vegetatio 83:35–47
Avolio S, Ciancio O (1985) I giganti della Sila. Ann Ist Sper Selv Arezzo XVI:373–421
Bănărescu P, Boşcaiu N (1978) Biogeographie. Fischer, Jena
Barkman JJ (1973) Synusial approaches to classification. In: Whittaker RH (ed) Ordination and
    classification of communities, vol 5, Handbook of vegatation science. Junk, The Haag,
    pp 435–491
Barthlott W, Biedinger N, Braun G, Feig F, Kier G, Mutke J (1999) Terminological and methodo-
    logical analysis of the global biodiversity. Acta Bot Fennica 162:103–110
Beard JS (1979) Vegetation mapping in Western Australia. J R Soc West Aust 62(1–4):75–82
Beard JS (1981) Vegetation survey of Western Australia: Swan. University of Western Australia
    Press, Perth
Béguinot A (1933) Italia: flora e vegetazione. Enciclopedia Italiana 19:729–736
Belov AV, Ljamkin VF, Sokolova LP (2002) Cartographical study of biota. Publishing House
    Oblmashinform, Irkutsk
Bernert JA, Eilers JM, Sullivan TJ, Freemark KE, Rivic C (1997) A quantitative method for
    delineating regions: an example for the Western Corn belt plains ecoregion. Environ Manag
    21(3):405–420

Bertin J (1967) Sémiologie graphique. Gauthier-Villars, Paris

Bertin J (1977) La graphique et le traitement graphique d'informations. Flammarion, Paris

Bezoari G, Monti C, Selvini A (2002) Topografia generale con elementi di geodesia. UTET, Turin

Biondi E (ed) (1999) Ricerche di geobotanica ed ecologia vegetale di Campo Imperatore (Gran Sasso d'Italia). Braun-Blanquetia 16:1–247

Biondi E, Baldoni M (1995) The climate and vegetation of Peninsular Italy. Coll Phytosoc XXIII:675–721

Biondi F, Pedrotti F, Tomasi G (1981) Relitti di antiche foreste sul fondo di alcuni laghi del Trentino. St Trent Sc Nat 58:93–117

Biondi F, Taffetani F, Allegrezza M, Ballelli S (1990) La cartografia della vegetazione del Foglio Cagli. Atti Ist Bot Lab Critt Univ Pavia 9:51–74

Biondi E, Casavecchia S, Bianchelli M, Pesaresi S, Pinzi M (2007) SCI IT5320007 (AB25) Monte Conero phytosociological and habitat maps. SCI IT5320007 (AB25) Monte Conero plant landscape map. Regione Marche, Ancona

Bioret F, Fichaut B, Gourmelon F (1995) Cartographie de la végétation de la partie terrestre de l'Archipel de Molène (Réserve de biosphère de la Mer d'Iroise). Coll Phytosoc XXIII:169–187

Blasi C (1994) Clima e fitoclima. In: Pignatti S (ed) I boschi d'Italia. UTET, Turin, pp 33–71

Blasi C (ed) (2005) Stato della biodiversità in Italia. Palombi, Rome

Blasi C (ed) (2010) La vegetazione d'Italia. Palombi, Rome

Blasi C, Michetti L (2005) Biodiversità e clima. In: Blasi C, Boitani L, La Posta S, Manes F, Marchetti M (eds) Stato della biodiversità in Italia. Palombi, Rome, pp 57–66

Blasi C, Scoppola A, Abbate G, Michetti L, Scagliusi E, Kuzminsky E, Antinori F (1989) Carta della naturalità della Caldera del Lago di Vico. Dipart Biol. Veg, Rome

Bohn U, Gollub G, Hettwer C (2000a) Karte der natürlichen Vegetation Europas. Massstab 1:2,500,000. Legende. Bundesamt f. Naturschutz, Bad Godesberg

Bohn U, Gollub G, Hettwer C (2000b) Karte der natürlchen Vegetation Europas. Massstab 1:2,500,000. Karten. Bundesamt f. Naturschutz, Bad Godesberg

Bohn U, Gollub G, Hettwer C, Neuhäuslova Z, Schlüter H, Weber H (2003) Karte der natürlchen Vegetation Europas. Massstab 1:2,500,000. Erläuterungstext. Bundesamt f. Naturschutz, Bad Godesberg

Bolognini G, Nimis PL (1993) Phytogeography of italian deciduous oak woods based on numerical classification of plant distribution ranges. J Veg Sci 4:847–860

Borfecchia F, De Cecco L, Dibari C, Iannetta M, Martini S, Pedrotti F, Schino G (2003) Carta della vegetazione reale della "Riserva naturale di Torricchio" ottenuta mediante elaborazione di immagini satellitari multispettrali. La Riserva Naturale di Torricchio 11(4):359–369

Borza A, Boşcaiu N (1965) Introducere in studiul covorului vegetal. Acad. Rep. Pop. Romậne, Bukarest

Boullet V, Géhu JM (1988a) Cartes des risques d'incendie méditerranéen. Commune de Figanières. Cahiers Phytosoc 1:1–20

Boullet V, Géhu JM (1988b) Cartes des risques d'incendie méditerranéen. Commune de Cogolin. Cahiers Phytosoc 2:1–16

Boullet V, Géhu JM (1988c) Cartes des risques d'incendie méditerranéen. Commune de Cassis. Cahiers Phytosoc 3:1–20

Box EO (1978) Geographical dimensions of terrestrial net and gross primary productivity. Radiat Environ Biophys 15:305–322

Box EO (1979a) Use of synagraphic computer mapping in geoecology. In: Computer mapping in education, research, and medicine, Harvard library of computer mapping, vol 5. Harvard University Press, Cambridge, pp 11–27

Box EO (1979b) The MAPCOUNT series of programs for analysis and manipulation of SYMAP maps. In: Mapping software and cartographic data-bases, Harvard library of computer mapping, vol 2. Harvard University Press, Cambridge, MA, pp 21–29

Box EO (1979c) Quantitative cartographic analysis: a summary (with geoenvironmental applications) of SYMAP auxiliary programs developed at the Jülich Nuclear Research Center. Jülich (Germany). Reports of the Kernforschungsanlage Jülich GmbH, n. 1582

Box EO (1981) Macroclimate and plant forms: an introduction to predictive modeling in phytogeography, vol 1, Tasks for vegetation science. Junk, The Haag, 25 world maps

Box EO (1988) Estimating the seasonal carbon source-sink geography of a natural steady-state terrestrial biosphere. J Appl Meteorol 27:1109–1124

Box EO (1995) Global potential natural vegetation: dynamic benchmark in the era of disruption. In: Murai Sh (ed) Toward global planning of sustainable use of the Earth: development of global eco-engineering. Elsevier, Amsterdam, pp 77–95

Box EO, Bai X-M (1993) A satellite-based world map of current terrestrial net primary productivity. Seisan Kenkyuu (Tokyo) 45(9):666–672

Box EO, Meentemeyer V (1991) Geographic modeling and modern ecology. In: Esser G, Overdieck D (eds) Modern ecology basic and applied aspects. Elsevier, Amsterdam, pp 773–804

Box EO, Holben BN, Kalb V (1989) Accuracy of the AVHRR vegetation index as a predictor of biomass, primary productivity, and net CO_2 flux. Vegetatio 80:71–89

Braun-Blanquet J (1928) Pflanzensoziologie. Springer, Berlin

Braun-Blanquet J (1936) La chênaie d'yeuse méditerranéenne (Quercion ilicis). Comm SIGMA 45:1–147

Braun-Blanquet J (1937–1943) Carte des groupements végétaux de la France. Région N.O. de Montpellier

Braun-Blanquet J (1951) Pflanzensoziologie. Springer, Wien

Braun-Blanquet J (1958) Lagunenverlandung und Vegetationsentwicklung an der französischen Mittelmeerküste bei Palavas, ein Sukzessionsexperiment. Comm SIGMA 141:9–32

Braun-Blanquet J (1964) Pflanzensoziologie. Springer, Berlin/Wien/New York

Bredenkamp G, Chytrý M, Fischer HS, Neuhäuslova Z, van der Maarel E (eds) (1998) Vegetation mapping: theory, methods and case studies. Appl Veg Sci 1:161–266

Brockmann-Jerosch H (1919) Baumgrenze und Klimacharakter. Beitr Geobot Landesausfnahme 6:1–255

Brockmann-Jerosch H (1930) Klimatisch bedingte Formationsklassen der Erde. In: Rübel E (ed) Die Pfanzengesellschaften der Erde. Huber, Bern

Bruno F, Bazzichelli G (1966) Note illustrative alla carta della vegetazione del Parco Nazionale d'Abruzzo (scala 1:25,000): Progetto conservazionale geobotanico. Ann Bot 28(3):739–778

Bruno F, Giacomini V, Pirola A (1976) Realizazioni di cartografia vegetazionale in Italia. Giorn Bot It 113(5–6):451–455

Bruno F, Petriccione B, Attorre F (2003) La cartografia della vegetazione in Italia. Braun-Blanquetia 26:1–27

Buffa M, Venturella G, Raimondo FM (1986) Carta della vegetazione del Parco della Favorita (Palermo). Il Naturalista Siciliano X (ser. IV)(suppl):1–90

Bulgarian Academy Science (1984) Red data book of the people's Republic of Bulgaria. Publishing House of Bulgarian Academy, Sofia

Cain SA (1951) Fundamentos de Fitogeografia. ACME Agency, Buenos Aires

Campetella G, Canullo R, Angelini G (2002) Lo stato delle querce camporili in un territorio del bacino del fiume Chienti (Macerata). Monti e Boschi 5:4–11

Canullo R (1991a) The map of vegetation structure of a coppice wood in the nature reserve of the Fiastra Abbey (Central Italy). Phytocoenosis (Suppl Cart Geobot) 2:195–198

Canullo R (1991b) Vegetation structure and regeneration process in a mixed deciduous coppice in the Central Apennines (Italy). Phytocoenosis (Suppl Cart Geobot) 2:199–207

Canullo R (1993a) L'évolution de la végétation vers la forêt: études des populations. Coll Phytosoc XX:121–140

Canullo R (1993b) Lo studio popolazionistico degli arbusteti nelle successioni second concezioni, esempi ed ipotesi di lavoro. Ann Bot 51(suppl 10):379–394

Canullo R, Campetella D (2010) Dinamica, struttura e diversità funzionale delle comunità vegetali. In: Pedrotti F (ed) La Riserva Naturale di Torricchio 1970–2010. TEMI, Trento, pp 265–285

Canullo R, Falińska K (2003) Ecologia vegetale. Liguori, Naples

Canullo R, Pedrotti F (1993) The cartographic representation of the dynamical tendencies in the vegetation: a case study from the Abruzzo National Park, Italy. Oecologia Montana 2:13–18

Canullo R, Tavolini A (1999) Population spatial dynamics of Anemone nemorosa in the Białowieza clearing (PL): a cartographic approach. Phytocoenosis 11 (N.S.), (Suppl Cartogr Geobot 11):213–224

Canullo R, Pedrotti F, Venanzoni R (1990) Les processus dynamiques dans la végétation de la tourbière de Fiavé (Italie du Nord). Phytocoenosis (Suppl Cart Geobot) 2:189–194

Canullo R, Campetella G, Manzi A, Nola P, Pierdominici MG (1993) Popolamento arboreo e sottobosco nell'area permanente Gariglione (Parco Nazionale della Calabria, Sila Piccola). Atti workshop progetto strategico Clima Ambiente e Territorio nel Mezzogiorno, Amalfi, 28–30 aprile 1993, pp 421–436

Canullo R, Pedrotti F, Venanzoni R (1994) La torbiera di Fiavé. In: Pedrotti F (ed) Guida all'escursione della socioetà Italiana di fitosociologia in Trentino (1–5 luglio 1994). Dipartimento di Botanica ed Ecologia, Camerino, pp 78–110

Cassola F (1985) Parchi e aree protette regionali, provinciali, locali e di altri enti. In: Parchi e aree protette in Italia. Atti Conv Lincei 66:121–166

Catorci A, Gatti R (eds) (2007) Le praterie montane dell'Appennino Maceratese. Braun-Blanquetia 42:1–272

Ceruti G (1993) Aree naturali protette. Domus, Milan (2nd edn, 1996)

Chorley RI, Kennedy BA (1971) Physical geography: a systems approach. Prentice-Hall, London

Chrisman NR, Lester MK (1991) A diagnostic test for categorical maps. Technical papers. ACSM-ASPRS annual convention 6:330–348

Chytrý M (1998) Potential replacement vegetation: an approach to vegetation mapping of cultural landscapes. Appl Veg Sc 1:177–188

Chytrý M, Grulich V, Tichy L, Kouril M (1999) Phytogeographical boundary between the Pannonicum and Hercynicum: a multivariate analysis of landscape in the Podyi/Thayatal National Park, Czech Republic/Austria. Preslia 71:1–19

Chytrý M, Pysek P, Wild J, Pino J, Maskell C, Vila M (2009) European map of alien plant invasion based on the quantitative assesment across habitats. Divers Distribut 15:98–107

Ciferri R (1936) Studio geobotanico dell'Isola Hispaniola (Antille). Atti R Ist Bot Lab Critt Ital R Univ Pavia VIII(IV):3–336

Clements EF (1912) Plant succession: an analysis of the development of vegetation. Carn Inst Wash Publ 242:1–512

Clements EF (1928) Plant succession and indicators: a definitive edition of plant succession and plant indicators. Wilson, New York

Commission European Communities (1991) Corine biotopes. ECSC-EEC-EAEC, Bruxelles

Consiglio Nazionale Ricerche (1982) Progetto finalizzato "Promozione della qualità dell'ambiente". Risultati delle ricerche. Catalogo delle pubblicazioni. Mostra dei risultati. CNR, Rome

Conti F, Manzi A, Pedrotti F (1992) Libro Rosso delle piante d'Italia. Ministero Ambiente-W.W.F., Rome

Cornelius JM, Reynolds JF (1991) On determining the statistical significance of discontinuities within ordered ecological data. Ecology 72:2057–2070

Cortini Pedrotti C (1989) La flore bryologique de la ville de Camerino (Italie centrale). Braun-Blanquetia 3:241–246

Cortini Pedrotti C (1992) Le Briofite quale componente strutturale e funzionale degli ecosistemi forestali. Ann Acc Ital Sc For 41:163–190

Cortini Pedrotti C (1996) Aperçu sur la bryogéographie de l'Italie. Bocconea 5:301–318

Cortini Pedrotti C (2001) New check-list of the mosses of Italy. Flora Mediterranea 11:23–107

Cortini Pedrotti C, Orsomando E, Pedrotti F, Sanesi G (1973) La vegetazione e i suoli del Pian Grande di Castelluccio di Norcia (Appennino centrale). Att Ist Bot Lab Critt Univ Pavia IX:155–249

Costa JC, Aguiar C, Capelo JH, Lousa M, Neto C (1998) Biogeografia de Portugal Continental. Quercetea 0:5–56

Cristea V, Gafta D, Pedrotti F (2004) Fitosociologie. Presa Universitaria Clujana, Cluj-Napoca

Currie DJ (1991) Energy and large-sale pattern of animal- and plant-species richness. Am Nat 137(1):27–49

Cwiklinski E, Glowacki Z (2000) Atlas florystyczny doliny Bugu. Phytocoenosis 12 (Suppl Cart Geobot 12):75–299

Dansereau P (1951) Description and recording of vegetation upon a structural basis. Ecology 32:172–229

Dansereau P (1957) Biogeography: an ecological perspective. Ronald Press, New York

de Agart PM, De Pablo CL, Pineda FD (1995) Mapping the ecological structure of a territory: a case study in Madrid (central Spain). Environ Manag 19(3):345–357

de Laubenfels DJ (1975) Mapping the world's vegetation. Syracuse University Press, Syracuse

de Philippis A (1937) Classificazioni ed indici del clima in rapporto alla vegetazione forestale italiana. Nuovo Giorn Bot It 44:1–169

Didukh YP, Eremenko LP, Kukovitsa GS, Shelag-Sosonko YR (1984) Large-scale geobotanical map as a model for the study of antropogenic succession in vegetation (using as an example the vegetation map of urochistche "Goroditsche" Western Podolia). Geobotaniceskoe Kartografirovanie, pp 25–33

Diels L (1929) Pflanzengeographie. Grunter, Berlin/Leipzig

Dierschke H (1994) Pflanzensoziologie. Ulmer, Stuttgart

Donița N (1979) Vegetația R.S. România. Principalele unități zonale. In: Ivan D (ed) 1979 – Fitocenologie și vegetația Republicii Socialiste România. Didactică și pedagogică, Bukarest

Donița N, Roman N (1979) Vegetația. In: Atlas Republica socialista România. Academia României, Bukarest

Donița N, Ivan D, Pedrotti F (2003) Struttura e produttività delle praterie delle Viotte del Monte Bondone. Report C.E.A. 32:1–36

Drude O (1890) Handbuch der Pflanzengeographie. Engelhorn, Stuttgart

Drummond J (1990) A framework for handling error in geographic data manipulation. In: Ripple WJ (ed) Fundamentals of geographical information systems: a compendium. ASPRS, Falls Church, pp 109–118

Dulloo ME, Maxted N, Guarino L, Florens D, Newbury HJ, Ford Lloyd BW (1999) Ecogeographic survey of the genus Coffea in Mascarene Islands. Bot J Linn Soc 131(3):263–284

Dunn R, Harrison AR, White JC (1990) Positional accuracy and measurements error in digital databases of land use: an empirical study. Int J Geogr Info Syst 4:385–398

Dupias G, Gaussen H, Izard M, Rey P (1965) Carte de la végétation de la France, vol 80–81, Corse. CNRS, Paris

Ehrlich PR, Mooney HA (1983) Extinction, substitution and ecosystem services. BioScience 33:248–254

Ellenberg H, Klötzli F (1967) Vegetation und Bewirtschaftung des Vogelreservates Neeracher Riet. Ber Geobot Inst ETH Stiftung Rübel 37:88–103

Ellenberg H, Müller-Dombois D (1967) Tentative physionomic-ecological classification of plant formation on the earth. Ber Geobot Inst ETH Rübel 37:21–55

Elzinga CL, Valzer DW, Willoughby JW (1998) Measuring and monitoring plant populations. U.S. Department of the Interior Bureau of Land Management National Applied Resource Sciences Center, Denver

Engler A, Gilg E (1924) Syllabus der Pflanzenfamilien. Borntraeger, Berlin

Falińska K (1984) Problemi di demografia nelle piante. Info Bot Ital 16(1):19–37

Falińska K (1993) The influence of disturbances in foreste communities upon the spatial organisation of Cyclamen hederifolium populations on the promontorio del Gargano (Italy). Fragm Flor Geobot 2(2):681–698

Falińska K (1998) Plant population biology and vegetation processes. W. Szafer Institute of Botany, Polish Academy of Sciences, Kraków

Falińska K (1999) Space patterns of plant populations on the San Domino island (Tremiti, S-Italy). Phytocoenosis II (N.S.), (Suppl Cart Geobot 11):231–236

Falińska K (2002) Przewodnik do badan biologii populacii roslin. PWN, Warsaw

Falińska K (2003) Alternative pathways of succession: species turnover patterns in meadows abandoned for 30 years. Phytocoenosis Archiv Geobot 9:1–104

Faliński JB (1975) Anthropogenic changes of the vegetation of Poland. Phytocoenosis 4(2):97–115

Faliński JB (1976) Windwürfe als Faktor der Differenzierung und der Veränderung der Urwaldbiotopes im Licht der Forschungen auf Dauerflächen. Phytocoenosis 5(2):85–106

Faliński JB (1978) Uprooted trees distribution and influence in the primeval forest biotope. Vegetatio 38(3):175–183

Faliński JB (1986) Vegetation dynamics in temperate lowland primeval forests. Ecological studies in Białowieza forest. Geobotany 8:1–537

Faliński JB (1990–1991) Kartografia geobotaniczna. PPWK, Warsaw

Faliński JB (1991) Vegetation processes ad subject of geobotanical map. In: Proceedings of the XXXIII symposium of the international association for vegetation science, Warsaw, 8–12 Apr 1990. Phytocoenosis 3, (Suppl Cart Geobot 2):1–383

Faliński JB (1993) Applied geobotany and "ecologization" of geobotanical map. Fragm Flor Geobot Suppl 2(2):501–512

Faliński JB (1994) Vegetation under the diverse anthropogenic impact as object of basic phyto-sociological map. Results of the international cartographical experiment organized in the Białowieza Forest. Phytocoenosis 6 (Suppl Geobot 4):1–134

Faliński JB (1998a) Dioecious woody pioneer species (Juniperus communis, Populus tremula, Salix spp. div.) in the secondary succession and regeneration. Phytocoenosis 10 (Suppl Cart Geobot 8):1–156

Faliński JB (1998b) Maps a anthropogenic transformations of plant cover (maps of synanthro-pisation). Phytocoenosis 10:15–54

Faliński JB (1999) Geobotanical cartography: subject, source basis, transformation and application fundamentals of maps. Phytocoenosis 11 (Suppl Cart Geobot 11):43–65

Faliński JB, Mulenko W (eds) (1995) Cryptogamous plants in the forests communities of Białowieza National Park (Project CRYPO). Phytocoenosis 7 (Archiv Geobot 4):1–176

Faliński JB, Pedrotti F (1990) The vegetation and dynamical tendencies in the vegetation of Bosco Quarto, Promontorio del Gargano, Italy. Braun-Blanquetia 5:1–31

Faliński JB, Venanzoni R (1991) Large-scale ecological map of synusial structure of the forest communities in the Piktovka study area. Phytocoenosis 3 (N.S.) (Archiv Geobot 2):79–82

Famiglietti A, Schmid E (1969) Fitocenosi forestali e fasce di vegetazione dell'Appennino Lucano centrale (Gruppo del Vulturino e zone contermini). Ann Centro Economia Montana Venezie 7:3–180

Fanelli G (2002) Analisi fitosociologica dell'area metropolitana di Roma. Braun-Blanquetia 27:1–269

Farina A (2001) Ecologia del paesaggio. UTET, Turin

Farinelli F (1989) "Certezza del rappresentare": la questione cartografica. Urbanistica 97:7–16

Fenaroli L (1970) Note illustrative della carta della vegetazione reale d'Italia. Collana Verde 28:1–125

Ferrari C (1977) La vegetazione attuale. In: Bertolani Marchetti D, Studi ecologici e paleoecologici nella palude della Chioggiola presso Pavullo nel Frignano. "Pavullo e il Medio Frignano", Aedes Muratoriana, Biblioteca n.s., 38:10–14

Ferrari C (2001) Biodiversità. Dall'analisi alla gestione. Zanichelli, Bologna

Ferrari C, Rossi G (1990) La cartografia della vegetazione con il metodo fitosociologico in Italia. Boll AIC 78–79:109–120

Ferrari C, Pirola A, Piccoli F (1972) Saggio cartografico della vegetazione delle Valli di Comacchio. Ann Univ Ferrara Sez I, I(2):35–54

Ferrari C, Mandrioli P, Rinaldi A (1978) Integrazioni tra il rilevamento fotoaereo a bassa quota ed il rilevamento fitosociologico per la cartografia vegetazionale di un biotopo palustre. Not Fitosoc 13:1–11

Ferrarini E (1972) Carta della vegetazione delle Alpi Apuane e zone limitrofe. Note illustrative. Webbia 27:551–582

Ferrarini E (1988) Carta della vegetazione dell'Appennino settentrionale dalla Cisa al Gottero e alle Cinque Terre. Mem. Accad. Lunigianese Sc. "Giovanni Capellini". LI–LIII:173–192

Fiori A (1939) Formazioni vegetali. In: Consociazione Turistica Italiana, Atlante fisico-economico d'Italia. Esperia, Milan, pp 42–43 e tav. 21

Fiori A (1908) Prodromo di una geografia botanica dell'Italia. In: Fiori A, Paoletti G (eds) Flora analitica d'Italia, Ith edn. Tip. Seminario, Padova, pp I–LXXXVI

Forman RTT, Godron M (1986) Landscape ecology. Wiley, New York

Fosberg FR (1961) A classification of vegetation for general purposes. Trop Ecol 2:1–28

Francini E (1936) Ricerche sulla vegetazione dell'Etruria marittima. II. La vegetazione del Laghetto di Sibolla (Valdarno inferiore). Nuovo Giorn Bot Ital 43:62–130

Froment A (1987) L'écologie et le paysage. Notes Recherches Soc Géogr Liège 8:37–48

Fujiwara K (ed) (2008) Integrated vegetation mapping of Asia. Research report for grants-in-aid of scientific research program A (International Scientific Research) no. 16255003 in 2004–2007 of the Japanese Society for the Promotion of Science (JSPS), Yokohama

Fujiwara K, Miyawaki A, Terada J (1991) Functional vegetation map of Oiso-machi, 1:10,000. In: Vegetation of Oiso Town. Bulletin Yokohama Phytosociological Society 61

Gafta D (1994) Tipologia, sinecologia e sincorologia delle abetine nelle Alpi del Trentino. Braun-Blanquetia 12:1–69

Gafta D (2002) Influenœa antropo-zoogena asupra pădurilor periurbane. In: Cristea V, Baciu C, Gafta D (eds) Municipiul Cluj-Napoca æi zona periurbana, Studii ambientale. Editura Accent, Cluj-Napoca, pp 241–274

Gafta D (2006) Vegetation landscape patterns in the Torricchio Natural Reserve: implication for biological conservation. La Riserva Naturale di Torricchio 12:61–80

Gafta D, Canullo R (1992) The role of Alnus glutinosa (L.) Gaertner in the secondary succession on wet meadows in the Piné high plain (North Italy). Studia Geobotanica 12:105–120

Gafta D, Pedrotti F (1994) Phytosociological and ecological research in a protected area as basis for its management: the example of Loppio Lake (North Italy). Applied vegetation ecology. In: Proceedings of 35th symposium I.A.V.S. Shangai, East China. Normal University Press, pp 31–40

Gafta D, Pedrotti F (1997) Environmental units of the Stelvio National Park as basis for its planning. Oecologia Montana 6:17–22

Gafta D, Pedrotti F (1998) Fitoclima del Trentino-Alto Adige. St Trent Sc Nat 73:55–111

Gambino R (1991) I parchi naturali. La nuova Italia scientifica, Rome

Gams H (1918) Prinzipienfragen der Vegetationsforschung. Ein Beitrag zur Begriffsklärung und Methodik der Biocoenologie. Vierteljahrschrift Naturf Ges Zürich 63:293–493

Gaussen H (1954) Géographie des plantes. A. Colin, Paris

Gaussen H (ed) (1961a) Méthodes de la cartographie de la végétation. CNRS, Paris

Gaussen H (1961b) L'emploie des couleurs dans la cartographie de la végétation. In: Méthodes cartographie végétation. CNRS, Paris, pp 137–145

Gavioli O (1936) Ricerche sulla distribuzione altimetrica della vegetazione in Italia. III. Limiti altimetrici delle formazioni vegetali nel Gruppo del Pollino (Appennino Calabro-Lucano). Nuovo Giorn Bot Ital 43(3):636–706

Géhu JM (1984) La cartographie en réseau et l'analyse de la végétation. In: Knapp R (ed) Sampling methods and taxon analysis in vegetation science. Junk, The Haag, pp 121–128

Géhu JM (1987) Des complexes de groupements végétaux à la phytosociologie paysagère contemporaine. Inform Bot Ital 18(1–2–3):53–83

Géhu JM (1991a) L'analyse symphytosociologique et géographique de l'espace. Théorie et méthodologie. Coll Phytosoc 17:11–46

Géhu JM (1991b) Livre rouge des phytocoenoses terrestres du littoral français. Centre Reg. Phytosoc, Bailleul

Géhu JM (2006) Dictionnaire de Sociologie et Synecologie végétales. J. Cramer, Berlin/Stuttgart

Gentile S (1968) Memoria illustrativa della carta della vegetazione naturale potenziale della Sicilia (Prima approssimazione). Quaderni Ist Bot Univ Lab Critt Pavia 40:1–114

Giacomini V (1954) Per la conoscenza geobotanica dei pascoli valtellinesi. Valtellina e Valchiavenna VII(11):3–11

Giacomini V (1966) Italy. In: Küchler AW (ed) International bibliography of vegetation maps. 2. Europe. University of Kansas, Lawrence, pp 400–420

Giacomini V, Fenaroli L (1958) La flora. TCI, Milan

Giacomini V, Pignatti S (1955) Flora e vegetazione dell'alta Valle del Braulio con speciale riferimento ai pascoli di altitudine. Mem Soc Ital Sc Nat Mus Civ Milan 11(2–3):1–194

Gianguzzi L (1999) Vegetazione e bioclimatologia dell'Isola di Pantelleria (Canale di Sicilia). Braun-Blanquetia 22:1–70

Gillet F (1986) Analyse concrète et theorique des relations à differents niveaux de perception phytoecologique entre végétation forestière et géomorphologie dans le Jura nord-occidentale. Coll Phytosoc XIII:101–127

Gillet F (1988) L'approche synusiale intégrée des phytocoenoses forestières. Application aux forêts du Jura. Coll Phytosoc 14:81–92

Gillet F, Gallandat J-D (1996) Integrated synusial phytosociology: some notes on a new, multiscalar approach to vegetation analysis. J Veg Sci 7(1):13–18

Gillet F, de Foucault B, Julve Ph (1991) La phytosociologie synusiale intégrée: objets et concepts. Candollea 46:315–340

Ginanni F (1774) Istoria civile e naturale delle pinete ravennati. Salomone, Rome

Giordano A, Mondino GP, Palenzona M, Rota L, Salandin R (1970) Ecologia e utilizzazioni colturali prevedibili dell'alta Val Pellice. Ann Ist Sper Selvicoltura Arezzo 1:423–539

Giordano A, Mondino GP, Palenzona M, Rota L, Salandin R (1972) Ecologia e utilizzazioni prevedibili nel comune di Chiomonte. Ann Ist Sper Selvicoltura Arezzo III:81–188

Giordano A, Mondino GP, Palenzona M, Rota L, Salandin R (1974) Ecologia e utilizzazioni prevedibili della Valle di Susa. Ann Ist Sper Selvicoltura Arezzo V:83–196

Giurgiu V, Doniţă N, Băndiu C, Radus S, Cenusa R, Disescu R, Stoicolescu C, Biriş I-A (2001) Les forêts vierges de Roumanie. Forêt Wallonne, Lovain-la-Neuve

Gomez C, Dessert F, Hamon P, Hamon S, de Kochko A, Poncet V (2009) Current genetic differentiation of Coffea canephora Pierre ex A. Froehn in the Guineo-Congolian Afrian zone: cumulative impact of ancient climatic changes and recent human activities. Evol Biol 9:1–167

Goodchild MBF (1993) Data models and data quality: problems and properties. In: Goodchild MF, Parks BO, Steyaert LT (eds) Environmental modeling with GIS. Oxford University Press, New York, pp 94–104

Gratani L, Rossi A, Crescente MF, Frattaroli AR (1999) Ecologia dei pascoli di Campo Imperatore (Gran Sasso d'Italia) e carta della biomassa vegetale. In: Biondi E (ed) 1999 – Ricerche di geobotanica ed ecologia vegetale di Campo Imperatore (Gran Sasso d'Italia). Braun-Blanquetia 16:227–247

Greco S, Petriccione B (1988–1989) La cartografia della vegetazione nella definizione della qualità ambientale: il caso di Cocullo (AQ). Not Fitosoc 24:63–98

Greco S, Persia G, Petriccione B, Pezzotti E (1991) Il valore di qualità ambientale degli ecosistemi forestali a partire da indici e funzioni di correlazione floristici e vegetazionali. Atti SiTE 12:749–759

Greco S, Petriccione B, Pignatti F (1994) Vegetation mapping: a numerical comparative study of six maps of Białowieza Forest (Poland). Phytocoenosis 6 (Suppl Cart Geobot 4):105–113

Green DR, Hartley S (2000) Integrating photointerpretation and GIS for vegetation mapping: some issues of error. In: Alexander R, Millington AC (eds) Vegetation mapping. Wiley, Chichester/London, pp 103–134

Gribova AA, Samarina GD (1963) Detailed large-scale mapping of the dinamic of vegetational cover. Geobotaniceskoe Kartografirovanie, pp 15–25

Grisebach A (1872) Die Vegetation der Erde nach ihrer klimatischen Anordnung. Engelmann, Leipzig

Guinochet M (1973) Phytosociologie. Masson, Paris

Haber W (1990) Basic concept of landscape ecology and their application in land management. In: Ecology for tomorrow. Physiol Ecol Jpn, pp 131–146

Habiyareme MK (2000) Interprétation syntaxonomique du dynamisme des groupements végétaux de la dorsale orientale du Lac Kiwu. Coll Phytosoc XXVII:443–464

Hanganu J, Gridin M, Drost HJ, Chifu T, Stefan N, Sårbu I (1993) Romanian Danube delta biosphere reserve. Flevobericht 356:1–68

Hegg O, Béguin C, Zoller H (1993) Atlas de la végétation à protéger en Suisse. Office fédéral environnement, forêts, paysage, Bern

Hofmann G (1985) Potentielle natürliche Nettoprimärproduktion an oberirdischer Pflanzentrockenmasse. Akademie Verlag, Berlin

Holland MM, Whigam D, Gopal B (1990) The characteristics of wetland ecotones. In: Naiman RJ, Decamps H (eds) The ecology and management of aquatic-terrestrial ecotones, vol 4, Man and the biosphere. The Parthenon Publishing Group, London/New York, pp 171–198

Hooper PT, Sallaway MM, Latz PK, Maconochie JR, Hyde KW, Corbett LK (1973) Ayers Rock-Mt Olga National Park environmental study, 1972. Australian Government Publishing Service, Canberra

Hou H-Y (1979) The legend to the "vegetation map of China". Chinese Academy Sciences, Pekin

Hou H-Y (1983) Vegetation of China with reference to its geographical distribution. Ann Mo Bot Gard 70:509–548

Howard JA (1970) Aerial photo-ecology. Faber & Faber, London

Hruska K (1991) The mapping of urban flora and vegetation in Central Italy. Phytocoenosis (Suppl Cart Geobot 2):381–383

Hruska K (1998) Carta della vegetazione della città di Camerino. SELCA, Florence

Hueck K, Seibert P (1981) Vegetationskarte von Südamerika, 2nd edn. Springer, New York/Berlin/Heidelberg

Hultén E (1964) The circumpolar plants. I. Vascular cryptogams, conifers, monocotyledons. Almqvist & Wiksell, Stockholm

Hultén E (1971) Atlas of the distribution of vascular plants in northwestern Europe. Generalstabens Litografiska Anstalts, Stockholm

Ingegnoli V (2002) Landscape ecology: a widening foundation. Springer, New York/Berlin/Heidelberg

Isachenko TI (1962) Principles and methods of generalization in geobotanical mapping in large, medium and small scale. In: Sochava V (ed) Principles and methods of vegetation mapping. Akad. Nauk U.R.S.S, Moscow/Leningrad, pp 28–46

Isachenko TI, Lavrenko EM (1979) Karta rastitelnosti URSS. Akad. Nauk. U.R.S.S, Moscow

Ivan D (1979) Fitocenologie şi vegetaţia Republicii Socialiste România. Didactică şi pedagogică, Bukarest

Ivan D, Doniţă N (1975) Metode practice pentru studiul ecologic æi fitogeografic al vegetaœiei. Universitatea din Bucureæti – Facultatea Biologie, Bukarest

Ivan D, Doniţă N, Coldea G, Sanda V, Popescu A, Chifu T, Boşcaiu N, Mititelu D, Paucă-Comanescu M (1993) Végétation potentielle de la Roumanie. Braun-Blanquetia 9:1–79

Jalas J (1955) Hemerobe und hemerochore Pflanzenarten ein terminologischer Reformversuch. Acta Soc Fauna Flora Fenn 72:1–15

Jalas J, Suominen J (1972) Atlas Florae Europaeae. Helsinki

Janssen J (2001) Monitoring of salt-marsh vegetation by sequential mapping. Rijkswaterstaat Meetkundige Dienst, Delft

Journaux A (1975) Légende pour une carte de l'environnement et de sa dynamique. Faculté Lettres Sciences humaines, Caen

Julve P, Gillet F (1994) Experiences of french Authors. In: Falinski JB (ed) Vegetation under the diverse anthropogenic impact as object of basic phytosociological map. Results of the international cartographical experiment organized in the Białowieza Forest. Part 2. Conceptions and methods of the individual vegetation maps. Phytocoenosis 6 (Suppl Geobot 4):45–66

Kaluzny SP, Vega SC, Cardoso TP, Shelly AA. (1998) S+ Spatial Stats: User's Manual for Windows and UNIX. Springer-Verlag. New York

Kamagata N, Hara K, Mori M, Akamatsu Y, Li Y, Hoshino Y (2006) A new method of vegetation mapping by object-based classification using high resolution satellite data. J Jpn Soc Photogram Remote Sens 54(1):43–49

Karamysheva ZV, Khramtsov VN (1995) The steppes of Mongolia. Braun-Blanquetia 17:1–79

Karamysheva ZV, Rachkowskaya EI (1975) Maps of subdivision. In: Exhibition guide "Geobotanical mapping in USSR", Leningrad. XII Int Bot Congr II:3–15

Kent M, Coker P (1994) Vegetation description and analysis. A practical approach. Wiley, Chichester/London

Kepczynski K (1960) Plant groups of the lake district of Skepe and the surrounding peat-bogs. Studia Soc Sc Torunensis Suppl VI:1–244

Kirkpatrick J (1990) A synusia-based mapping system for the conservation management of natural vegetation, with an example from Tasmania. Aust Biol Conserv 53(2):93–104

Kokaly RF, Depain DG, Clark RN, Livo KE (2003) Mapping vegetation in Yellowstone National Park using spectral feature analysis of AVIRIS data. Remote Sens Environ 84:437–456

Kondō H, Sakai A (2011) Distribution of tree species in relation to the micro-topography in riparian area, sub-alpine zone, central Japan. In: 54th symposium of the international association for vegetation science, 20–24 June 2011, Abstracts. Université Claude Bernard, Lyon, p 187

Kowarik I (1987) Kritische Anmerkungen zum theoretischen Konzept der potentiellen natürlichen Vegetation mit Anregungen zu einer zeitgemässen Modifikation. Tüxenia 7:53–67

Krebs CJ (2001) Ecology. Benjamin Cummings, San Francisco

Küchler AW (1964) Manual accompany the map potential natural vegetation of the conterminous United States. American Geographical Society, New York

Küchler AW (1967) Vegetation mapping. The Ronald Press Company, New York

Küchler AW, Zonneveld OS (1988) Vegetation mapping. Kluwer, Dordrecht

Ladle RJ, Whittaker RJ (2011) Conservation biogeography. Wiley-Blackwell, Oxford

Lavrenko EM (1950) Botanical-geographic regions of the Palearctic. In: Problems of botany, I

Lavrenko EM (1964) Botanical-geographic dominions. In: Physical-geographic atlas of the world

Lavrenko EM, Sochava VB (1954) Geobotaniceskaia Karta CCCR. Akademia Nauk SSSR, Botanicheskii Institut V.L. Komarova, Leningrad

Lavrenko EM, Sochava VB (1956) The legend to the "geobotanical map of the USSR", scale 1:4,000,000. Acad. Sc USSR Press, Leningrad

Leibundgut H (1959) Über Zweck und Methodik der Struktur und Zuvachsanalyse von Urwälder. Schweiz Zeitschr Fortswesen 110:3

Leibundgut H (1978) Über die Dynamik europäischer Urwälder. Allg Zeitsch Forstwesen 24:686–690

Leuschner C (1997) Das Konzept der potentiellen natürlichen Vegetation (PNV): Schwachstellen und Entwicklungsperspektiven. Flora 192:379–391

Liberman Cruz M (1986) Microclima y distribuciòn de Polylepis tarapacana en el Parque nacional del Nevado Sajama. Doc Phytosoc 10(II):235–272

Liberman Cruz M, Pedrotti F (2006) Woody formations in a mesothermic valley of Tarija Province, Boliva. In: Gafta D, Akeroyd J (eds) Nature conservation. Concepts and practice. Springer, Berlin/Heidelberg, pp 75–86

Liberman CM, Pedrotti F, Venanzoni R (1988) Le associazioni della classe Lemnetea del Lago Titicaca (Bolivia). Riv Idrobiol 27(2–3):377–388

Liberman CM, Gafta D, Pedrotti F (1997) Estructura de la poblaciòn de Polylepis tarapacana en el Nevado Sajama, Bolivia. In: Liberman CM, Baied C (eds) Desarrollo sostenible de ecosistemas de montana: Manejo de areas fragiles en los Andes. The United Nations University, La Paz, pp 59–70

Libermann CM, Pedrotti F, Venanzoni R (1995) La Isla del Sol en el Lago Titicaca (Bolivia). II Simposio int. desarrollo sostenible ecosistemas de montana: manejo areas fragiles en Los Andes (Huarina–Bolivia, 2–12, 1995). La Paz, pp 71–78

Lieth H (1975) Modeling the primary productivity of the world. In: Lieth H, Whittaker RH (eds) Primary productivity of the biosphere. Springer, New York, pp 237–263

Lieth H, Box EO (1972) Evapotranspiration and primary productivity; C. W. Thornthwaite memorial model. Publications in Climatology (University of Delaware, Elmer, New Jersey) 25(3):37–46

Lieth H, Box EO (1977) The gross primary productivity pattern of the land vegetation: a first attempt. Trop Ecol 18:109–115

Li-Kuo Fu (1992) China plant red data book. Science Press, Pekin/New York

Lillesand T, Raulth R, Kieser RW (1999) Remote sensing and image interpretation. Wiley, New York

Lobo GM, Castro I, Moreno JC (2001) Spatial and environmental determinants of vascular plant species richness distribution in the Iberian Peninsula and Balearic Islands. Biol J Linn Soc 73:233–253

Loidi J, Bascones JC (1995) Mapa de series de vegetaciòn de Navarra. Gobierno de Navarra, Departamento Ordenaciòn, Territorio, Medio Ambiente, Navarra

Lüdi W (1921) Die Pflanzengesellschaften des Lauterbrunnentales und ihre Sukzession. Beitr Geobot Landesaufnahme 9:1–364

Lüdi W (1935) Beitrag zur regionalen Vegetationsgliederung der Apenninhalbinsel. In: Rübel E (ed) Ergebnisse der Internat. Pflanzengeogr. Eskursion durch Mittelitalien 1934. Veröff Geobot Inst Rübel Zürich 12:212–239

Lüdi W (1946) Die Gliederung der Vegetation auf der Apenninenhalbinsel, insbesondere der montanen und alpinen Höhenstufen. In: Rikli M (ed) Das Pflanzenkleid der Mittelmeerländer. II. Huber, Bern, pp 573–596

Magurran AE (1988) Ecological diversity and its measurement. Cambridge University Press, Cambridge

Mahito K, Takeshi O (1998) Vegetation mapping with the aid of low-altitude aerial photography. Appl Veg Sci 1:211–218

Marchesoni V (1954) Il Lago di Molveno e la foresta riaffiorata in seguito allo svaso. St Trent Sc Nat 31:9–24

Marchesoni V (1958) Aspetti mediterranei lungo il margine meridionale delle Alpi con particolare riguardo al settore prealpino antistante al bacino atesino. St Trent Sc Nat 34(2–3):47–69

Marchesoni V (1959) Il cembro, l'albero più espressivamente alpino. Natura Alpina X(4):117–128

Martinelli M (1990) Impatto ambientale nel territorio di Camerino (Italia centrale). Dipart. Bot. Ecol, Camerino

Martinelli M (1999) La cartographie environnementale: une cartographie de synthèse. Phytocenosis 11 (Suppl Cart Gobot 11):23–129

Martinelli M, Pedrotti F (2001) A cartografia das unidades de paisagem: questoes metodologicas. Rev Depart Geografia Univ São Paulo 14:39–46

Mason F (2002) Dinamica di una foresta della Pianura Padana. Bosco della Fontana. Centro Nazionale Studio Conservazione Biodiversità Forestale – Corpo Forestale Stato, Verona

Matuszkiewicz JM (1972) Analysis of the spatial variation of the field layer in the contact zone of two phytocoenoses. Phytocoenosis 1(2):121–150

Matuszkiewicz W (1984) Die Karte der potentiellen natürlichen Vegetation von Polen. Braun-Blanquetia 1:1–99

Mccoy RM (2005) Field methods in remote sensing. The Guilford Press, New York/London

Mcintosh RP (1967) The continuum concept of vegetation. Bot Rev 33(2):130–187

Merloni N, Piccoli F (2001) La vegetazione del complesso Punte Alberete e Valle Mandriole (Parco regionale del delta del Po – Italia). Braun-Blanquetia 29:1–17

Meusel H, Jäger E (1992) Vergleichende Chorologie der Zentraleuropäischen Flora, III. Fischer, Jena

Meusel H, Jäger E, Weinert E (1965) Vergleichende Chorologie der centraleuropäischen Flora. Fischer, Jena

Meusel H, Jäger E, Rauschert S, Weinert E (1978) Vergleichende Chorologie der Zentraleuropäischen Flora, II. Fischer, Jena

Michalet R, Pautou G (1998) Végétation et sols de montagnes. Diversité, fonctionnement et évolution. Écologie 29(1–2):1–440

Minghetti P (2003) Le pinete a Pinus sylvestris del Trentino-Alto Adige (Alpi italiane). Braun-Blanquetia 33:1–95

Minghetti P, Pedrotti F (2000) La vegetazione del laghetto delle Regole di Castelfondo (Trento). St Trent Sc Nat 74:175–189

Ministero Agricoltura Foreste (1976) Carta della montagna. Geotecneco, Urbino

Mittermeier RA, Gil PR, Mittermeier CG (1999) Hotspots, Earth's richest and most endangered terrestrial ecoregions. Agrupaciòn Sierra Madre, Mexico City

Miyawaki A (1979) Vegetation und Vegetationskarten auf den Japanischen Inseln. In: Miyawaki A, Okuda S (eds) Vegetation und Landschaft Japans. Yokohama Phytosociological Society, Yokohama, pp 49–70

Miyawaki A, Box EO (2006) The healing power of forests. Kôsei Publishing, Tokyo

Miyawaki A, Fujiwara K (1970) Vegetationskundliche Untersuchungen im Ozegahara-Moor, Mittel-Japan. The National Parks Association, Tokyo

Miyawaki A, Fujiwara K, Okuda Sh (1987) The status of nature and re-creation of green environments in Japan. In: Miyawaki A et al (eds) Vegetation ecology and creation of new environments. Tokai University Press, Tokyo, pp 357–376

Moggi G (1969) Some reflections on the phytogeographical subdivisions of Italy. Public University of Sevilla, Separata V Simposio Flora Europaea (20–30 de mayo de 1967), pp 229–234

Moggi G (1992) Conservazione e documentazione: gli erbari come archivi floristici di situazioni pregresse. In: Pedrotti F (a cura di) La Società Botanica Italiana per la protezione della natura (1888–1990). L'Uomo e l'Ambiente 14:162–175

Morandini R (1969) Abies nebrodensis (Lojac.) Mattei. Inventario 1968. Pubbl Ist Sper Selvicolt Arezzo 18:1–93

Morandini R, Ducci F, Menguzzato G (1994) Abies nebrodensis (Lojac.) Mattei. Inventario 1992. Ann Ist Sper Selvicolt Arezzo 22:5–51

Moser D, Dullinger S, Englisch T, Niklfeld H, Plutzar C, Sauberer N, Zechmeister G, Grabherr G (2005) Environmental determinants of vascular plant species richness in the Austrian Alps. J Biogeogr 32:1117–1127

Mucina L, Rutherford MC, Powrie LW (2006) The vegetation of South Africa, Lesotho and Zwaziland. SA, South African Biodiversity Institute, Pretoria

Mutke J, Barthlott W (2005) Patterns of vascular plant diversity at continental to global scales. Biol Skr 55:521–537

Myers N, Mittermeier RA, Mittermeier CG, da Fonseca CG, Kent GABJ (2000) Biodioversity hotspots for conservation priorities. Nature 403:853–858

Naveh Z, Lieberman A (1984) Landscape ecology: theory and application. Springer, New York/Berlin/Heidelberg

Negri G (1934) Ricerche sulla distribuzione altimetrica della vegetazione in Italia. Introduzione. Nuovo Giorn Bot Ital 41(2):327–364

Negri G (1947) Considerazioni sulla classificazione dei piani altimetrici della vegetazione in Italia. Riv Geogr Ital 54:1–28

Negri G (1951) Geografia botanica. In: Gola G, Negri G, Cappelletti C (eds) Trattato di Botanica. UTET, Turin, pp 1042–1136

Negri G (1954) Interpretazione individualistica del paesaggio vegetale. Nuovo Giorn Bot Ital 61:579–694

Neshataev YN (1971) The selective-statistical method in distinguishing of plant associations. In: Alexandrova VD (ed) Methods for distinguishing of plant associations. Academy Sc, Leningrad, pp 181–205

Neuhäuslova Z (2001) Potential natural vegetation of the Czech Republic. Braun-Blanquetia 30:1–80

Nimis PL, Bolognini G (1990) The use of chorograms in quantitative phytogeography and in phytosociological syntaxonomy. Fitosociologia 25:69–87

Nimis PL, Bolognini G (1993) Quantitative phytogeography of the Italian beech forests. Vegetatio 109:125–143

Noirfalise A (1987) Carte de la végétation naturelle des états membres des Communautées européennes et du Conseil de l'Europe. Council of Europe, Strasbourg

Ogureeva G (1999) Zoni i tipi poiasnosti rastitelnosti Rossii i sopredelnih territorii, Geographiceskoii Fakultet MGU, Moscow

Ogureeva G, Kotova TU, Emelianova LG (2010) Biogeograficeskoe Kartografirovanie. Geograpiceskoii Fakultet MGU, Moscow

Olaczek R (1974) Kierunki degeneracji fitocenoz lesnych i metody ich badania. Phytocenosis 3(3/4):179–190

Oldeman RAA (1990) Forests: elements of silvology. Springer, New York/Berlin/Heidelberg

Olivier L, Galland JP, Mauron H (1995) Livre rouge de la flore de France. I. Espèces prioritaires. Museum National d'Histoire Naturelle, Paris

Orloci L (1972) On obiective functions of phytosociological resemblance. Am Mildl Nat 88:28–55

Orsomando E (1969) Areale italiano di Ephedra nebrodensis (Tin.). Mitt ostalp-din pflanzensoz Arbeitsgem 9:341–348

Orsomando E (1993) Carte della vegetazione dei fogli Passignano sul Trasimeno e Foligno. Braun-Blanquetia 10:1–46

Orsomando E, Pedrotti F (1986) Le praterie galleggianti a Carex pseudocyperus L. di alcuni laghi dell'Italia centrale. Riv Idrobiol XXV(1–3):87–103

Orsomando E, Pedrotti F (1992) Il programma di cartografia della vegetazione delle Regioni Marche ed Umbria in scala 1:50,000. Boll AIC 84–85:139–142

Orsomando E, Catorci A, Beranzoli N, Ferranti G, Ciarapica A, Segatori R, Grohmann F (1998) Regione Umbria. Carta geobotanica con principali classi di utilizzazione del suolo. SELCA, Florence

Orsomando E, Catorci A, Pitzalis M, Raponi M (1999) Carta fitoclimatica dell'Umbria. SELCA, Florence

Ozenda P (1964) Biogéographie végétale. Doin, Paris

Ozenda P (1974) De la carte de la végétation à une carte de l'environnement. Doc Cart Écol 13:1–18

Ozenda P (1975) La cartographie écologique. Courrier du C.N.R.S. 24 (hors série):1–11

Ozenda P (1979) Carte de la végétation des États membres du Conseil de l'Europe. Council of Europe, Strasbourg

Ozenda P (ed) (1980–1982) Mitteilungen der 16. Tagung Vegetationskartierung im Gebirge, Klagenfurt, 10–13 Sept 1979. Doc Cart Écol XXIII:1–72 e XXV:89–95

Ozenda P (ed) (1981) Colloque Internat. Cartographie végétation à petite échelle, Grenoble, France, 24–27 Sept 1980. Doc Cart Écol XXIV:1–134

Ozenda P (1982) Les végétaux dans la biosphère. Doin, Paris

Ozenda P (1984) La végétation de l'Arc alpin. Council of Europe, Strasbourg

Ozenda P (1985) La végétation de la châine alpine dans l'espace montagnard européen. Masson, Paris

Ozenda P (1986) La cartographie écologique et ses applications. Masson, Paris

Ozenda P (1994) Végétation du continent européen. Delachaux & Niestlé, Lausanne

Ozenda P (2002) Perspectives pour une géobiologie des montagnes. Presses Polytech Univ Romandes, Lausanne

Ozenda P, Borel J-L (2006) La végétation des Alpes occidentales. Un sommet de la biodiversité. Braun-Blanquetia 41:1–45

Ozenda P, Lucas M-J (1987) Esquisse d'une carte de la végétation potentielle de la France à 1/1.500.000. Doc Cart Ecol 30:49–80

Pampanini R (1903) Essai sur la géographie botanique des Alpes et en particulier des Alpes sudorientales. Fragnière, Fribourg

Pasquale GA (1863) Relazione sullo stato fisico-economico-agrario della Prima Calabria Ulteriore. Mem R Ist Incoraggiamento Sc Nat Naples 11:1–432

Pedrotti F (1965–1968) Carta fitosociologica della vegetazione della media Val di Sole. Soc Geogr, Florence

Pedrotti F (1967) Carta fitosociologica (1:3,000) della vegetazione dei Piani di Montelago (Camerino). Notiz Soc Ital Fiosoc 4:1–8

Pedrotti F (1976) Vegetazione e ambiente delle Marche e relativi problemi di salvaguardia. Giorn Bot It 110(6):383–399

Pedrotti F (1978) Einige Bemerkungen über die Entwicklung der Vegetation im Naturreservat von Torricchio. Phytocoenosis 7(1-2-3-4):11–19

Pedrotti F (1979) La conservazione della vegetazione negli ambienti umidi. CNR, Rome, AC/1/103:63–80

Pedrotti F (1980) La végétation de la tourbière du Vedes (Trento). Coll Phytosoc VII:231–250

Pedrotti F (1982) Les "piani" de Montelago (Camerino). In: Guide-Itinéraire Escursion Internationale de Phytosociologie en Italie centrale, 2–11 juillet 1982. Università Studi, Camerino, pp 242–249

Pedrotti F (1983) Cartografia geobotanica e sue applicazioni. Ann Accad It Sc Forestali 32:317–363

Pedrotti F (1985) Géomorphologie et repartition de la végétation dans les bassins karstiques des Apennins. Coll Phytosoc XIII:507–539

Pedrotti F (1987) Presenza e diffusione di Bromus inermis Leyss in. Trentino-Alto Adige. Inf Bot It 19(1):60–66

Pedrotti F (1988a) Carta della vegetazione del Foglio Borgo Valsugana (1:50,000). Giorn Bot Ital 122(3–4):59

Pedrotti F (1988b) La cartografia geobotanica in Italia. In: Pedrotti F (ed) 100 anni di ricerche botaniche in Italia 1888–1988. SBI, Florence, pp 731–761

Pedrotti F (1988c) La flora e la vegetazione del Lago di Loppio (Trentino). Giorn Bot Ital 122(3–4):105–147

Pedrotti F (1988d) über das Vorkommen von Fels- und Mauer-Chasmophyten in Monte S. Angelo (Gargano, Italien). Flora 180:145–152

Pedrotti F (1989) La vegetazione. In: Ambiente M (ed) Relazione sullo stato dell'ambiente. Ist Poligrafico Stato, Rome, pp 69–77

Pedrotti F (1990) Exhibition guide "Geobotanical mapping in Italy". In: Fifth international congress of ecology, Yokohama, 23–30 Aug 1990, pp 1–24

Pedrotti F (1991) La carta della vegetazione d'Europa. In: Attualità biologiche, Atti I Conv Sez Ital IUBS:31–41

Pedrotti F (1992) La vegetazione. Carta della vegetazione naturale attuale d'Italia. In: Ambiente M (ed) Relazione sullo stato dell'ambiente. Ist Poligrafico Stato, Rome, pp 94–100

Pedrotti F (1993) Vegetation mapping in Italy. Vegetatio 109:187–190

Pedrotti F (1995) La vegetazione forestale italiana. Atti Conv Lincei 115:39–78

Pedrotti F (1996a) Suddivisioni botaniche dell'Italia. Giorn Bot Ital 130(1):214–225

Pedrotti F (1996b) Il pioppo tremulo (*Populus tremula* L.) nella colonizzazione dei terreni abbandonati del Parco Nazionale d'Abruzzo. Coll Phytosoc XXIV:111–121

Pedrotti F (1997a) Les données de la phytosociologie pour la cartographie de la végétation. Coll Phytosoc 27:503–541

Pedrotti F (1997b) Geobotanik und Landschaftskartierung – Beispiele aus Italien. Ber Reinh-Tüxen-Ges 9:123–137

Pedrotti F (1997c) Sintesi geobotanica della Valle di Tovel (Trentino). L'Uomo e l'Ambiente, 2007 46:1–39

Pedrotti F (1998) La cartographie géobotanique des biotopes du Trentin (Italie). Écologie 29(1–2):105–110

Pedrotti F (1999a) Carta delle unità ambientali dei Monti Sibillini. SELCA, Florence

Pedrotti F (1999b) Cartografia della vegetazione e qualità dell'ambiente. Natura Alpina 4:21–41

Pedrotti F (2001) Biotopkartierung im Trentino: Methoden und Resultate. Sauteria 11:61–74

Pedrotti F (2004a) Ricerche geobotaniche al Laghestel di Piné (Trento) (1967–2001). Braun-Blanquetia 35:1–55

Pedrotti F (2004b) Cartografia Geobotanica. Pitagora, Bologna

Pedrotti F (2004c) Vegetation mapping in wetlands. Ann Bot iv:29–36

Pedrotti F (2008) La vegetazione del Monte Pennino e dei piani carsici di Montelago (Sefro, Macerata). In: Di Martino V, Pedrotti F, Valeriani P (eds) Per l'istituzione del Parco Naturale Regionale dell'area Monte Pennino, Valle Scurosa e Montelago. TEMI, Trento, pp 63–72

Pedrotti F (2010) Le serie di vegetazione della Regione Trentino-Alto Adige. In: Blasi C (ed) La vegetazione d'Italia. Palombi, Rome, pp 83–109

Pedrotti F, Cortini Pedrotti C (1976) The vegetation map of the nature reserve of Burano (Central Italy). Geobotaniceskoe Kartographirovanie, pp 68–69

Pedrotti F, Faliński JB (2002) Real vegetation of Foresta Umbra, Promontorio del Gargano, Italy. SELCA, Florence

Pedrotti F, Gafta D (1994) La palude di Roncegno. In: Pedrotti F (ed) Guida all'escursione della Società Italiana di Fitosociologia in Trentino, 1–5 luglio 1994. Dipart Bot Ecol, Camerino, pp 123–129

Pedrotti F, Gafta D (1996) Ecologia delle foreste ripariali e paludose dell'Italia. L'Uomo e l'Ambiente 23:1–165

Pedrotti F, Gafta D (2003) Approccio fitogeografico alla distinzione di megageoserie di vegetazione nelle Alpi del Trentino-Alto Adige (con carta 1:250,000). Centro Ecologia Alpina, Report 30, pp 1–18

Pedrotti F, Minghetti P (1994) Le marocche di Dro. In: Pedrotti F (ed) Guida all'escursione della Società Italiana di fitosociologia in Trentino, 1–5 luglio 1994. Dipart Bot Ecol, Camerino, pp 29–65

Pedrotti F, Minghetti P (1997) Carta della naturalità della vegetazione della Regione Trentino-Alto Adige. In: Centro Ecologia Alpina, 1999, Report 20

Pedrotti F, Orsomando E (1977) Studio per la tutela e la valorizzazione del patrimonio naturalistico del bacino del Trasimeno: flora e vegetazione, vol 3. Ministero Agricoltura Foreste, Rome, pp 1–66

Pedrotti F, Venanzoni R (1994a) Experiences of italian authors. In: Falinski JB (ed) Vegetation under the diverse anthropogenic impact as object of basic phytosociological map. Results of the international cartographical experiment organized in the Białowieza Forest. Part 2. Conceptions and methods of the individual vegetation maps. Phytocoenosis 6 (Suppl Geobot 4):31–37

Pedrotti F, Venanzoni R (1994b) Carta della vegetazione del Bosco dell'Incoronata (Foggia). SELCA, Florence

Pedrotti F, Pratesi F, Patella A (1969) La conservazione della natura attraverso la pianificazione territoriale. In: Studi per la valorizzazione naturalistica del Parco nazionale dello Stelvio, II, Ed. Mevio, Sondrio, pp 565–656

Pedrotti F, Orsomando E, Cortini PC (1975) Carta della vegetazione del Lago di Burano e della Duna di Capalbio (Grosseto). LAC, Florence

Pedrotti F, Orsomando E, Cortini PC (1979) The phytosociological map of Burano (Tuscany). Webbia 34:529–531

Pedrotti F, Gafta D, Minghetti P (1994) Il Monte Bondone. In: Pedrotti F (ed) Guida all'escursione della Società Italiana di fitosociologia in Trentino, 1–5 luglio 1994. Dipart Bot Ecol, Camerino, pp 5–19

Pedrotti F, Minghetti P, Sartori G (1996) Evoluzione della vegetazione e del suolo delle marocche di Dro (Trento, Italia). Coll Phytosoc XXIV:203–222

Pedrotti F, Gafta D, Martinelli M, Patella SA, Barbieri F (1997) Le unità ambientali del Parco Nazionale dello Stelvio. L'Uomo e l'Ambiente 28:1–103

Perring FH, Walters SM (1962) Atlas of the British flora. Nelson, London

Petriccione B, Claroni N (1996) The dynamical tendencies in the vegetation of Velino Massif (Abruzzo, Italy). Doc Phytosoc XVI:365–373

Pettit C, Cartwright W, Bishop I, Lowell K, Pulsar D, Duncan D (eds) (2008) Landascape analysis and visualisation. Springer, New York/Berlin/Heidelberg

Pignatti S (1959) Fitogeografia. In: Cappelletti C (ed) Trattato di Botanica. UTET, Rurin, pp 681–811

Pignatti S (1976) Geobotanica. In: Cappelletti C (ed) Trattato di Botanica. UTET, Turin, pp 801–977

Pignatti S (1978) Dieci anni di cartografia floristica nell'Italia di Nord-Est. Inf Bot Ital 10(2):210–219

Pignatti S (1982a) Flora d'Italia. Edagricole, Bologna

Pignatti S (1982b) The origins of the flora of central Italy. In: Pedrotti F (ed) Guide-Itinéraire Excursion Internationale Phytosociologie Italie centrale, 2–11 juillet 1982. Università Studi, Camerino, pp 75–90

Pignatti S (1988a) Phytogeography and chorology – definitions and problems. Ann Bot 46:7–23

Pignatti S (1988b) Le ricerche dei botanici italiani in campo fitogeografico (1888–1988). In: Pedrotti F (ed) 100 anni di ricerche botaniche in Italia (1888–1988). SBI, Florence, pp 681–697

Pignatti S (1994) Ecologia del paesaggio. UTET, Turin

Pignatti S (2004) La biodiversità. Un concetto complesso. In: Ecoregioni e reti ecologiche. La protezione incontra la conservazione. Atti Conv Naz, Roma, 27–28 maggio 2004. WWF, Rome, pp 10–14

Pirola A (1982) Attualità e applicazione della cartografia della vegetazione. Monti e Boschi 33(3–5):15–19

Pirola A (1988) Gli studi vegetazionali e lo sviluppo della fitosociologia in Italia. In: Pedrotti F (ed) 100 anni di ricerche botaniche in Italia (1888–1988). SBI, Florence, pp 699–729

Pirola A, Vianello G (1992) Cartografia tematica ambientale. Suolo, vegetazione, fauna. La Nuova Italia Scentifica, Rome

Poldini L (1991) Atlante corologico delle piante vascolari nel Friuli-Venezia Giulia. Inventario floristico regionale. Regione Autonoma Friuli-Venezia Giulia – Università Studi Trieste, Udine

Poldini L (2002) Nuovo atlante corologico delle piante vascolari nel Friuli Venezia Giulia. Regione autonoma Friuli-Venezia Giulia – Università Studi Trieste, Udine

Poldini L (2009) La diversità vegetale del Carso fra Trieste e Gorizia. Lo stato dell'ambiente. Ed. Goliardiche, Trieste

Poli E (1965) La vegetazione altomontana dell'Etna. Gianasso, Sondrio

Poli E (1973) Indagine geobotanica e pianificazione territoriale nel Parco dell'Etna. Atti III Simp Naz Conservaz. Natura, Bari, 2–6 maggio 1973, II, pp 309–320

Poli ME, Patti G (2000) Carta della vegetazione dell'Etna. SELCA, Florence

Poli E, Maugeri G, D'Urso A (1974) La Celtis tournefortii Lam. sull'Etna. Arch Bot Biogeogr Ital XX(1–2):27–50

Poli E, Maugeri G, Ronsisvalle G (1981) Note illustrative della carta della vegetazione dell'Etna. Collana programma finalizzato "Promozione qualità ambiente". CNR, Rome. AQ/1/131, pp 1–29

Polish Academy Science (2001) Polish red data book of plants. Inst. Bot. W. Szafera, Polska Akad. Nauk, Kraków

Polli M, Sala B (2003) La cartografia oggi in Provincia di Trento e suoi futuri sviluppi. Boll AIC 117-118-119:104–110

Polunin N (1967) Éléments de géographie botanique. Gauthier-Villars, Paris

Pop I (1977–1979) Biogeografie ecologica. Dacia, Cluj-Napoca

Pott R (1995) Die Pflanzengesellschaften Deutschlands. Ulmer, Stuttgart

Pratesi F (a cura di) (1968) Piano di riassetto del Parco Nazionale d'Abruzzo. Ass. Italia Nostra, Rome

Prentice IC, Cramer W, Harrison SP, Leemans R, Monserud RA, Solomon AM (1992) A global biome model based on plant physiology and dominance, soil properties and climate. J Biogeogr 19:117–134

Primack RB (2002) Essentials of conservation biology. Sinauer Associates, Sunderland

Provasi T (1926) Osservazioni e ricerche sulla vegetazione di alcuni laghetti dell'Appennino Tosco-Emiliano. Nuovo Giorn Bot Ital 33:681–725

Puppi BG, Zanotti AL (1989) Methods in phenological mapping. Aerobiologia 5:44–54

Puppi G, Speranza M, Pirola A (1980) Carta della vegetazione dei dintorni del Lago Brasimone, Emilia Romagna. Programma finalizzato Qualità Ambiente. CNR, Roma, AQ/1/1974:1–29

Puscaru-Soroceanu E, Popova-Cucu A (1966) Geobotanica. Ed. Stiințifică, Bukarest

Quezel P, Barbero M (1985) Carte de la végétation potentielle de la Ragion Méditerranéene. N. 1: Feuille Méditerranée orientale. CNRS, Paris

Radford AE, Ahles HE, Bell CR (1968) Manual of the vascular flora of the Carolinas. The University of North Carolina Press, Chapel Hill

Raimondo FM (2000) Carta del paesaggio e della biodiversità vegetale della Provincia di Palermo. Quad Bot Amb Appl 9:1–160

Raimondo FM, Schicchi R (1998) Il popolamento vegetale della Riserva naturale dello Zingaro (Sicilia). Regione Siciliana – Azienda Foreste Demaniali, Palermo

Regione Abruzzo (2000) Carta dell'uso del suolo 1:250,000. Note illustrative, 2nd edn. SELCA, Florence

Regione Lazio (2003) Carta dell'uso del suolo. SELCA, Florence

Regione Toscana (1998a) Carta della vegetazione forestale (scala 1:250,000). Note illustrative. SELCA, Florence

Regione Toscana (1998b) Carta della vegetazione forestale potenziale (scala 1:250,000). Note illustrative. SELCA, Florence

Rey P (1988) Notions générales d'utilisation des cartes de la végétation. CNRS, Paris

Ricotta C, Carranza ML, Avena G, Blasi C (2002) Are potential natural vegetation maps a meaningful alternative to neutral landscape models? Appl Veg Sci 5:271–275

Rivas Martínez S (1985) Biogeografia y vegetaciòn. Real Acad. Ciencias Exactas,Fisicas y Naturales, Madrid, pp 1–103

Rivas Martínez S (1987) Memoria del mapa de series de vegetaciòn de España. ICONA, Madrid

Rivas Martínez S (1996) Bioclimatic map of Europe. Cartographic Service, University of Leon. In: Rivas Martinez S (ed) Geobotanica y climatologia. Discursos pronunciados en el acto de investidura de Doctor "honoris causa" del exc.mo señor D. Salvador Rivas Martínez. Universidad de Granada, Granada, pp 25–98

Rivas Martínez S (2005a) Avances en Geobotánica. Discurso de Apertura Curso Académico Real Academia Nacional Farmacia año:1–128

Rivas Martínez S (2005b) Notions of dynamic-catenal phytosociology as a basis of landscape science. Plant Biosyst 139(2):135–144

Rivas Martínez S (2007–2011) Mapa de series, geoseries y geopermaseries de vegetaciòn de España. Parte I, Parte II. Itinera Geobot 17:5–436; 18(1):1–424; 18(2):425–800

Rivas Martínez S, Sànchez-Mata D, Costa M (1999) North american boreal and western temperate forest vegetation. Itinera Geobot 12:5–316

Rivas-Martínez S (2008) Global bioclimatics (Clasificaciòn Bioclimatica de la Tierra), version01-12-2008. http://www.globalbioclimatics.org/book/bioc/global_bioclimatics-2008_00.htm

Roig FA, Faggi AM (1985) Transecta botànica de la Patagonia austral. C.O.N.I.C.E.T. (Argentina), I. P. (Chile), R.S. (Great Britain), Buenos Aires

Roig FA, Anchorena J, Dollenz O, Faggi AM, Mendez E (1985) Las comunidades vegetales de la Transecta botànica de la Patagonia austral. Primera parte: le vegetaciòn del àrea continental. In: Boelcke O, Moore DM, Roig FA (ed) Transecta botànica de la Patagonia austral. Buenos Aires, C.O.N.I.C.E.T. (Argentina), I.P. (Cile), R.S. (Great Britain), pp 350–591

Rosi D (2005) Serie storiche e successioni secondarie negli incolti del Monte Cardosa (Parco Nazionale dei Monti Sibillini). L'Uomo e l'Ambiente 44:1–53

Rübel E (1912a) Vorschläge zur geobotanischen Kartographie. Beitr Geobot Landesaufn 1:1–14

Rübel E (1912b) Pflanzengeographische Monographie des Berninagebietes. Engelmann, Leipzig

Rübel E (1930) Die Pflanzengesellschaften der Erde. Huber, Bern

Safronova I (2004) On phytocoenotical mapping of the caspian desert region. Ann Bot IV:83–93

Sanesi G (1982) I suoli dei Piani di Montelago. In: Guide-Itinéraire Escursion Int. Phytosoc. Italie centrale, 2–11 juillet 1982. Università Studi, Camerino, pp 249–257

Sappa F (1955) Carta della vegetazione forestale delle Langhe (1:50,000). Allionia 2:269–292

Sappa F, Charrier G (1949) Saggio della vegetazione forestale della Val Sangone (Alpi Cozie). Nuovo Giorn Bot Ital 56(1–2):106–187

Schino G, Borfecchia F, de Cecco L, Dibari C, Iannetta M, Martini S, Pedrotti F (2003) Satellite estimate of grass biomass in a mountainous range in central Italy. Agrofor Syst 59:157–162

Schmid E (1961) Erläuterungen zur Vegetationskarte der Schweiz. Pflanzengeogr Kommission Schweiz Naturforsch Gesell 39:1–52

Sochava V (1954) Printipi i zadaci gheobotaniceskoi kartografii. Vopr Botaniki, Moscow/Leningrad

Sochava V (1962) Principles and methods of vegetation mapping. Akad. Nauk U.R.S.S., Moscow/Leningrad

Sochava V (1975) The content of vegetation maps and how to enriche it. In: XIIth international botanical congress, Leningrad, pp 1–7

Sochava V (1979) Vegetation on tematic maps. Nauka, Novosibirsk

Sochava V, Isachenko TI (1976) Materiali XII mezdunarodnogo botaniceskogo kongressa. Geobotaniceskoe Kartografirovanie, pp 1–100

Stohlgren TJ (2007) Measuring plant diversity. Oxford University Press, Oxford

Stropp J (2011) Towards an understanding of tree diversity in Amazonia forests. University of Utrecht, CNPq, IEB, Gordon & Betty Moore Foundation, Utrecht

Stropp J, ter Steege H, Mahli Y (2009) Disentangling regional and local tree diversity in the Amazon. Ecography 32(1):46–54

Sukachev VN, Zonn SV (1961) Metodiceskie ukazanua k izuceniu tipov lesa, Moscow

Sukopp H (1972) Wandel von Flora und Vegetation in Mitteleuropa unter des Menschen. Ber Landwirtsch 50:112–139

Takhtajan A (1986) Floristic regions of the world. University of California Press, Los Angeles/London/Berkeley

Tassi F (1983) La situazione in Italia dei parchi e delle riserve. La sfida del 10% per gli anni '80. In: Atti Convegno Nazionale "Strategia 80 per i parchi e le riserve d'Italia", Camerino, 28–30 ottobre 1980. L'Uomo e l'Ambiente, 4:65–78

Theurillat JP (1992) étude et cartographie du paysage végétal (Symphytocoenologie) dans la Région d'Aletsch (Valais, Suisse). Matér Levé Géobot Suisse 68:1–384

Theurillat JP (1994) Symphytocoenologie: du paysage végétal aux divisions phytogéographiques. Rev Valdôtaine Hist Nat Suppl 48:317–333

Thuiller W, Midgley GF, Rouget M, Bowling RM (2006) Predicting patterns of plant richness in megadiverse South Africa. Ecography 29:733–744

Tomaselli R (1955) Osservazioni sulla Primula tyrolensis Scott. Arch Bot 15:162–177

Tomaselli R (1956a) Introduzione allo studio della fitosociologia. Industria Poligrafica Lombarda, Milan

Tomaselli R (1956b) Note sulla vegetazione dei prati e dei pascoli dell'alta Valle di Scalve sulla sinistra del fiume Dezzo (Bergamo). Ann Sper Agr 10(suppl 3):XLVII–LXIX; (suppl 4): I–XXXI

Tomaselli R (1970a) Note illustrative della carta della vegetazione naturale potenziale d'Italia. Collana Verde 27:1–63

Tomaselli R (1970b) Tipologia ecologico-strutturale della vegetazione del mondo. Att Ist Bot Lab Critt Univ Pavia VI:1–232

Tomaselli R (1973a) La vegetazione forestale d'Italia. Collana Verde 33:25–60

Tomaselli R (1973b) Carta della vegetazione forestale potenziale d'Italia. In: Tecneco, Prima relazione sulla situazione ambientale del paese, II, pp 61–62

Tomaselli R (1977) Gli aspetti fondamentali della vegetazione del mondo (Ecologia e corologia). Parte I. Tipologia ecologico-strutturale della vegetazione. Collana Verde 48:1–290

Tomaselli R (1981) Gli aspetti fondamentali della vegetazione del mondo (Ecologia e corologia). Parte II. La vegetazione che caratterizza i paesaggi naturali. Collana Verde 58:1–301

Töpfer F (1979) Kartographische Generalisierung. Geogr.-Kart. Anstalt Gotha, Leipzig

Trautmann W (1966) Erläuterungen zur Karte der potentiellen natürlichen Vegetation der Bundesrepublik Deutschland 1:200,000. Blatt 85 Minden. Schriftenr Vegetationsk 1:1–137

Trautmann W (1973) Vegetationskarte der Bundesrepublik Deutschland 1:200,000 – Potentielle natürliche Vegetation – Blatt CC 5502 Köln. Schriftenr Vegetationsk 6:1–172

Troll C (1964) Karte der Jahreszeiten-Klimate der Erde. Erdkunde 18:5–28

Tucker CJ (1978) A comparison of satellite sensor bands for vegetation monitoring. Photogramm Eng Remote Sens 44(11):1169–1180

Tucker CJ (1979) Red and photographic infra-red linear combinations for monitoring vegetation. Remote Sens Environ 8(2):127–150

Tucker CJ, Townshend JRG, Goff TE (1985) African land-cover classification using satellite data. Science 227:369–375

Tüxen R (1956) Die heutige potentielle natürliche Vegetation als Gegenstand der Vegetationskartierung. Angew Pflanzensoz 13:1–55

Tüxen R (ed) (1963) Bericht über das Internationale Symposion für Vegetationskartierung vom. 26.3.1959 im Stolzenau/Weser. Cramer, Weinheim

Tüxen R (1979) Sigmeten und Geosigmeten, ihre Ordnung und ihre Bedeutung für Wissenschaft, Naturschutz und Planung. Biogeographica 16:79–92

UNESCO-FAO (1963) Carte bioclimatique der la zone méditerranéenne. Notice explicative. UNESCO, Paris

UNESCO-FAO (1970) Carte de la végétation de la région méditerranéenne. Notice explicative. UNESCO, Paris

Ustin SL, Di Pietro D, Olmstead K, Underwood E, Scheer GJ (2002) Hyperspectral remote sensing for invasive species detection and mapping. Geoscience and remote sensing symposium, Toronto, Ontario, Canada pp, 1658–1660

van der Maarel E, Westhoff V (1964) The vegetation of the dunes near Oostvorne (the Netherlands) with a vegetation map. Wentia XII:1–61

van der Maarel E, Westhoff V (1973) The Braun-Blanquet approach. In: Whittaker RH (ed) Classification and ordination of plant communities. Junk, The Haag, pp 617–726

Velasquez A, Duran E, Ramírez I, Mas JF, Ramírez G, Bocco G, Palacio JL (2003) Land use-cover change processes in highly biodiverse areas: the case of Oaxaca, Mexico. Global Environ Change 3(12):8–24

Venables WN, Ripley BD (1997) Modern applied statistics with S-Plus. Springer, New York/Berlin/Heidelberg

Venanzoni R, Kwiatkowski W, Kaomocki A (1999) The geographic information system as a tool to archive field and cartographic data in the "Montagna di Torricchio" naturale reserve (Central Italy). Phytocoenosis (Suppl Cart Geobot) 11:191–197

Vianello G (1998) Cartografia e fotointerpretazione. CLUEB, Bologna

Videsott R (1955) Problemi di organizzazione e di vita dei parchi nazionali d'Italia e particolarmente del Gran Paradiso. Atti I Conv. Internaz. Amministratori Direttori Parchi Nazionali, Cogne, 27 agosto 1955. Parco Naz. Gran Paradiso, Turin, pp 69–102

Vigo J (1998) Some reflections on geobotany and vegetation mapping. Acta Bot Barc 45:535–556

Virgilio F, Schicchi R, La Mela Vaca DS (2000) Aggiornamento dell'inventario della popolazione relitta di Abies nebrodensis (Lojac.) Mattei. Naturalista Sicil 24(1–2):13–54

von Bertalanffy L (1950) The theory of open systems in physics and biology. Science 111:23–29

Vos W, Stortelder A (1992) Vanishing tuscan landscapes. Landscape ecology of a submediterranean area (Solano Basin, Tuscany, Italy). Pudoc, Wageningen

Walker DA, Gould WA, Maier HA, Raynolds MK (2002) The Circumpolar Arctic Vegetation Map: AVHRR-derived base maps, environmental controls, and integrated mapping procedures. Int J Remote Sens 23(21):4551–4570

Walsby JC (1995) The causes and effects of manual digitising on error creation in data in data input to GIS. In: Fisher P (ed) Innovations in GIS 2. Taylor & Francis, London, pp 113–122

Walter H (1954) Grundlagen der Pflanzenverbreitung. II. Teil. Arealkunde. Ulmer, Stuttgart

Walter H (1977) Vegetationszonen und Klima, 3rd edn. Ulmer, Stuttgart

Walter H, Box EO (1976) Global classification of natural terrestrial ecosystems. Vegetatio 32:72–81

Welten M, Sutter R (1982) Verbreitungsatlas der Farn- und Blütenpflanzen der Schweiz. Birkhäuser, Basel

Whittaker H (1960) Vegetation of Siskiyou Mountains, Oregon and California. Ecol Monogr 30:279–338

Whittaker H (1962) Classification of natural communities. Bot Rev 28(1):1–239

Williams JW, Seablom EW, Slayback D, Stoms DM, Joshua HV (2004) Anthropogenic impacts upon plant species richness and net primary productivity in California. Ecol Lett 8:127–137

Wilson EO (1992) The diversity of life. Harward University Press, Cambridge

Wohlgemuth T (1998) Modelling floristic species richness on a regional scale: a case study in Switzerland. Biodivers Conserv 7:159–177

Xie Y, Sha Z, Yu M (2008) Remote sensing imagery in vegetation mapping: a review. J Plant Ecol 1(1):9–23

Zangheri P (1936) Flora e vegetazione delle pinete di Ravenna e dei territori limitrofi tra queste e il mare. Valbonesi, Forlì

Zangheri P (1959) Flora e vegetazione della fascia gessoso-calcarea del basso Appennino Romagnolo. Webbia 14(2):243–595

Zenari S (1925) I caratteri della vegetazione della Val Cellina. Arch Bot 1:101–140; 2:149–169

Zenari S (1941) La vegetazione del Comelico (Alto Cadore). Nuovo Giorn Bot Ital 48(1):1–388

Zenari S (1950) Fitogeografia. Liviana, Padova

Zonneveld IS (1989) The land unit – a fundamental concept in landscape ecology and its applications. Landsc Ecol 3(2):67–86

Zunino M, Zullini A (2004) Biogeografia. Ambrosiana, Milan

Printed by Publishers' Graphics LLC
LSI20121211.19.20.128